EUROPEAN COMMISSION

The Single Market Review

AGGREGATE AND REGIONAL IMPACT

EMPLOYMENT, TRADE AND LABOUR COSTS IN MANUFACTURING

The Single Market Review

SUBSERIES VI: VOLUME 4

OFFICE FOR OFFICIAL PUBLICATIONS
OF THE EUROPEAN COMMUNITIES

KOGAN PAGE . EARTHSCAN

The Single Market Review series

This report is part of a series of 39 studies commissioned from independent consultants in the context of a major review of the Single Market. The 1996 Single Market Review responds to a 1992 Council of Ministers Resolution calling on the European Commission to present an overall analysis of the effectiveness of measures taken in creating the Single Market. This review, which assesses the progress made in implementing the Single Market Programme, was coordinated by the Directorate-General 'Internal Market and Financial Services' (DG XV) and the Directorate-General 'Economic and Financial Affairs' (DG II) of the European Commission.

This document was prepared for the European Commission

by

Cambridge Econometrics

It does not, however, express the Commission's official views. Whilst every reasonable effort has been made to provide accurate information in regard to the subject matter covered, the Consultants are not responsible for any remaining errors. All recommendations are made by the Consultants for the purpose of discussion. Neither the Commission nor the Consultants accept liability for the consequences of actions taken on the basis of the information contained herein.

The European Commission would like to express thanks to the external experts and representatives of firms and industry bodies for their contribution to the 1996 Single Market Review, and to this report in particular.

Office for Official Publications of the European Communities
2 rue Mercier, L-2985 Luxembourg
ISBN 92-827-8809-1   Catalogue number: C1-72-96-004-EN-C

Kogan Page . Earthscan
120 Pentonville Road, London N1 9JN
ISBN 0 7494 2341 2

Table of contents

List of tables

List of figures

List of abbreviations

CGE	computable general equilibrium
ECM	error correction model
Eurostat	Statistical Office of the European Communities
IMF	International Monetary Fund
IO	input-output
IV	Instrumental Variables
LTU	long-term unemployment
NIC	newly industrialized country
OECD	Organization for Economic Co-operation and Development
OLS	Ordinary Least Squares
OPEC	Organization of Petroleum Exporting Countries
PA	per annum
PP	percentage points
R&D	Research and Development
SEM	Single European Market
SMP	single market programme

Country Codes

BE	Belgium
DK	Denmark
DO	East Germany
DW	West Germany
EL	Greece
ES	Spain
FR	France
IR	Ireland
IT	Italy (north)
IS	Italy (south)
LX	Luxembourg
NL	Netherlands
PO	Portugal
UK	United Kingdom

E3ME Manufacturing Sector Codes

08	Ferrous & Non-ferrous Metals
09	Non-metallic Mineral Products
10	Chemicals
11	Metal Products
12	Agricultural & Industrial Machinery
13	Office Machines
14	Electrical Goods
15	Transport Equipment
16	Food, Drink & Tobacco
17	Textiles, Clothing & Footwear
18	Paper & Printing Products
19	Rubber & Plastic Products
20	Recycling/Emissions Abatement *
21	Other Manufactures

* This sector is not included in the tables as data is not available to identify it sufficiently for modelling purposes. Instead the sector is distributed among the residual manufacturing sectors of the model, namely Other Manufactures.

1. Summary

1.1. Scope of the report

This study evaluates the impact of the single market programme (SMP) on manufacturing employment and labour costs, through its effects on intra-EU trade. The study is primarily concerned with the SMP effects on the manufacturing sectors, as defined by the NACE-CLIO 25 sector disaggregation, in each of the EUR-12 Member States up to the period 1993.

The study begins in Chapter 2 with a critical review of the academic literature concerning trade and employment linkages. Chapter 3 then describes the off-model analysis necessary for the study, i.e. the empirical work which does not require a full simulation of the structural model used in the analysis. In Chapter 4 the study reports the detailed simulation results by Member State, manufacturing sector and type of labour. The model results are compared with the conclusions from other studies on manufacturing to check for both similarities and differences. A variety of model simulations are described in Section 4.5: these serve to test the robustness of the results. Finally, Section 4.6 discusses the policy issues raised by the report and draws conclusions from these.

The following limitations of the study should be noted. Data quality and availability, particularly at the sectoral level, restricted the methodologies used in the study and renders some results uncertain. The short period (1987–93) that the SMP measures have had to take effect makes it unwise to draw definitive conclusions, particularly with regard to the long-term impact. Finally, a number of contemporaneous events may not be adequately captured by the model, so that their influence is incorrectly attributed to the SMP. These are discussed in Section 4.4 and include the expansion of spending under the Structural Funds, the accession of Spain and Portugal to the EU and the re-unification of Germany.

1.2. Methodological approach

The main tool used for the analysis is the model, **E3ME**, described in detail in Appendix B. E3ME is disaggregated by region (the Member States of the EUR-12, excluding Greece at present due to data limitations), and by sector (NACE-CLIO 25 sector disaggregation, including 14 manufacturing sectors). It is an econometric, sectoral model and distinguishes intra-EU and extra-EU imports and exports by sector and Member State.

The underlying causality implied by the study is that the SMP has affected trade in manufactures and services in the EU, and that this in turn affected employment and labour costs, as well as other aspects of the economy. The basic methodological approach, therefore, involves including a variable to represent the SMP in E3ME's trade equations, estimating its coefficient (by sector and Member State) using econometric techniques, and then simulating several 'antimondes', under different assumptions, to represent what would have happened if the SMP had not been implemented. To represent the SMP it was decided to construct a synthetic variable which took the value of zero up until 1985 and then rose gradually to reach unity in 1993 (see Section 3.4 for more details). Different forms of this variable have been devised and a comparison is made of the effects of the specification on the estimates of the SMP effects on trade. Model runs with the SMP variable set to zero in different sets of trade equations are then compared with one in which the SMP variable is as originally constructed.

1.3. Review of theoretical and empirical literature

Section 2.1 describes in detail how traditional neoclassical theories of trade suggest that a reduction in barriers to trade will lead countries to specialize in sectors for which they have a comparative advantage: thus the low-wage countries would attract (low-skill) labour-intensive sectors. However, these theories provided no explanation of intra-industry trade which now dominates intra-EU trade. Their predictions may therefore be unreliable for analysing the impact of the SMP. More recent theories of trade offer a better explanation of intra-industry trade, but their implications for the impact of a reduction in barriers to trade are less certain. Moreover, the employment effects of this kind of trade usually occur between individual firms within an industry so that what, at the industrial level, may appear to be small changes could be disguising larger flows in opposing directions.

The empirical literature on trade and employment linkages can be divided in similar fashion to the academic studies by their focus on either neoclassical or so-called 'new-trade' methodologies. The choice of method depends on the nature of the trade being studied, for example most neoclassical methodologies focus on trade between the US and developing countries, while trade within the EU or OECD is typically analysed using intra-industry techniques. The studies of EU trade tend to rely more on sectoral characteristics, both in terms of the type of product produced and the industry structure through which production takes place, for undertaking empirical analysis.

1.4. Effects of the single market programme on trade

The long-run partial impact of the SMP on trade is estimated to be a volume increase in intra-EU exports and imports of the order of 19 to 30% depending on the form of the SMP synthetic variable chosen (see Table 3.10b). The effects are very uneven between sectors and Member States, although the impact on imports is more diverse than for exports. The effects for Spain and Portugal may have been exaggerated due to the difficulty in isolating the impact of the SMP from the effects due to accession to the EU.

A SMP impact of similar magnitude – a long-run increase of 25 to 33% – is estimated for extra-EU imports. It does not appear, therefore, that the SMP has had a protectionist effect on trade with the EU economies. The boost to imports could reflect the impact of the 1992 advertising on non-EU producers, the sourcing of inputs from outside the EU by plants set up under foreign direct investment, or the fact that lower internal barriers to trade reduce the costs of non-EU producers to service EU markets (as they do for EU producers). If these explanations are correct, the impact on extra-EU imports could be a step effect, rather than one which would be expected to increase over time.

The impact of the SMP on extra-EU exports in the estimated equations is also positive overall (averaged across all sectors and Member States), but of much smaller magnitude (of the order of 4 to 10%). However, for many manufacturing industries, the effect appears to be negative, suggesting that trade has been diverted towards the single market. It is likely that the improvements to competitiveness that are the expected result from the SMP (for example, as industries rationalize and economies of scale are secured) are not evident by 1993, and there could be a different impact on extra-EU trade in the long term.

1.5. The SMP simulations and the effects on trade

Two main sets of simulations were performed, testing for the effects of the SMP through intra-EU trade only, and through both intra-EU and extra-EU trade.

1.5.1. Intra-EU trade only

The first set of simulations used a version of the model having the SMP variable included in the intra-EU trade equations for both imports and exports (by sector). The extra-EU trade equations in this simulation were estimated without any SMP effect. The antimonde simulation was performed by setting the value of the synthetic SMP variable to zero over the whole period of the simulation. This causes intra-EU imports and intra-EU exports to change, by the amount of the estimated SMP effect. This then feeds through to output, employment and the other variables in the model including, via indirect influences such as relative labour costs, extra-EU trade. A simulation with the SMP effect included was then compared with this *antimonde* to provide the results shown in Table 1.1. The effect on intra-EU trade in manufactures builds up over the period from 1986, rising to about 5% above base by 1993.

1.5.2. Intra-EU and extra-EU trade

The second simulation was with a version of the model having the SMP variable included in both in the intra-EU trade equations and the extra-EU trade equations. The SMP variable was then set to zero for both the intra- and extra-EU trade equations. This is model simulation 2 with the results giving the direct effects of the SMP on both intra- and extra-EU trade. Total imports are about 5% above base by 1993; but the effect of the SMP is to reduce extra-EU exports (interpreted as trade diversion), with EU exporters redirecting exports to the SM, and total exports of manufactures are about 2% below base by 1993.

Table 1.1.   Model simulation 2: impact of the SMP on EU manufacturing employment by sector (%)

		1985	1986	1987	1988	1989	1990	1991	1992	1993
08	Ferrous & Non-f. Metals	0.00	0.00	-0.02	-0.06	-0.06	0.13	0.72	2.61	6.72
09	Non-metallic Min.Pr.	0.00	0.00	0.00	0.01	-0.03	-0.06	-0.07	-0.22	-0.48
10	Chemicals	0.00	0.00	0.00	-0.01	-0.02	-0.12	-0.14	0.04	-0.64
11	Metal Products	0.00	0.00	0.02	0.07	0.20	0.58	0.86	1.26	1.84
12	Agri. & Indust. Mach.	0.00	0.00	-0.03	-0.11	-0.28	-0.60	-3.03	-3.45	-3.24
13	Office Machines	0.00	0.00	0.00	0.02	0.06	0.25	-0.52	0.23	0.64
14	Electrical Goods	0.00	0.00	-0.01	-0.03	-0.04	-4.37	-2.84	-1.30	-1.01
15	Transport Equipment	0.00	0.00	-0.01	-0.03	-0.07	-0.16	-0.31	-0.57	-0.86
16	Food, Drink & Tobacco	0.00	0.00	0.00	0.00	0.01	0.26	-0.96	-0.35	-0.61
17	Tex., Cloth. & Footw.	0.00	0.00	0.00	0.00	0.02	0.10	0.42	0.62	0.78
18	Paper & Printing Pr.	0.00	0.00	0.00	-0.01	-0.04	-0.15	0.39	0.30	-0.63
19	Rubber & Plastic Pr.	0.00	0.00	0.00	0.00	-0.01	-1.23	-2.42	-3.07	-4.16
21	Other Manufactures	0.00	0.00	0.00	-0.01	-0.04	-0.10	-0.33	-0.62	-0.91
	Total	0.00	0.00	-0.01	-0.01	-0.03	-0.55	-0.77	-0.52	-0.53

Source: Cambridge Econometrics Forecast C51F1V-C51F1W/da1/F1WA.

1.6. Effects of the single market programme on employment

The overall impact on manufacturing employment averaged across Member States is very small in both simulations. When only intra-EU trade is assumed to respond directly to the SMP (simulation 1), manufacturing employment rises 0.5% above base by 1993; when intra-

and extra-EU trade respond (simulation 2) manufacturing employment is down 0.5%. A set of results for EUR-12 sectors, aggregated across countries, is shown in Table 1.1. The differing effects by sector are apparent, as is the tendency for the effects to accumulate in most cases.

In general, the impact on employment across all sectors (including services) in the higher cost economies is more positive than it is for manufacturing alone, suggesting that the SMP has had the effect of accelerating structural change towards services. The results by sector suggest evidence of rationalization in Rubber and Plastic Products, Agricultural and Industrial Machinery and to a lesser extent in Electrical Goods, Other Manufactures and Transport Equipment. Job losses in these sectors are offset by gains in Ferrous and Non-ferrous Metals, Metal Products, and Textiles, Clothing and Footwear.

Examination of the detailed results in Appendix D shows evidence of a shift in employment in Textiles, Clothing and Footwear away from the higher cost economies and towards Spain and Ireland. There are few other such clear examples of sectors commonly characterized as 'low-skill, labour-intensive' shifting towards the lower cost economies, nor are there consistent examples of a reverse movement for 'skill-intensive' sectors.

The interpretation is that even with the detailed sectoral breakdown adopted for this study, the sectors cannot be readily characterized as low or high skill. Higher cost economies can find niches even in the clothing sector where high skill and R&D are important for competitive advantage.

There is no evidence for the EU as a whole that the SMP has had a major effect on overall employment (all sectors). The results suggests a small increase by 1993 of the order of 0.4%. However, there is some indication that the impact is accumulating, and so the effects could be larger in the longer term. Also, the effects vary substantially by sector and Member State.

1.7. Effects of the single market programme on labour costs

The results for unit labour costs were mixed, and no evidence was found for a general interpretation of a kind similar to that outlined above for employment. In simulation 1 there was a slight overall fall of 0.2% in manufacturing unit labour costs, but in simulation 2 there was no overall change. In particular, there was no evidence of any tendency towards equalization of unit labour costs among Member States.

Table 1.2 shows the summarized results for simulation 2 by EU sector. Initially there is a slight increase in unit labour costs for the manufacturing sectors, but by 1993 various countries show a decrease in unit labour costs for these sectors. On average across the EU there has been a small increase in unit labour costs of 0.05% in 1993 compared to costs if the SMP had not occurred. A second measure of labour costs, namely labour costs per employee, is also considered. Overall, the impact of the SMP on these costs appears to have been very small. On average across the EU, the largest falls in labour costs per employee were experienced by the Electrical Goods, Transport Equipment and Metal Products industries, while Other Manufactures experienced the largest increases (see Table 1.3 for details).

Table 1.2. **Model simulation 2: impact of the SMP on EU manufacturing unit labour costs by sector (%)**

		1985	1986	1987	1988	1989	1990	1991	1992	1993
8	Ferrous & Non-f. Metals	0.00	0.00	0.04	0.11	-0.11	-0.16	0.34	0.75	1.53
9	Non-metallic Min.Pr.	0.00	0.01	0.10	0.44	0.80	1.13	1.59	1.78	0.54
10	Chemicals	0.00	0.02	0.14	0.72	0.73	1.55	2.44	5.95	5.31
11	Metal Products	0.00	0.00	0.03	0.12	0.39	0.28	0.83	1.84	3.45
12	Agri. & Indust. Mach.	0.00	-0.01	0.00	-0.02	-0.11	-1.50	-2.74	-4.34	-1.85
13	Office Machines	0.00	-0.01	0.02	0.12	0.27	0.26	0.89	1.69	-0.09
14	Electrical Goods	0.00	0.02	0.13	0.47	1.40	-1.37	0.27	1.31	1.35
15	Transport Equipment	0.00	0.01	0.21	0.14	0.13	-0.83	-0.60	-1.15	-3.70
16	Food, Drink & Tobacco	0.00	0.00	0.01	0.08	0.25	0.58	0.55	1.13	2.00
17	Tex., Cloth. & Footw.	0.00	-0.03	-0.38	-0.77	-0.89	-1.01	-4.04	-7.38	-3.29
18	Paper & Printing Pr.	0.00	0.00	0.03	0.13	0.29	1.05	0.47	1.17	-0.04
19	Rubber & Plastic Pr.	0.00	0.00	0.02	0.08	0.22	-0.54	-0.44	-0.32	-0.47
21	Other Manufactures	0.00	0.00	0.02	0.04	0.08	-0.17	0.34	0.76	0.39
	Total	0.00	0.00	0.03	0.11	0.27	-0.14	-0.20	-0.19	0.05

Source: Cambridge Econometrics Forecast C51F1V-C51F1W/da1/F1WA.

Table 1.3. **Model simulation 2: impact of the SMP on EU manufacturing labour costs per employee by sector (%)**

		1985	1986	1987	1988	1989	1990	1991	1992	1993
8	Ferrous & Non-f. Metals	0.00	-0.01	-0.01	0.01	0.07	0.27	-0.04	0.13	-0.86
9	Non-metallic Min.Pr.	0.00	0.00	-0.02	-0.06	-0.08	-0.32	-0.31	-0.42	0.10
10	Chemicals	0.00	0.00	0.00	0.00	-0.02	0.17	0.16	0.06	-0.11
11	Metal Products	0.00	-0.01	-0.02	-0.06	-0.19	-0.82	-1.07	-1.33	-1.40
12	Agri. & Indust. Mach.	0.00	-0.01	0.00	-0.01	-0.03	-0.56	0.05	-0.06	0.42
13	Office Machines	0.00	-0.01	0.00	0.04	0.05	0.19	0.57	0.89	1.03
14	Electrical Goods	0.00	0.01	0.02	0.02	0.15	-1.45	-0.99	-1.38	-1.26
15	Transport Equipment	0.00	0.00	-0.02	-0.05	-0.15	-1.37	-0.61	-1.09	-1.26
16	Food, Drink & Tobacco	0.00	0.00	-0.01	0.05	0.10	0.16	0.19	0.52	0.94
17	Tex., Cloth. & Footw.	0.00	0.00	-0.01	0.00	-0.01	-0.19	-0.20	0.12	0.45
18	Paper & Printing Pr.	0.00	0.00	0.00	0.00	0.02	0.47	-0.16	0.46	0.59
19	Rubber & Plastic Pr.	0.00	0.00	-0.01	0.02	0.03	-0.30	0.18	0.51	0.84
21	Other Manufactures	0.00	0.00	0.01	0.06	0.16	-0.15	0.60	1.33	1.64
	Total	0.00	0.00	0.00	0.00	0.01	-0.28	-0.12	-0.04	0.01

Source: Cambridge Econometrics Forecast C51F1V-C51F1W/da1/F1WA.

1.8. The implications for skills

The implications for skills were analysed by conducting off-model analysis on employment by occupation, although data quality and consistency are poor in this area. The technical details behind this analysis are described in Appendix C, while Section 4.2 focuses on the results. The data suggest that there are large variations in occupational structure among Member States. The change in employment due to the SMP is in general spread across all occupations, with the smallest proportional decrease registered by Service Workers, Craft and Related Trades Workers and Plant and Machinery Operators. Technicians & Associate Professionals suffer the largest proportional decline in employment from the SMP. Craft and Related Trade Workers show the biggest variation in occupational impacts between Member States.

1.9. Policy conclusions

On the basis of evidence to 1993, the SMP appears to have been a major factor neither in creating jobs in manufacturing nor in destroying them, for the EU as a whole. There is evidence of an acceleration of restructuring, and the economies that have lost jobs in manufacturing have generally gained jobs in services.

There is no evidence that the SMP has reinforced disparities between rich and poor Member States: if anything, there are indications that Ireland, Spain and Portugal have seen larger proportionate gains in employment, although this finding must be qualified by the difficulty of distinguishing SMP effects from those of EU accession and supply of structural funds.

There are indications that the SMP has enabled non-EU producers to increase their exports to the EU to a greater extent, so far, than EU producers have been able to increase their extra-EU exports. The main policy implications therefore relate to actions that can ease the adjustment to restructuring, and to the monitoring of extra-EU trade performance to check whether the hoped-for improvements to competitiveness emerge in the longer term.

2. Review of trade and employment linkages

2.1. Trade and the labour market

Within the academic literature on empirical linkages between trade and the labour market there are several strands of development, which can be broadly described as follows.

2.1.1. Heckscher–Ohlin–Samuelson (HOS) theory

The basic proposition of the theory is that trade follows relative factor endowments, such that factor prices will tend to equalize between countries, and is typically described in a two-good, two-factor, two-country framework. In the context of the EU, which is rich in physical and human capital, the model would imply that exports of goods and services from the EU will tend to embody these factors, while there will be a tendency to import raw materials and those goods which have a relatively higher labour intensity.

The theory was developed in the 1950s, when the heavily industrialized northern hemisphere was exporting highly capital-intensive goods to a developing southern hemisphere. Since then there has been a rapid increase in intra-industry trade (import and export of similar goods) as opposed to inter-industry trade (one-way trade in different products with different factor intensities). As a result, intra-EU exports accounted for 88% of total EU exports in 1992, with the equivalent import figure at 84%.

Thus, inter-industry trade at best explains only a minor share of overall trade patterns, although according to OECD trade data it tends to dominate trade between areas such as the Dynamic Asian Economies (DAEs) and China. EU trade is dominated by manufactures, which is in turn dominated by intra-EU trade, and this in turn is dominated by intra-industry trade. Hence, the HOS theory is largely irrelevant in accounting for import and export patterns.

The HOS theory also has little to say about linkages with the labour market. The main benefit from trade will occur through lower prices of imported goods and higher prices of exported goods, and is therefore channelled through consumption. The theory is difficult to test because of the strong nature of its assumptions (perfect competition, constant returns to scale, homogeneous goods) and largely ignores income and activity effects. Several features of production and trade are not well represented by the theory, e.g. increasing returns to specialization and scale, complete specialization by countries in some goods and services and diversity in technology and tastes.

The model is one of factor displacement (e.g. labour, capital) and changes in relative rewards. The effect of increased trade is to favour some industries at the expense of others, and the effects on employment follow from this.

2.1.2. Factor content analysis

As the name suggests, this analysis uses the factor content of exporting and importing industries to estimate the implicit change in factor endowments due to trade. For example, if imports are heavily low-skill-intensive while exports are heavily high-skill-intensive, it is possible to calculate the implied employment that would be required to produce the same output in a no-trade (or reduced trade) situation, given fixed wages. Given the employment effect, estimates of wage elasticities from the supporting literature are used to work out the effect on wages. The analysis typically focuses on trade between advanced, high-income countries and developing, low-income countries. Most empirical studies have taken the US as the example of a high-income country, and have used the analysis to try and explain the stylized facts of:

(a)   widening wage inequality between high-skill and low-skill workers;
(b)   falling employment share of low-skill workers relative to high-skill workers.

The analysis can be regarded as an extension of the HOS theory because a similar argument is applied on factor endowments, except that the factors are no longer labour and capital but infrastructure capital and human capital, i.e. the skill content of labour. There are difficulties in measuring skill levels, but the analogous argument is that the high-income EU countries would specialize in producing goods with a relatively high skill content, while lower income countries would specialize in goods with a low skill content, and a similar tendency to factor price equalization would take place. Where labour markets were not flexible, a rise in unemployment would occur.

Some empirical evidence has been undertaken using this analysis, notably by Wood (1994 and 1995), who extends earlier work by taking into account factors such as differing mixes of products within industries, non-competing manufactured imports, intermediate (inter-industry) demand effects and labour-saving technical change induced by trade competition (so-called 'defensive innovation'); this latter effect is analysed empirically in Chapter 3. Using adjustments like this, Wood (1994) claims to explain most of the decline in employment of unskilled workers in advanced countries by changing trade patterns, in contrast to earlier studies which identified trade as a cause, but not a particularly major one.

The criticisms of the factor content approach can be listed as follows:

(a)   The assumption that employment adjusts while wages are fixed is a strong one, although in the EU, where wages can be held steady by institutional factors, there may be some relevance.
(b)   No adjustment is made for how demand for output may respond to changes in prices.
(c)   The increase in trade is taken as exogenous, i.e. the approach is only valid if the change in trade is due to reduced trade barriers, increased skills in developed countries or the spread of technology, whereas other factors such as domestic labour market forces or macroeconomic expansion could also play a part. The reduction of trade barriers (tariff or non-tariff) are undoubtedly a factor in increased imports of manufactured products from low-income countries, but changes in the strategies of these countries from import substitution to export promotion must also have played a role.
(d)   The model seems to ignore the possibility of movement of labour between the tradable and non-tradable sectors of the economy, the latter being where many low skill-intensive jobs occur.
(e)   Trade with low-skill countries accounts for such a small percentage of GDP that at best this type of trade could have only a minor impact on wages and employment levels. Growth of trade with China and the DAEs is accelerating rapidly, but this is from a very small base.

2.1.3. New trade theory

Given that the majority of intra-EU trade is of the intra-industry type, a theory of intra-industry trade would seem to be the most appropriate representation to use in a modelling exercise.

Several different intra-industry trade theories have been developed, and most have at their core imperfectly competitive producers with scale economies. Reasons offered for imperfect competition include barriers to entry, differentiated access to technology, a consumer demand for differentiated products, and economies of scale. A reduction of barriers to trade leads to more trade because consumers have access to a greater variety of products, and costs fall in response to economies of scale.

Helpman (1981) explains intra-industry trade as resulting from the interaction of demand for variety with increasing returns to scale technology. The model is cast in a general equilibrium framework in a two-industry world: one produces a standard good, the other produces differentiated products. Factors of production in both industries are capital and labour. The standard good production function is characterized by constant returns, whereas the industry producing the differentiated products has increasing returns to scale. With increasing returns to scale, the number of products which can be viably produced will depend on the characteristics of individual demand functions, the technology behind production and the size of the market.

In this model, the possibility of trade gives firms access to a larger market so that more differentiated products can be viably produced. Hence, the new equilibrium will involve a greater variety of products, some of which will be produced domestically, the remainder produced abroad. Moreover, the expansion of the market allows each producer to produce higher volumes, thereby enabling prices to fall, resulting in an additional welfare gain for consumers, in addition to the increase in variety.

Helpman also shows how the share of intra-industry trade in total trade, and the volume of trade, is related to such factors as income per capita and relative country size. In particular, Helpman suggests that there is a relationship between the composition of total trade (whether inter- or intra-industry) and the dispersion of relative factor supplies between countries. Similarity of factor supply ratios is seen to stimulate intra-industry trade, while unequal factor endowment ratios stimulate inter-industry, or inter-sectoral, trade: the share of intra-industry trade in the total is a declining function of the absolute difference in the capital–labour ratios.

Thus the Helpman model allows inter- and intra-industry trade to coexist. If production of the differentiated products is relatively capital-intensive, the capital-abundant country can be shown to export more of these products than it imports, i.e. run a trade surplus in differentiated products. On the other hand, the country well endowed with labour will tend to run a deficit in differentiated products, which is financed by the surplus it runs on its trade in the standard product. The volume and similarity in composition of trade are predicted to be largest where countries are roughly equal in their economic fundamentals such as per capita income levels.

The scope for product differentiation is much greater in the case of manufactured goods, as the variety of internationally traded car models shows. It can be said that liberalizing trade creates and expands potential markets for a greater variety of products; and this would seem to provide a significant stimulus to certain manufacturing sectors' output and employment. The increase of technological knowledge is another benefit of specialization and increased product innovation. Problems might arise in cases where the removal of regional protection left domestic firms unable to compete with imports which possess greater variety and lower prices, particularly if there is not a symmetric removal of trade barriers. This may specifically affect smaller producers, in smaller economies, who cannot reach a minimum efficient scale of production.

While the theory is plausible, there does not seem to be a generally accepted standard model for empirical analysis as yet. Nevertheless, the process by which gains from trade are dispersed within the economy can be followed quite well, i.e. increased specialization leads to lower production costs and hence to lower prices. There is, however, no clear message emerging on how this type of trade affects the demand for different types of labour, as this depends on how production of different goods changes. Unlike HOS, there is no definite tendency to equalize factor prices and/or product prices. As the changes are occurring within the industry, the longer term effects on a sector are less pronounced as workers can relocate across similar firms whereas for inter-industry trade whole sectors could face terminal decline with derived demand falling accordingly.

Work undertaken by the OECD in *The Jobs Study* (1994) looked explicitly at the link between trade and employment within this type of framework. The methodologies used to look at the trade-employment linkage and the trade-wages linkage were very similar:

(a) A pooled time-series cross-section regression was performed on the periods 1973–79, 1979–85 and 1985–89.
(b) A sectoral analysis was possible using the STAN database.
(c) The dependent variable (employment or wages) was constructed relative to total manufacturing.
(d) Fixed effect dummy variables were used to adjust for industry-specific variables.

The specification of the equations was as follows:

Relative employment = f(change in relative import penetration, change in relative export intensity, change in per capita value added).

Relative wages per employee = f(change in relative import penetration, change in relative export intensity, non-trade-related variables).

The a priori reasoning behind the equation specifications and expected coefficient signs is as follows:

(a) Coefficient of import penetration in employment equation: negative.
(b) Coefficient of export intensity in employment equation: positive (direct effect on output effect) or negative (trade expansion leads to relocation of parts of the production process in different countries such that effect of decline in local production is greater than the export boost).
(c) Coefficient of import penetration in wage equation: positive (rising relative wages induce out-sourcing or bring in imports) or negative (firms hold down costs to compete on the international market).
(d) Coefficient of export intensity in wage equation: positive (employ more skilled workers to compete effectively) or negative (hold down costs to compete on international market).

This inability to predict from theory even the appropriate sign for the coefficients of the 'independent' variables highlights the problems in identifying unambiguous effects in empirical work, and makes it more likely that any measured effect is net (i.e. the sum of two opposing forces) rather than gross.

The *OECD Jobs Study* summarizes its findings thus:

The traditional HOS model predicts well what sorts of goods will be traded between countries with different resource endowments... An increase in intensity of trade with non-OECD countries will thus tend to depress the demand for unskilled labour in the tradeables sector of OECD countries. Various studies...find that there is indeed such an impact and that it is negative but also very small (at the macro level) relative to other influences, especially technical progress.' (OECD, 1994a, p.108).

New trade theories give a plausible explanation of trade in similar products between countries with similar resource endowments. Changes in this kind of trade could affect the demand for labour (skilled or unskilled) in individual firms in either direction in a given industry.

The main criticism to which the OECD's analysis is open is that the regressions were not part of a structural model, implying the following limitations:

(a)   no secondary (intermediate) linkage to demand for services; and
(b)   no endogenous effects from the independent variables are accounted for, i.e. changes in trade intensities are taken as exogenous, by virtue of the single-equation methodology.

Also, no distinction was made between skilled and unskilled labour due to lack of time-series data.

The majority of empirical studies analysing trade–employment linkages adopt a partial equilibrium approach, which is incomplete and therefore potentially misleading for two reasons. Firstly, partial equilibrium analysis assumes that resources drawn into the industry under study are available at prices equal to social opportunity cost. If one imperfectly competitive industry's expansion is at the expense of another's because of overall resource constraints, then this assumption is invalid, and the analysis may overestimate the resulting welfare gains. Secondly, partial equilibrium studies assume that input supply curves are horizontal so that resources are available to the industry at a constant price. If input supply curves to each industry are in fact upward sloping, partial equilibrium studies overestimate the quantity effects of the policy.

For these reasons Gasiorek, Smith and Venables (1992) adopt a general equilibrium approach to trade and welfare in which production uses intermediate goods and five primary factors of production (capital and labour disaggregated into four skill types). The model consists of eight countries (seven from the EU and one rest-of-the-world), 13 manufacturing industries and one financial services sector, all of which are assumed to be imperfectly competitive. The industries operate under increasing returns to scale and support an equilibrium with intra-industry trade. Gasiorek *et al.*. report experimental results for both a 'segmented' (i.e. price-taking) market, where the costs of trade are reduced by an amount equivalent to 2.5% of the value of trade, and an 'integrated' (i.e. price-making) market which has similar cost reductions alongside a switch to a segmented market equilibrium. The sectoral impact depends on a number of factors, including the relative share of intra-EU trade in the production of each industry, the extent of economies of scale in the industry, the degree of concentration in the industry, and the elasticity of demand for the individual product varieties. The effects on wages are also investigated.

A key finding is that the boost to welfare from reduced trade costs is very sensitive to assumptions as to which sectors see the greatest cost reduction. Two criticisms of the study are firstly that the rest of the world is modelled in a rudimentary fashion, so that results on external trade effects should be treated with caution. The second criticism, which applies to most existing CGE models, is that since the model parameters have not been derived by econometric estimation, there is no statistical basis for assessing their reliability. Hence, the model has not been validated empirically.

2.1.4. New developments linking geography and economics

The work summarized here represents a recent extension of new trade theory, focusing on how industry agglomeration and regional differentiation can arise as a consequence of transport costs, market sizes and trade policy regimes.

The theories were developed in response to a perceived gap in the current literature, namely the failure to explain the emergence of multinational corporations and foreign direct investment (FDI). New trade theory focuses on production and trade between firms resident in different countries; the theory has little to say about the ownership and control of firms.

The rapid growth of FDI during the 1980s has been dominated by two-way investment flows between developed countries (more generally between countries with similar factor endowments and per capita incomes) and in horizontal form, i.e. producing similar products to those manufactured at home.

Multinational corporations (the companies undertaking the FDI) tend to be found in industries with the following characteristics:

(a)   high R&D intensities,
(b)   a high share of professional and technical workers,
(c)   new or technically complex products,
(d)   high levels of product differentiation and advertising.

Dunning's OLI (Ownership, Location and Internationalization) framework provides a theory for the existence of multinationals. Related to this is Markusen's Convergence hypothesis (Markusen, 1995), which states that as countries converge in terms of relative factor endowments and per capita incomes then international activity starts to become dominated by multinational activity. These organizations actually displace trade, provided that transport costs are not small. This proposition contrasts with that of new trade theory, which suggests intra-industry trade will rise. Models incorporating geography and economics in this way predict that intra-industry trade will rise at first, but at the extreme level of integration will fall to zero as multinationals completely displace trade. Thus, a non-linear relation is posited between the volume of intra-industry trade and the degree of integration between economies.

A key assumption of this approach, which appears to make it less relevant for the current study, is that it relies on transport costs being prohibitively high so as to favour domestic production (by a multinational plant) over imports. Within the EU however, transport costs should be coming down as a result of the SMP, so a reverse effect (increased trade) may well occur. No evidence is available on this theory for the EU and only circumstantial evidence is available for EU–US trade, which is arguably a more plausible direction for the theory.

Indeed, the theory suggests that trade within an integrated economy, such as that within each of the Member States of the EU, or that within the US, would wither away. The evidence, however, is quite to the contrary: trade between regions or localities within integrated economies appears to continue to increase until particular localities export virtually all the goods they produce and import virtually all the goods they consume.

2.2. Trade and investment

Krugman and Lawrence (1994) conclude that technological change is the underlying cause of the fall in demand for unskilled labour in developed countries, and that international trade explains only a small part of the decline in the importance of the manufacturing economy. They suggest that a shift in the composition of domestic spending away from manufactured goods to services is the main reason for decline in the US industrial sector. They also note that many of the accompanying predictions of the HOS model have not been borne out, namely that the ratio of skilled to unskilled employment should decline in most industries, and that employment should increase more rapidly in skill-intensive industries than in those that employ more unskilled labour. Between 1979 and 1989 the ratio of real wage of skilled to that of unskilled workers rose, but most industries employed an increasing share of skilled workers. In summary, Krugman and Lawrence (1994, p. 27) state that,

'The share of manufacturing in GDP is declining because people are buying relatively fewer goods; manufacturing employment is falling because companies are replacing workers with machines and making more efficient use of those they retain. Wages have stagnated because the rate of productivity growth in the economy as a whole has slowed, and less skilled workers in particular are suffering because a high-technology economy has less and less demand for their services. Our trade with the rest of the world plays at best a small role in each case.'

Wood (1994) argues that adoption of labour-saving technologies is in part a response to fierce competition from low-wage countries. He also argues that there is much econometric evidence to demonstrate the complementarity between skill and capital, e.g. computers require skilled workers, and trade may have pushed up the skill mix through a shift in the composition of production towards more capital-intensive industries, and the adoption of defensive innovation strategies. Hence, the rise in skill intensity associated with the innovation is trade-induced, i.e. trade precedes innovation. It is therefore very difficult to isolate the autonomous effects of innovation.

Sachs and Shatz (1994, p. 33) and Lawrence and Slaughter (1993, Fig. 10) are quoted by Wood as providing some empirical support to these statements. Sachs and Shatz note that they might expect to find higher total factor productivity (TFP) in sectors most open to competition, the reasoning being that firms will let go of low productivity workers first, thus raising overall productivity. In their US study, using a sample over the period 1960–78, TFP is found to be lower (on average) in low-skill sectors, while in 1978–89 TFP is slightly higher in low-skill sectors (relative to high-skill). This latter period coincides with the period of rapid trade growth with developing countries.

Lawrence and Slaughter (1993) plot the percentage change in the TFP of certain US manufacturing sectors during the 1980s against the ratio of non-production to production labour employed. No link is established between TFP and the employment ratio, implying that Hicks-neutral technical progress did not raise the relative wage of non-production labour. However, when the variable constructed to represent technology is weighted by the shares of production and non-production labour, Lawrence and Slaughter find that technological change is concentrated in industries using non-production labour intensively. Hence, they argue that technological progress is concentrated in skilled labour-intensive industries, and this raises the wages of skilled labour and adds to the divergence relative to unskilled labour.

Certain criticisms have been raised about the issue of causality between trade and investment. Firstly, if producers can lower costs by the introduction of a certain technology, why wait for trade competition to do it? Secondly, no causality has been established empirically between labour-saving technical progress and degree of openness to trade, i.e. that highly open sectors innovate most rapidly and vice versa. In the OECD *Jobs Study* (1994a), agriculture is given as a prime example of a protected sector that has labour productivity rates rivalling those of unprotected manufacturing sectors.

Wood (1995) responds to the first criticism by arguing that the limited knowledge and information available to firms prevents them from exploiting all their production possibilities, given the current state of technology. In a market of costly information, R&D is necessary to investigate alternative production techniques. Further, the various search techniques available are generally subjected to a cost-benefit analysis before proceeding, and this analysis itself must be related to the market conditions which the firm finds itself in. Thus the prospect of low-wage competition (or new competition in general) which results in a dismantling of trade barriers or a change in policy that facilitates trade flows represents a change in market conditions that may influence search behaviour and hence innovative activity.

Given the lack of empirical evidence for causality in trade-investment linkages, econometric tests of this proposition are reported in Chapter 3.

2.3. Partial equilibrium studies

This section is split into three parts, each looking at papers which focus on different dimensions (typically geographical) for their analysis. Section 2.3.1 looks at studies focusing on a particular sector, e.g. textiles, or a particular EU Member State, e.g. the UK. Section 2.3.2 investigates sectoral studies at an area level – most of these have occurred using the OECD's STAN database. The final section relates to studies of US manufacturing, an area that is rich in the analysis of trade and employment linkage. While the US economy differs from that of the EU in many respects, not least of all its labour market structure, it is interesting to compare the kind of results obtained here with those which use EU data.

2.3.1. Specific sector or country studies

Driver, Kilpatrick and Nesbitt (1986)

This paper uses an input–output framework to consider the employment effects in the UK of a balanced trade expansion with the newly industrializing countries (NICs) and the EU. Both the overall net effects and the results by industry are presented. They are subdivided into the direct effects on an industry due to the trade change in that industry and the indirect effects arising from that part of the industry's production which is sold as an intermediate input to other industries. The exercise, carried out for 1979, finds small negative overall employment effects in each case examined, but important differences in the inter-industry pattern of employment changes are reported.

The methodology used can be explained by the following expression,

$$e = E(I - A)^{-1}t$$

where
e = the resulting employment change (by sector),
E = a diagonal matrix of sectoral output-employment ratios,
A = the matrix of input-output coefficients,
t = the vector of proposed changes in trade.

Such a methodology is restrictive in the sense that it ignores relative price movements or the effect of real income changes on the pattern of trade. The assumption of balanced trade is also unrealistic if one considers the pattern of UK manufacturing trade balances in recent years. Nevertheless, the study provides a useful benchmark analysis providing the limitations of the study are realized.

For a balanced trade expansion with NICs, mechanical engineering has the greatest expected employment gain, while textiles and clothing suffer the biggest job losses. Agriculture and the service sector also show negative employment results. In contrast, a similar experiment for trade with the EU showed the motor vehicles sector and food production to be the biggest losers of jobs, while the 'residual' sector, other manufacturing, had the largest individual gain. Employment in agriculture and services was again shown to be adversely affected by the balanced trade expansion.

Guillaume, Meulders and Plasman (1989)

This somewhat *ex ante* study looks at the potential sectoral impact of the SMP on manufacturing in Belgium. The repercussions on employment of the creation of the single market is generally summarized in terms of an overall macro-economic growth forecast of 1,800,000 jobs in the medium term. Concerning the types of jobs that will be created, the authors have no information on their occupational or sectoral distribution. The purpose of the paper is to provide an evaluation of the repercussions of the creation of the single market on Belgian employment. For this purpose a macrosectoral econometric model of the Belgian economy (DRY) was simulated. For the simulations, the same assumptions as those of the Cecchini report were used. However, despite using the same assumptions, quite different and less optimistic results are obtained.

Haskel and Jukes (1995)

The authors investigate changes in skilled and unskilled employment over the 1980s using UK data skills from the Labour Force Survey and the New Earnings Survey Panel Data set, matched with industrial data from the Census of Production. The major findings are that there was a rise in non-manual wage and employment shares over the 1980s but only a slight rise in that of the skilled, and that in industries where foreign competition intensified, the manual wage share fell, but there was no significant effect on the unskilled.

These results provide some conflicting outcomes. The UK is widely recognized as having one of the most flexible labour markets in the EU, and as such one would expect more wage adjustment (maintaining levels of employment where possible) in the face of increasing competition from abroad. This seems to be the case for manual workers, but not for the unskilled – both groups might be expected to see wages (in absolute and relative terms) fall. A similar dichotomy is true of the results for skilled and non-manual labour, which might be expected to benefit as firms specialize in high quality niche markets.

It is difficult to read too much into the results in the context of this study, however, as there is no distinction between the effects of foreign competition within or outside the EU.

Henderson and Sanford (1991)

In this study, an elasticity of textile employment with respect to textile imports is estimated for the nation, two regions, and four local areas. A theoretical labour demand equation is derived which endogenizes textile imports and facilitates direct estimation of import–employment elasticities. The results indicate the textile imports only partially substitute for domestic employment and that the estimated elasticity varies by region. The estimated elasticity was -0.27 for the nation and ranged from -0.22 to -0.31 at the regional level and from -0.43 to -0.19 for the local analysis. The elasticities are used in an impact analysis to demonstrate the over and underestimation of direct employment loss which occurs when estimates from spatial averages are used.

2.3.2. Area studies

OECD (1994c)

This part of the OECD *Jobs Study* examined the possible links between foreign trade, employment and relative wages by industrial sector and type of labour. Two sets of cross-section regression analysis for selected OECD areas were undertaken, the first relating changes in relative employment to trade intensity, the second relating changes in relative wages to trade intensity.

For example, it was generally believed that low-skill industries would be more vulnerable to changes in trade competition. However, the OECD found the most significant relationships between trade volumes and employment changes (as well as between trade and relative wages) to have been in the high-skill industries such as Chemicals, Electrical Machines and Aircraft.

A priori, it was also expected that increased import penetration in some industries or industry groupings would be related to a fall in employment in those industries relative to the average for all industries. The OECD study confirmed that there was a positive correlation between relative wages and import penetration in Europe in high-skill industries. This suggests that the adjustment to rising import penetration is not through wage changes, but rather via relative employment changes.

Firms in high-skill industries were believed to be holding down labour costs so they could compete on foreign markets, particularly in the case of Europe. Finally, the impact of trade with non-OECD countries was found to be statistically significant but small in magnitude. In conclusion, the impact of changing trade patterns on labour market conditions appeared to be significant but generally small relative to other factors. However, the OECD empirical analysis did not set out to measure the impact of changing trade patterns on aggregate employment or unemployment. This would depend on the number of workers displaced from a given sector due to foreign competition, and the ease with which workers can find jobs in expanding sectors. This in turn depends on the transferability of skills and flexibility of the labour market.

Oliveira Martins (1994)

As a precursor to the work undertaken for the OECD *Jobs Study*, this paper introduces the concept of market structure into the explanation of the relationship between international trade and industry relative wages. In order to identify the market structure prototypes, the paper employs a four-way classification where each industry is grouped according to its product differentiation (homogeneous and differentiated) and market concentration (fragmented and segmented) characteristics.

Martins considers empirical evidence for 12 OECD countries but the results from individual sectoral analysis do not support the hypothesis leading us to believe that the sign of the relation between relative wage rates and the trade variables is rather ambiguous. However, stronger results are obtained the results are considered at the level of the four industry sub-groupings; see Table 2.2 for an explanation of how the E3ME sectors fit into these categories. The estimated impact of import penetration on industry relative wages appears to be largely negative in industries with low product differentiation and/or low market power. This result may be expected on the basis of the traditional Stopler–Samuelson theorem which suggests that foreign competition from countries with relatively abundant low-skill labour forces may induce an absolute fall in the real wage of OECD unskilled workers. As a consequence, increasing import penetration in these industries may lead to income distribution conflicts among the owners of relatively scarce production factors (such as low-skilled labour) in the importing country.

On the other hand, in industries with high product differentiation and large scale economies there is no evidence that increasing openness to trade leads to reduced industry relative wages. In fact, industries can be open to competition in both domestic and foreign markets and still have above-average wages, a result which can be found in the trade theory models with imperfect competition. On the basis of this latter result, there is no reason to expect a potential income distribution conflict between the owners of production factors that are intensively used in the production of differentiated products. This is because the impact of trade on income distribution in the presence of strong product differentiation and scale economies may be very different from the one predicted by the Heckscher–Ohlin–Samuelson paradigm which predicts the equalization of relative factor prices and an absolute fall in the real wage rate of OECD unskilled workers due to foreign competition from countries with relatively abundant low-skill labour forces (the Stopler–Samuelson theorem). In trade models with imperfect competition, a reverse Stopler–Samuelson result can also occur when products are sufficiently differentiated. If this is the case, then the arguments related to income distribution may not be valid. As Martins points out, trade theory suggests exactly the opposite, namely that import penetration may be vital in order to benefit from increasing returns to scale or the welfare effects of greater product variety in all sectors.

Martin's results have two important policy implications. First, they lend support to the suggestion that the intensification of foreign competition may lower wages in those sectors where typically OECD countries have low or have lost market power, relative to the rest of manufacturing industry. As a result, in some instances policy measures may be necessary to ensure that the owners of relative scarce production factors (such as unskilled labour) receive their share of the overall benefits from increased openness to international trade. Second, from the analysis of trade patterns and Martin's econometric results it emerges that, in those industries where OECD countries have a strong market position, relative wages are not driven down by import penetration. Indeed, the results show that openness to trade may even increase relative wages in these sectors.

Saeger (1995)

Using panel data for OECD countries, Saeger finds that there is a statistically and economically significant link between imports from the South and the manufacturing share of employment and real value added, even after allowing for other possible causes of de-industrialization. Saeger suggests that increased trade with the South may explain up to one-third of de-industrialization in the OECD between 1970 and 1990. However, there are two reasons for not interpreting Saeger's results as evidence that North–South trade is the primary cause of the decline in the share of manufacturing employment. First, there is also a strong relationship between de-industrialization and differential productivity growth. A simple counterfactual suggests that the expansion of trade with the South can explain approximately one-third of the decline in the share of manufacturing employment in OECD countries.

Second, we should also consider the problem of simultaneity. As Saeger points out, given the endogeneity of trade flows, it was not possible to establish a causal link running from trade to de-industrialization. Saeger's regressions indicate that North–South trade and intra-OECD trade have fundamentally different effects on the patterns of production and employment in developed economies. Intra-OECD trade appears to have had only a minor impact on the structure of employment and production at the aggregate level.

The results of Saeger's pooled regressions indicate that across the OECD, countries with deficits in intra-OECD manufacturing trade tend to have lower shares of manufacturing employment. However, this does not imply that trade with other OECD countries leads to de-industrialization. For example, if a country such as Germany is heavily dependent on manufacturing production, then we would expect net imports of manufactured goods from both other OECD and non-OECD countries to be low compared to other countries. Therefore, Saeger's results do not lend support to the hypothesis that trade is linked to de-industrialization in the OECD. However, his paper does suggest that the expansion of North–South trade has played a role in determining recent changes in the structure of developed economies.

2.3.3. US studies

Armah (1992)

The paper is a reaction to the persistent trade deficits in the United States since 1971, and growing foreign competition, which have revived debate about the net effects of trade on the domestic economy. Focusing on the employment consequences of trade, this study examines the demographic and industrial characteristics of trade sensitive manufacturing industries in the United States. The findings reveal two significant trends. Firstly, there has been a decline in the importance of high-tech manufacturing as a source of trade-related employment opportunities since 1975. Secondly, while trade-enhanced industries still employ relatively fewer women and minorities than industries adversely affected by trade, the gap has narrowed, especially for women. Since 1975, the proportion of all women employed in adversely affected industries actually declined; for every 1% decline in female employment in an adversely affected manufacturing industry there was a corresponding 0.6% gain in employment in a trade-enhanced manufacturing industry.

Revenga (1992)

This paper investigates the effect of increased import competition on US manufacturing employment and wages using data for a panel of manufacturing industries over the 1977–87 period. The empirical analysis uses previously unavailable industry import price data and an instrumental variables estimation strategy. The estimates suggest that changes in import prices have a significant effect on both employment and wages. The dramatic appreciation of the dollar between 1980 and 1985 is estimated to have reduced wages by 2% and employment by 4.5–7.5% on average in this sample of trade-impacted industries.

The equations estimated are first-differenced and of a quasi-reduced form, i.e. they do not aim to explicitly describe a structural economic relationship. Three equations are estimated, for employment, wages and hours-worked with the same independent variables used in each case, i.e. unemployment, import price, alternative wage, materials price and sectoral dummies. The sectors identified by the data are grouped for the panel estimation according to size of import share, i.e. there are three groups: high, medium and low import share. Unfortunately, the paper does not say which sectors are allocated into which groups, so no industries can be identified with a particular set of results.

The range of elasticities estimated in the study are as follows:
Employment: 0.24 to 0.39.
Wages: 0.06 to 0.09.

Thus, it can be seen that the wage impact is quite small, with most of the effect coming through employment. It should be noted that the estimates are obtained through panel averages, not individual sectors, and so the results may hide large industry variations. There is also no way of measuring within-industry labour movements, a feature which is of more importance for EU industries, many of which are characterized by high levels of intra-industry trade.

Part of the study looks at the differences between OLS and 2SLS estimates. The conclusion reached is that a simultaneous relationship exists between import prices and employment, and import prices and wages. The estimates seem to be biased in a downwards direction, perhaps helping to explain why previous studies (that used OLS) reported only weak employment effects.

Sachs and Shatz (1994)

This paper emphasizes that US trade with developing countries has increased substantially over the past 15 years and that such trade is generally characterized by the patterns proposed by the Heckscher–Ohlin–Samuelson theory. In other words, the US exports skill-intensive goods and imports less skill-intensive goods. Moreover, increased trade with East Asia, Brazil, and Mexico since 1978 has bolstered these trends. As a result of increased specialization, employment has declined sharply in low-skill sectors and has increased in high-skill sectors. According to Sachs and Shatz's estimates, the increase in net imports between 1978 and 1990 in the US is associated with a decline of 7.2% in production jobs in manufacturing and a decline of 2.1% in non-production jobs in manufacturing. In addition, since production jobs, overall, tend to be less highly skilled than non-production jobs, such trends may have contributed to the widening of wage inequalities between skilled and unskilled workers.

Sachs and Shatz calculate their results by comparing actual employment change among a range of two-digit manufacturing industries with the employment change from a counterfactual scenario. The counterfactual is calculated by assuming a 1990 trade structure equal to that of 1978, with imports increased by factors for cif tariffs and tariff equivalents of quotas. In other words, the study is a classic example of factor content analysis, and as such is subject to the criticisms outlined in Section 2.1. Nevertheless, the results are worthy of mention, notably that the largest employment falls (explained by trade) are in the textiles and clothing sectors, and also in miscellaneous manufacturing. In most other cases the employment change is small or even in the opposite direction to that suggested by Sachs and Shatz's model.

Increased trade can only explain the trends if it not only shifted labour out of low-skill jobs but also reduced the relative prices of products made mainly by low-skilled labour such as clothing, footwear and textiles imported by the US from low-wage countries. If so, then according to the Stopler–Samuelson proposition, the wages of low-skilled labour would be driven down relative to the wages of high-skilled labour. Sachs and Shatz suggest that relative product prices have moved in the indicated direction, but not markedly so.

They also emphasize that increased internationalization cannot, on its own, explain most of the observed labour market trends. Technological change, for example, has a role which is independent of internationalization although we cannot measure technological change with any degree of precision. In addition, another reason for the large fall in US manufacturing employment could be the slow growth of non-manufacturing productivity in the 1980s. Manufacturing sector employment tends to contract when manufacturing productivity grows faster than service sector productivity. This, in turn, can be the result of rapid growth of manufacturing productivity, or a slowdown in the growth of service sector productivity.

Table 2.1 provides a brief summary of the partial equilibrium studies. The table outlines the methodology used and the key findings of the reports.

2.4.   Background sectoral analysis

This section provides a brief review of the characteristics of manufacturing sectors across the EU Member States, in order to identify what effects might be expected on an a priori basis, against which the model results reported below can be compared.

The study by Oliveira Martins (1994), mentioned in Section 2.3, makes use of a pooling methodology which groups industries into certain categories depending on their market and product characteristics. The similarity between the industrial classifications of the OECD STAN database, which Oliveira Martins used, and the E3ME model (which is based on the NACE 25 sector breakdown) makes this a potentially useful way in which to analyse E3ME's sectors so as to guide interpretation of the modelling results.

Table 2.1. Summary of partial equilibrium studies

Study	Area	Description of methodology	Employment impacts
Driver et al. (1986)	UK	Input–output analysis to examine the employment effects of a balanced trade expansion, carried out for 1979.	Mechanical engineering positive impact. Agriculture, textiles, clothing and service sectors: negative impact.
Guillaume et al. (1989)	BE	Ex ante study of SMP on manufacturing employment using the same assumptions as the Cecchini report.	Positive impact on employment, but the results were not as optimistic as the Cecchini report.
Haskel and Jukes (1995)	UK	Investigates changes in skilled and unskilled employment structure over the 1980s.	Positive impact for non-manual and skilled labour, decrease in wages where foreign competition intensified.
Henderson and Sanford (1991)	OECD	Estimated effect of increased textiles imports on employment. Textile imports are endogenized enabling direct estimation of import-employment elasticities.	Estimated textile employment elasticity with respect to textile imports of -0.2 to -0.5.
OECD (1994c)	OECD	Examines the possible links between foreign trade, employment and relative wages by industrial sector and type of labour for the EU.	Macroeconomic models predict employment gains of 2 to 5 million. Low skilled more vulnerable with high skilled less vulnerable.
Oliveira Martins (1994)	OECD	Investigates the relationship between trade and wages using market structure. Four classifications are used, fragmented, segmented, low-differentiation, high-differentiation.	Results ambiguous. The estimated impact of import penetration on industry relative wages for fragmented, low-differentiation sectors is -0.02 to -0.60.
Saeger (1995)	OECD	Using panel data for OECD countries investigating the impact of trade on the manufacturing employment share dividing trade between North and South.	Employment share: imports from North coefficient of -0.18 to -0.45. Employment share: exports to North coefficient of 0.03 to 0.19.
Armah (1992)	US	Examines the employment consequences of trade by investigating the demographic and industrial characteristics of trade sensitive manufacturing industries.	Decrease in high-tech trade-related employment.
Revenga (1992)	US	Examines the effect of increased manufacturing import competition on employment.	Between 1980 and 1985 the appreciation of the dollar is estimated to have reduced employment by 4.5 to 7.5%.
Sachs and Shatz (1994)	US	Investigates the impact of trade on employment distinguishing between developed/developing countries and low-skilled/high-skilled employment.	Positive impact on employment for high skilled, negative impact for low skilled. Estimated impact on employment of trade with developed countries: -0.2%.

According to Oliveira Martins, the net impact of international trade on wages (and the labour market in general) is identified as depending on three industrial characteristics:

(a)   extent of economies of scale,
(b)   elasticity of product substitution,
(c)   similarity of factor endowments between trading countries.

Oliveira Martins classifies industries by nature of market concentration and by degree of product differentiation. Under market concentration, industries are classified as either 'fragmented' or 'segmented', with the following characteristics:

Fragmented market structure:

(a)   industrial concentration moves inversely to output,
(b)   low set-up costs,
(c)   wide product variety,
(d)   highly competitive.

Segmented market structure:

(a)   industry concentration stable,
(b)   high set-up costs,
(c)   non-price competition.

Under product differentiation, industries are classified either as producing homogeneous products (HP) or differentiated products (DP). Homogeneous products are very substitutable, while the predominance of differentiated products includes both horizontal (taste for variety) and vertical (quality, R&D investment) kinds of differentiation.

The grouping according to the E3ME industrial disaggregation is shown in Table 2.2. Those sectors typified by segmented markets and differentiated products are most likely to have the characteristics required for intra-industry trade, while those sectors identified as having fragmented markets and homogeneous products are most likely to be dominated by inter-industry trade. Sectors located in the other two combinations (fragmented market/differentiated product and segmented market/homogeneous product) are likely to fall somewhere between the two extremes. The resulting predictions from the literature review of trade–employment linkages could thus be expected to hold true on the basis of this analysis. However, the study is fairly broad in its categorization of industries across the whole OECD area, and does not look at the details on a country-by-country basis.

Tables 2.5a–2.16b provide the dimension of the analysis missing in Oliveira Martins (1994) by looking, for each Member State in turn, at the sectoral characteristics surrounding them. Where data permit, descriptive statistics are constructed for all the E3ME regions and are relative to the equivalent figure for total manufacturing (except for the measure of inter-industry trade), i.e. a value of unity indicates a performance in line with the EU average. The tables cover the periods 1980 and 1990 in order to provide a snapshot of data before and after the implementation of the single market programme.

Table 2.2.   E3ME sectoral groupings

Market type	E3ME code	Industry description
HP/FM	09	Non-metallic Mineral Products
	17	Textiles, Clothing and Footwear
	21	Other Manufactures
HP/SM	08	Ferrous and Non-ferrous Metals
	11	Metal Products
	18	Paper and Printing Products
	19	Rubber and Plastic Products
DP/FM	12	Agricultural and Industrial Machinery
	13	Office Machines
	14	Electrical Goods
DP/SM	10	Chemicals
	15	Transport Equipment
	16	Food, Drink and Tobacco

The construction of the statistics is as follows:

Share of output (%) = Qi/Qm

where

Qi = gross output of sector i,

Qm = total manufacturing gross output.

Balassa measures of inter-industry trade = $(Xi - Mi)/(Xi + Mi)$

where

Xi = exports of industry i,

Mi = imports of industry i.

The measure is constructed using value data on both extra- and intra-EU trade. The construction means the statistic is bounded by -1 and +1, with a value of 0 implying equal two-way trade, i.e. intra-industry trade.

Import penetration ratio = Mi/DDi

where

DDi = domestic demand for industry i.

Export penetration ratio = Xi/Qi

Wage rate = $RLCi/YEi$

where

$RLCi$ = real labour costs for industry i,

YEi = industry employment for industry i.

Unit labour costs = $YLCi/VYVMi$

where

YLCi = nominal labour costs for industry *i*,

VYVMi = nominal value added for industry *i*.

Capital intensity ratio = *Ki/YVMi*

where

Ki = Investment in industry *i*,

YVMi = value added in industry *i*.

R&D expenditure ratio = *RDi/YVMi*

where

RDi = R&D expenditure in industry *i*.

In the Technical Annex to this study, Tables A1a–A13b provide another way of viewing the data, by displaying data for each industry on a European basis, thus allowing direct comparison of the same sector in each Member State. Tables 2.3 and 2.4 show the Grubel–Lloyd Index of intra-industry trade for intra-EU trade and extra-EU trade respectively. The index is calculated as a weighted (across sectors) average of the Balassa measure described above, and is bounded between 0 and 1. A value of zero implies inter-industry trade, while a value of unity indicates entirely intra-industry trade.

The analysis also draws on the views expressed in Buigues *et al.* (1990) to indicate the expected sectoral effects of the SMP by Member State.

Table 2.3.   Grubel-Lloyd index of intra-industry trade – EUR-12

EUR-12

	1980	1981	1982	1983	1984	1985	1986	1987	1988	1989	1990	1991	1992	1993
Belgium	0.893	0.914	0.913	0.922	0.930	0.928	0.919	0.937	0.848	0.845	0.845	0.847	0.857	0.857
Denmark	0.614	0.623	0.589	0.614	0.617	0.609	0.590	0.637	0.672	0.672	0.677	0.703	0.792	0.786
Germany (East)	0.000	0.000	0.000	0.000	0.000	0.000	0.000	0.000	0.000	0.000	0.000	0.000	0.000	0.000
Germany (West)	0.824	0.823	0.817	0.828	0.829	0.817	0.793	0.785	0.784	0.794	0.826	0.826	0.809	0.816
Greece	0.403	0.405	0.430	0.473	0.479	0.459	0.464	0.484	0.462	0.528	0.500	0.491	0.487	0.490
Spain	0.717	0.720	0.725	0.699	0.704	0.708	0.714	0.702	0.697	0.670	0.684	0.691	0.675	0.667
France	0.857	0.870	0.869	0.880	0.886	0.876	0.869	0.866	0.890	0.895	0.888	0.905	0.930	0.935
Ireland	0.649	0.658	0.703	0.719	0.694	0.673	0.654	0.640	0.647	0.634	0.645	0.674	0.672	0.679
Italy (total)	0.666	0.664	0.661	0.671	0.678	0.670	0.701	0.710	0.711	0.704	0.707	0.707	0.697	0.693
Italy (south)	0.000	0.000	0.000	0.000	0.000	0.000	0.000	0.000	0.000	0.000	0.000	0.000	0.000	0.000
Luxembourg	0.444	0.468	0.536	0.525	0.541	0.578	0.555	0.576	0.579	0.582	0.594	0.603	0.609	0.612
Netherlands	0.775	0.816	0.798	0.785	0.777	0.789	0.774	0.750	0.756	0.756	0.752	0.738	0.748	0.743
Portugal	0.429	0.424	0.447	0.533	0.573	0.566	0.548	0.544	0.531	0.561	0.589	0.600	0.606	0.627
United Kingdom	0.804	0.795	0.759	0.709	0.716	0.736	0.724	0.752	0.718	0.768	0.808	0.867	0.860	0.858

Table 2.4.   Grubel-Lloyd index of intra-industry trade – rest of the world

Rest of the world

	1980	1981	1982	1983	1984	1985	1986	1987	1988	1989	1990	1991	1992	1993
Belgium	0.759	0.767	0.803	0.780	0.796	0.804	0.797	0.824	0.819	0.828	0.846	0.842	0.843	0.844
Denmark	0.797	0.777	0.757	0.767	0.760	0.761	0.751	0.767	0.796	0.762	0.761	0.755	0.719	0.723
Germany (East)	0.000	0.000	0.000	0.000	0.000	0.000	0.000	0.000	0.000	0.000	0.000	0.000	0.000	0.000
Germany (West)	0.597	0.623	0.595	0.610	0.621	0.620	0.620	0.622	0.647	0.664	0.679	0.692	0.680	0.684
Greece	0.533	0.533	0.579	0.501	0.476	0.516	0.574	0.548	0.531	0.513	0.537	0.486	0.522	0.513
Spain	0.780	0.704	0.728	0.697	0.640	0.669	0.789	0.805	0.757	0.714	0.701	0.672	0.699	0.683
France	0.693	0.667	0.671	0.666	0.640	0.664	0.714	0.739	0.770	0.756	0.768	0.775	0.742	0.738
Ireland	0.545	0.551	0.563	0.589	0.594	0.632	0.626	0.608	0.595	0.580	0.593	0.647	0.658	0.646
Italy (total)	0.599	0.575	0.567	0.554	0.591	0.615	0.619	0.643	0.664	0.656	0.661	0.694	0.709	0.712
Italy (south)	0.000	0.000	0.000	0.000	0.000	0.000	0.000	0.000	0.000	0.000	0.000	0.000	0.000	0.000
Luxembourg	0.594	0.574	0.635	0.591	0.600	0.642	0.645	0.634	0.676	0.681	0.699	0.681	0.712	0.711
Netherlands	0.774	0.679	0.749	0.761	0.751	0.781	0.763	0.743	0.748	0.730	0.727	0.698	0.726	0.721
Portugal	0.671	0.588	0.634	0.561	0.663	0.649	0.625	0.551	0.525	0.597	0.588	0.562	0.599	0.616
United Kingdom	0.723	0.740	0.749	0.786	0.792	0.788	0.758	0.781	0.794	0.809	0.797	0.801	0.811	0.809

2.4.1. Belgium (see Tables 2.5a and 2.5b)

Belgium is characterized by a number of features which were expected to work to its advantage following the onset of the SMP. Firstly, the economy is very open and already had experience of the pressures of international competition. For example, Tables 2.5a and 2.5b show that between 1980 and 1990 most sectors had relative import-penetration and export-intensity ratios close to 1. There is also evidence of a shift in the character of trade between 1980 and 1990 in the same direction that might be expected from the SMP measures. For trade with the rest of the EU in 1980, inter-industry trade was dominant in most industries. Metal Products, Agricultural and Industrial Machinery, Office Machines and Other Manufactures were orientated towards one-way trade (in deficit) while Ferrous and Non-ferrous Metals displayed an EU surplus. In terms of the rest of the world, the picture was often reversed, especially for Ferrous and Metal Products. In absolute terms, the figures tended to be larger with a greater tendency towards one-way trade. By 1990, Belgian EU trade was much more intra-industry orientated, except for Office Machines, Food, Drink and Tobacco and Other Manufactures. Rest of the world trade also gravitated towards intra-industry trade with the figures generally closer to 0.

The second reason why Belgium was expected to benefit from lower trade barriers is its relatively small market. It was expected that Belgian firms would benefit from the abolition of non-tariff barriers by reaping economies of scale and increasing their returns on R&D expenditures.

Thirdly, Belgium is characterized by its central location in Europe, good infrastructure and a skilled workforce. It was therefore hoped that the country would be a favourable location for production units of major European multinationals.

Belgium's strong sectors fall into two categories: 'traditional' and 'world-wide-reputation' industries. In the former category, Buigues et al. (1990) included Motor Vehicles, Textiles and Rubber Goods. Diamonds, Beer and Chocolate are examples of products where Belgium enjoys a world-wide reputation. It was expected that these sectors would withstand the increased competition following the completion of the SMP. In the past, trade in these products was often impeded by standards (Food, Drink and Tobacco, Chemicals) or regulations (Motor Vehicles and Textiles). These sectors would be boosted by the abolition of such trade barriers. Furthermore, by the end of the 1980s a number of these industries were already adopting strategies in readiness for the single market. For example, in the Chocolate industry integration into international groups was taking place.

However, the impact of the SMP was not expected to have a positive effect on all industries. We may distinguish between two potentially vulnerable types of sector: 'protected' and 'already exposed to competition' industries. In the 'protected' category Buigues *et al.* (1990) included Shipbuilding and Electrical Equipment. Both had experience of low or non-existent penetration of Community imports because non-tariff barriers protected them from foreign competition. The opening up of public procurement was expected to have a major impact upon these sectors. Therefore, a fall in market share in these industries may have had a negative effect upon employment. Finally, Buigues *et al.* emphasized the small size of Belgian firms, putting them in a weak position to offset foreign competitors. Restructuring appeared necessary and takeovers of Belgian firms were also a possibility.

Clothing, Footwear and Household Electrical Appliances were identified as sectors already exposed to Community competition. Office Machines and Telecommunications belonged to sectors which had experience of international competition. In the latter category, the single market was expected to bolster the trend towards cooperation and mergers/takeovers between European producers.

The group of exposed sectors also included the more traditional industries where the removal of frontiers was expected to stimulate intra-Community competition and accelerate the specialization process. A change in the geographical distribution of these activities within the EU was a possibility. For example, an industry such as Footwear was limited in size because of specialization towards its domestic market. Buigues *et al.* (1990) regarded Footwear and Clothing as vulnerable to greater trade.

In summary, though, the impact of the SMP was expected to have a beneficial effect upon Belgian industry as the proportion of strong sectors appeared to outweigh the vulnerable sectors. By the late 1980s, most firms were optimistic about Belgium being successful in the integration process.

Table 2.5a. Descriptive sectoral statistics (relative to manufacturing total) – Belgium, 1980

E3ME sector	Output share	Balassa (EU)	Balassa (non-EU)	Import penetration	Export intensity	Wage rate	Unit labour costs	Capital intensity	R&D intensity
Ferrous & Non-ferrous Metals	17.15	0.22	-0.24	0.80	0.89	1.37	1.24	1.41	0.65
Non-metallic Mineral Products	4.77	-0.08	0.03	0.58	0.57	1.11	0.10	1.00	0.26
Chemicals	10.98	0.09	0.30	1.53	1.54	1.22	0.91	1.53	2.33
Metal Products	3.90	-0.23	0.35	0.80	0.72	0.77	1.11	0.72	0.12
Agricultural & Industrial Machinery	5.44	-0.17	0.25	1.62	1.67	0.99	1.11	0.81	0.40
Office Machines	0.89	-0.30	-0.49	1.29	1.14	1.65	1.12	0.24	4.12
Electrical Goods	4.23	-0.10	0.32	1.44	1.47	0.86	1.12	0.78	2.36
Transport Equipment	11.65	-0.02	-0.26	1.27	1.27	1.30	1.13	1.15	4.46
Food, Drink & Tobacco	20.45	0.03	0.23	0.42	0.46	0.95	0.74	0.89	0.11
Textiles, Clothing & Footwear	8.57	-0.01	0.02	1.01	1.01	0.70	1.05	0.58	0.08
Paper & Printing Products	4.42	-0.02	-0.49	0.76	0.66	0.99	1.06	1.18	0.04
Rubber & Plastic Products	2.22	-0.07	0.18	1.10	1.08	0.90	1.01	1.55	1.05
Other Manufactures	5.27	-0.20	-0.21	1.43	1.46	0.66	0.73	0.78	0.38

Source: E3ME Database.

2.4.2. Denmark (see Tables 2.6a and 2.6b)

Given the international character of its industries and relative openness of its economy prior to the SMP, Denmark appeared less vulnerable to any negative effects which may have resulted from the SMP. For example, Table 2.3 illustrates the increase in intra-industry trade with the rest of the EU between 1980 and 1993, and Tables 2.6a and 2.6b show the high levels of import penetration of nearly all Danish industries. Within industries already exposed to international competition, it was expected that the direct impact of market integration would be negligible. Over time, however, the competitive environment would change as industries restructured in response to the enlarged market. Naturally, this was expected to have implications for a small-country participant such as Denmark. Whether the effects would be negative depended upon the competitive position and characteristics of Danish industries.

Table 2.5b   **Descriptive sectoral statistics (relative to manufacturing total) – Belgium, 1990**

E3ME sector	Output share	Balassa (EU)	Balassa (non-EU)	Import penetration	Export intensity	Wage rate	Unit labour costs	Capital intensity	R&D intensity
Ferrous & Non-ferrous Metals	15.50	0.12	-0.18	0.76	0.78	1.47	0.84	0.90	0.70
Non-metallic Mineral Products	3.94	-0.17	0.00	0.73	0.67	0.98	0.95	0.91	0.20
Chemicals	11.05	-0.10	0.15	1.46	1.50	1.35	1.01	1.90	1.56
Metal Products	4.28	-0.15	0.24	0.70	0.62	0.97	1.20	0.87	0.13
Agricultural & Industrial Machinery	5.97	-0.19	0.06	1.12	1.10	1.18	1.21	0.65	0.67
Office Machines	0.95	-0.31	-0.49	1.68	2.48	1.28	1.34	0.30	2.75
Electrical Goods	4.70	-0.17	0.14	1.21	1.20	1.18	1.13	0.83	3.05
Transport Equipment	12.55	0.17	0.09	1.22	1.29	0.96	1.16	0.72	4.05
Food, Drink & Tobacco	19.07	0.21	-0.01	0.49	0.63	0.85	0.82	0.77	0.16
Textiles, Clothing & Footwear	8.54	0.03	-0.01	1.10	1.11	0.72	1.17	0.65	0.08
Paper & Printing Products	5.15	-0.01	-0.43	0.73	0.68	1.04	1.07	1.19	0.04
Rubber & Plastic Products	3.77	-0.02	-0.05	0.92	0.93	0.89	0.74	0.61	0.47
Other Manufactures	4.47	-0.31	-0.37	1.40	1.49	0.59	0.79	1.44	0.36

Source: E3ME Database.

For example, the following industries experienced an increase in intra-industry trade within the EU and with the rest of the world between 1980 and 1990: Agricultural and Industrial Machinery, Transport Equipment, Food, Drink and Tobacco, and Rubber and Plastic Products. Tables 2.6a and 2.6b indicate that Transport Equipment and Rubber and Plastic Products shared common trends over the same period including a fall in their openness to trade, decreasing wage rates and unit labour costs and R&D expenditure (all relative to the Danish average for manufacturing as a whole). Buigues *et al.* (1990) identified Shipbuilding and Locomotives and Tramway (Transport Equipment) as highly sensitive but strongly competitive industries. Two large industries, Agricultural and Industrial Machinery and Food, Drink and Tobacco, accounted for only a small share of EUR-12 total production (around 2% in 1990), suggesting that, to the extent that scale economies were important, the industries might either be vulnerable or benefit from the greater opportunities offered by trade.

Office Machines appeared to be in a weak competitive position, characterized by very strong import penetration, relatively high wage rates and unit labour costs and falling relative R&D expenditure. By contrast, the overall position of Textiles, Clothing and Footwear was unclear. On the one hand, Buigues *et al.* considered Clothing and Accessories as having weak competitiveness but Household Textiles were believed to hold a strong competitive position. Tables 2.6a and 2.6b show that, relative to the average for Danish manufacturing, the whole industry experienced rising wage rates and unit labour costs and falling R&D expenditure between 1980 and 1990. However, these industries experienced an increase in globalization over this period and a rise in capital investment expenditure. Therefore, it was difficult to reach any firm conclusion about the potential strength of this industry.

At the time of the late 1980s most industrial firms expected the completion of the single market to have positive effects upon the Danish economy. For example, all large companies were generally confident of receiving net gains from market integration, but were less optimistic about sales potential. The greatest obstacle facing Danish industry appeared to be the absolute small size of firms and limited access to primary resources compared to their competitors elsewhere in the EU.

Table 2.6a   Descriptive sectoral statistics (relative to manufacturing total) – Denmark, 1980

E3ME sector	Output share	Balassa (EU)	Balassa (non-EU)	Import penetration	Export intensity	Wage rate	Unit labour costs	Capital intensity	R&D intensity
Ferrous & Non-ferrous Metals	1.18	-0.51	-0.50	2.05	1.58	1.26	1.32	0.00	0.60
Non-metallic Mineral Products	4.78	-0.22	-0.10	0.60	0.52	1.10	0.97	3.08	0.59
Chemicals	7.51	-0.39	0.03	1.50	1.29	1.27	0.89	0.00	3.03
Metal Products	6.71	-0.23	0.05	1.09	1.06	0.96	1.02	0.00	0.23
Agricultural & Industrial Machinery	11.31	-0.19	0.30	1.39	1.46	1.02	1.01	0.00	0.85
Office Machines	1.38	-0.03	0.26	1.70	1.88	1.04	1.11	0.00	2.31
Electrical Goods	4.05	-0.47	0.13	1.36	1.08	0.94	1.13	0.00	2.51
Transport Equipment	4.24	-0.38	-0.20	1.69	1.36	1.11	1.30	0.00	0.84
Food, Drink & Tobacco	36.15	0.63	0.32	0.41	0.90	0.99	0.84	2.23	0.33
Textiles, Clothing & Footwear	5.33	-0.43	-0.06	1.41	1.15	0.69	0.94	1.04	0.14
Paper & Printing Products	9.27	-0.23	-0.58	0.57	0.26	1.28	1.11	1.18	0.06
Rubber & Plastic Products	2.17	-0.29	0.10	1.15	1.03	0.96	0.97	0.00	0.64
Other Manufactures	5.93	0.01	0.01	1.05	1.04	0.79	1.02	1.86	3.54

Source: E3ME Database.

Table 2.6b.   Descriptive sectoral statistics (relative to manufacturing total) – Denmark, 1990

E3ME sector	Output share	Balassa (EU)	Balassa (non-EU)	Import penetration	Export intensity	Wage rate	Unit labour costs	Capital intensity	R&D intensity
Ferrous & Non-ferrous Metals	0.93	-0.50	-0.59	1.72	1.43	1.37	0.76	0.00	2.07
Non-metallic Mineral Products	3.88	-0.18	0.21	0.54	0.49	1.01	0.80	1.99	0.40
Chemicals	8.56	-0.39	0.24	1.36	1.30	1.23	0.68	0.00	3.31
Metal Products	7.89	-0.14	0.10	0.88	0.80	0.90	0.93	0.00	0.21
Agricultural & Industrial Machinery	11.64	-0.16	0.24	1.03	1.03	0.93	1.01	0.00	1.05
Office Machines	1.96	0.16	0.32	2.59	2.45	1.00	1.06	0.00	1.01
Electrical Goods	5.04	-0.35	0.13	1.38	1.24	1.04	1.25	0.00	2.44
Transport Equipment	4.65	-0.37	-0.18	1.50	1.31	0.93	0.98	0.00	0.77
Food, Drink & Tobacco	32.22	0.53	0.31	0.52	0.90	1.12	0.92	1.66	0.32
Textiles, Clothing & Footwear	4.79	-0.31	-0.12	1.52	1.34	0.73	1.15	2.25	0.11
Paper & Printing Products	8.89	-0.28	-0.43	0.55	0.29	1.11	1.30	2.65	0.06
Rubber & Plastic Products	2.70	-0.13	0.09	1.04	1.02	0.84	0.86	0.00	0.33
Other Manufactures	6.79	0.23	0.18	1.09	1.30	0.77	1.13	2.09	2.60

Source: E3ME Database.

2.4.3.   West Germany (see Tables 2.7a and 2.7b)

German manufacturing industry appeared to have a competitive advantage in Mechanical Engineering and Motor Vehicles. Tables 2.7a and 2.7b show that intra-industry trade (within the EU and with the rest of the world) in Transport Equipment increased between 1980 and 1990, and the industry's R&D expenditure also rose relative to the German average for manufacturing. Transport Equipment increased its share of manufacturing output in Germany from 10% to 14% over the same period (see Tables 2.7a, 2.7b, A8a and A8b). Buigues *et al.* (1990) also emphasized the strong competitive position of all subsectors of Mechanical Engineering, such as Agricultural and Industrial Machinery. For such strong sectors, competition from, say, Japan was expected to pose a greater threat than that from Germany's European counterparts. Weak points of German industry were identified in the 'traditional' industries such as Footwear and Wool and the Cotton industry, as well as parts of Food and Beverages.

Buigues *et al.* (1990) categorized German industry into three main groups: industries with a strongly protected domestic market, those with a better competitive position on the EU market, and industries with a better third-country position. According to Buigues *et al.*, Pharmaceuticals, Shipbuilding and Tobacco belonged to the first category. Tables A3a and A3b confirm the low levels of import penetration of the Transport Equipment and Food, Drink and Tobacco industries. Therefore, the dismantling of German barriers to market entry was expected to create problems for these industries.

Office Machines and Electrical Goods were identified as having a better competitive position on the EU market. Their products tend to be strongly internationally traded and Tables 2.7a and 2.7b highlight that Office Machines had the highest levels of import penetration and export-intensity relative to the German average for manufacturing (between 1980 and 1990). The future performance of such an industry would depend upon EU trade policy towards the US, Japan and the Asian NICs.

Buigues *et al.* (1990) included Chemicals, Soft Drinks, Rubber Products and Shipbuilding in those industries with a better third-country position. It was not clear though whether such industries could increase their market shares in a more open EU market. Tables 2.7a and 2.7b reveal the apparent strength of the Chemicals industry. It is characterized by high levels of intra-industry EU trade, openness, capital investment expenditure and R&D expenditure. Furthermore, the industry retained its 11% share of German manufacturing output between 1980 and 1990.

Overall, it was expected that the SMP would not result in pronounced losses in employment. The expectation was that German employment in manufacturing could exploit its comparative advantage in high-skilled labour and benefit from economies of scale associated with the SMP.

Table 2.7a. **Descriptive sectoral statistics (relative to manufacturing total) – West Germany, 1980**

E3ME sector	Output share	Balassa (EU)	Balassa (non-EU)	Import penetration	Export intensity	Wage rate	Unit labour costs	Capital intensity	R&D intensity
Ferrous & Non-ferrous Metals	10.99	-0.01	-0.15	1.07	0.80	1.20	1.05	1.23	0.46
Non-metallic Mineral Products	4.33	0.01	-0.11	0.74	0.58	1.03	0.87	1.12	0.20
Chemicals	11.02	0.04	0.41	1.22	1.35	1.40	1.00	1.25	2.79
Metal Products	8.53	0.24	0.45	0.47	0.68	0.98	1.04	0.72	0.23
Agricultural & Industrial Machinery	10.45	0.37	0.65	1.18	1.98	1.20	1.18	0.69	1.11
Office Machines	2.42	0.14	-0.14	1.88	1.44	0.93	1.10	1.41	0.82
Electrical Goods	8.03	0.10	0.35	1.23	1.20	1.00	1.07	0.75	2.28
Transport Equipment	10.73	0.27	0.62	0.96	1.55	1.25	1.10	1.30	1.87
Food, Drink & Tobacco	14.17	-0.17	-0.12	0.61	0.34	0.71	0.59	0.84	0.09
Textiles, Clothing & Footwear	5.16	-0.35	-0.33	1.95	0.94	0.62	1.04	0.61	0.10
Paper & Printing Products	6.28	0.12	-0.34	0.64	0.37	0.96	0.99	1.77	0.06
Rubber & Plastic Products	3.35	-0.03	0.41	0.88	0.85	0.90	1.00	1.19	0.43
Other Manufactures	4.48	-0.17	-0.32	0.87	0.52	0.79	1.00	0.66	0.56

Source: E3ME Database.

Table 2.7b.　Descriptive sectoral statistics (relative to manufacturing total) – West Germany, 1990

E3ME sector	Output share	Balassa (EU)	Balassa (non-EU)	Import penetration	Export intensity	Wage rate	Unit labour costs	Capital intensity	R&D intensity
Ferrous & Non-ferrous Metals	8.48	-0.01	-0.14	1.08	0.85	1.24	0.98	0.75	0.31
Non-metallic Mineral Products	3.86	0.03	0.11	0.60	0.60	1.03	0.87	1.14	0.22
Chemicals	11.10	0.06	0.35	1.34	1.45	1.53	0.99	1.08	2.21
Metal Products	8.23	0.18	0.27	0.48	0.65	0.88	1.00	0.81	0.28
Agricultural & Industrial Machinery	10.18	0.41	0.54	0.89	1.48	1.04	1.19	0.78	0.81
Office Machines	2.93	0.08	-0.32	2.28	1.88	1.07	1.25	1.38	1.28
Electrical Goods	10.52	0.13	0.16	1.01	1.00	1.19	1.10	0.72	1.94
Transport Equipment	14.34	0.24	0.44	0.71	1.05	1.27	1.09	1.23	2.02
Food, Drink & Tobacco	11.20	-0.21	-0.10	0.73	0.44	0.59	0.56	0.90	0.06
Textiles, Clothing & Footwear	4.39	-0.24	-0.32	2.16	1.51	0.65	0.99	0.70	0.09
Paper & Printing Products	7.33	0.19	-0.30	0.58	0.48	0.85	0.93	1.77	0.03
Rubber & Plastic Products	4.10	0.07	0.24	0.92	0.96	0.90	0.99	1.13	0.30
Other Manufactures	3.29	0.10	-0.09	1.17	1.01	0.66	0.98	0.82	0.53

Source: E3ME Database.

2.4.4.　Greece (see Tables 2.8a and 2.8b)

By 1990, the Greek economy was characterized by a trade deficit in manufactured products in its intra-EU trade relations, which, in both absolute and relative terms, was the highest among the 12 EU partners. In addition, as Buigues *et al.* emphasize, Greek exports to the EU in 1990 consisted of many consumer goods, fewer intermediate goods and a small quantity of capital goods. At this time, the following sectors accounted for 70% of intra-EU exports: Agricultural Products, Food Products and Textiles/Clothing. In contrast, the largest share of Greece's intra-EU imports was accounted for by capital goods.

In 1990, Food, Drink and Tobacco and Textiles, Clothing and Footwear were the dominant industries in the Greek economy. However, Tables 2.8a and 2.8b show that Food, Drink and Tobacco's share of Greek total manufacturing output fell from 25% in 1980 to 14% in 1990, while the output share for Textiles, Clothing and Footwear rose from 10% in 1980 to 22% in 1990. Indeed, the latter is the only industry in Greece which had a trade surplus both within the EU and with the rest of the world. However, Buigues *et al.* identified this industry as being relatively more protected in Greece than on average within the EU as a whole, with substantial export subsidies.

By 1990 both industries' export intensity ratios exceeded their import penetration ratios (both of which had risen strongly over the 1980s), and this characteristic was only shared by Ferrous and Non-ferrous Metals and Non-metallic Mineral Products (see Tables 2.8a and 2.8b). Buigues *et al.* identified the Non-ferrous Metals industry and the Insulated Copper Wire sector as the only intermediate goods activities which were performing relatively well in 1990.

Prior to the implementation of the SMP, Greece's other manufacturing industries were not expected to benefit much, since most of them had experienced vulnerable trade performances and displayed a strong tendency towards inter-industry trade. For example, the Chemicals industry was one of the key sectors within Greek manufacturing in 1990. Buigues *et al.* noted that Basic Chemical Products, Chemical Products for industrial-agricultural use, Paints-Varnishes-Inks and Pharmaceutical Products were activities in this industry which could have been sensitive to the SMP. All these industries had increased their production during the 1980s but their trade performance did not show such a marked improvement. Greece had tried to stimulate production in the Pharmaceuticals industry through protective measures such as restrictions on, or the taxation of, finished imports. By 1990, these import restrictions had led foreign companies to produce locally (97% of total production was accounted for by foreign-owned companies). Buigues *et al.* believed that the elimination of such protective measures could have had an adverse effect on this industry if foreign companies decided to shift their production to other Member States since Greek plants were often smaller than plants elsewhere.

In conclusion, Greece had a strong position in labour-intensive sectors with many industries expecting to expand their plants in order to take advantage of the economies of scale associated with the SMP. However, Greece was a weak performer in R&D-intensive sectors and industries were often more protected than the EU average so firms expected to find it difficult to compete with the more technologically advanced countries of Europe. Thus, overall the country was not expected to gain from the implementation of the SMP.

Table 2.8a. **Descriptive sectoral statistics (relative to manufacturing total) – Greece, 1980**

E3ME sector	Output share	Balassa (EU)	Balassa (non-EU)	Import penetration	Export intensity	Wage rate	Unit labour costs	Capital intensity	R&D intensity
Ferrous & Non-ferrous Metals	10.72	-0.25	-0.15	0.82	1.02	0.90	0.77	NA	NA
Non-metallic Mineral Products	6.54	0.16	0.66	0.58	2.33	1.03	0.55	NA	NA
Chemicals	8.61	-0.72	0.08	1.19	0.96	1.07	0.77	NA	1.90
Metal Products	6.75	-0.81	0.33	0.64	0.58	0.88	1.09	NA	NA
Agricultural & Industrial Machinery	3.85	-0.99	-0.87	2.18	0.13	0.85	1.37	NA	NA
Office Machines	0.55	-0.93	-0.98	0.80	0.03	0.00	0.00	NA	NA
Electrical Goods	4.43	-0.85	-0.15	0.81	0.41	0.89	0.85	NA	NA
Transport Equipment	8.92	-0.99	-0.94	1.92	0.09	0.90	0.89	NA	2.22
Food, Drink & Tobacco	25.79	-0.23	0.07	0.41	0.64	1.09	1.16	NA	0.04
Textiles, Clothing & Footwear	10.63	0.32	0.24	0.97	2.81	1.10	1.13	NA	NA
Paper & Printing Products	5.16	-0.87	-0.51	0.68	0.31	1.07	0.95	NA	0.01
Rubber & Plastic Products	3.14	-0.78	-0.05	1.50	0.82	0.98	0.82	NA	0.53
Other Manufactures	4.84	-0.42	0.19	1.12	1.57	0.98	2.47	NA	NA

Source: E3ME Database.

Table 2.8b.   Descriptive sectoral statistics (relative to manufacturing total) – Greece, 1990

E3ME sector	Output share	Balassa (EU)	Balassa (non-EU)	Import penetration	Export intensity	Wage rate	Unit labour costs	Capital intensity	R&D intensity
Ferrous & Non-ferrous Metals	8.24	-0.17	-0.46	0.97	1.38	1.02	NA	NA	NA
Non-metallic Mineral Products	5.79	-0.04	0.24	0.63	1.57	0.94	NA	NA	NA
Chemicals	10.81	-0.85	-0.38	1.02	0.46	1.29	0.28	NA	1.15
Metal Products	6.73	-0.86	0.03	0.67	0.44	0.92	NA	NA	NA
Agricultural & Industrial Machinery	3.70	-0.93	-0.70	1.62	0.41	0.85	NA	NA	NA
Office Machines	1.40	-0.96	-0.95	0.80	0.04	0.83	NA	NA	NA
Electrical Goods	5.43	-0.74	-0.50	0.85	0.42	1.04	NA	NA	NA
Transport Equipment	14.22	-0.98	-0.92	1.05	0.06	1.09	0.27	NA	2.98
Food, Drink & Tobacco	22.87	-0.37	0.06	0.66	0.88	0.94	0.46	NA	0.03
Textiles, Clothing & Footwear	8.18	0.03	0.27	1.88	4.53	0.98	NA	NA	NA
Paper & Printing Products	4.78	-0.86	-0.71	0.68	0.21	0.89	0.43	NA	0.00
Rubber & Plastic Products	3.30	-0.80	-0.50	1.25	0.59	0.90	0.39	NA	0.22
Other Manufactures	4.48	-0.57	-0.48	1.50	1.36	0.99	NA	NA	NA

Source: E3ME Database.

2.4.5. Spain (see Tables 2.9a and 2.9b)

There are major difficulties involved in separating the impact of the SMP upon the Spanish economy from the anticipatory and transition effects induced by entry to the EU along with Portugal in 1986. For example, Buigues *et al.* noted that in the late 1980s one notable feature of the Spanish economy was the sharp increase in manufacturing imports. Spain's manufacturing exports to the rest of the EU market had fallen, and drastically so, to other third countries. The manufacturing trade balance had also deteriorated, reflecting the high degree of protectionism prior to EU entry. Since 1980, a sharp fall in the degree of both intra- and extra-EU trade had also been experienced (see Tables 2.3 and 2.4). Intra-industry trade (intra- and extra-EU) had increased in Textiles, Clothing and Footwear and Rubber and Plastic Products. On the other hand, inter-industry trade (intra- and extra-EU) had risen in Chemicals, Agricultural and Industrial Machinery, Electrical Goods, Transport Equipment and Food, Drink and Tobacco.

Tables 2.9a and 2.9b indicate that globalization had occurred in the majority of the industries under consideration (relative to the Spanish manufacturing average). In only four out of the fourteen industries have import penetration and export-intensity not increased: Ferrous and Non-ferrous Metals, Textiles, Clothing and Footwear, Chemicals and Other Manufactures. The potential impact of the SMP on the Spanish economy depends, of course, on the competitive position of its firms. The Transport Equipment industry appeared to have the greatest potential to benefit from the programme for two main reasons. Firstly, it is characterized by very high R&D expenditure relative to the Spanish average for manufacturing. Secondly, it is strongly orientated towards intra-industry trade within the EU. By 1990, Transport Equipment accounted for 22% of manufacturing output in Spain (8% of the EU total). Electrical Goods and Office Machines had high R&D expenditure ratios, with the latter industry experiencing dramatic increases in investment expenditure since 1980. However, both industries remained characterized by high levels of inter-industry trade within the EU. They were, therefore, expected to be vulnerable to the competitive pressures which resulted from the removal of trade barriers.

Buigues *et al.* (1990) noted Textiles, Clothing and Footwear and Food, Drink and Tobacco industries as likely to increase their share in the EU market provided investment efforts were maintained for enhancing qualitatively both products and process. Large increases in R&D expenditure were also expected. Finally, an improvement in competitiveness was envisaged in the Chemicals industry which is characterized by high levels of foreign investment, and efforts had been made to modernize its productive structure. Despite an orientation towards intra-industry EU trade, below average performance was expected from the Non-Metallic Mineral Products and Paper and Printing Products industries.

One worrying feature of the Spanish economy was its poor technological capacity relative to the EU average. In the late 1980s, the poor trade performance of traditional sectors facing strong competition suggested that technological improvement of firms was required throughout industry. In a traditional sector such as Textiles, Clothing and Footwear, applying new technologies, specializing in upmarket product ranges and promoting brand policies appeared to be the best chance of offsetting the competitive threat from both the south-east Asian countries and other EU partners. In other sectors with an average and high technological content, developing new technologies was even more crucial, considering the importance of product and process innovation in determining international competitiveness.

The high degree of protectionism prior to the implementation of the SMP through both tariff and non-tariff barriers meant many sectors were vulnerable to the increased competition. Although the Spanish economy had a poor technological capacity relative to the EU average, the implementation of the SMP was expected to encourage investment in R&D, and improve the competitiveness of many industries.

Table 2.9a. **Descriptive sectoral statistics (relative to manufacturing total) – Spain, 1980**

E3ME sector	Output share	Balassa (EU)	Balassa (non-EU)	Import penetration	Export intensity	Wage rate	Unit labour costs	Capital intensity	R&D intensity
Ferrous & Non-ferrous Metals	10.72	-0.12	0.26	1.88	2.19	1.27	0.99	4.27	0.66
Non-metallic Mineral Products	6.54	-0.10	-0.01	0.62	0.48	1.11	0.98	0.24	0.10
Chemicals	8.61	-0.43	-0.05	2.08	1.14	1.48	0.81	1.42	2.95
Metal Products	6.75	-0.03	0.62	0.30	0.79	0.86	1.11	0.54	0.14
Agricultural & Industrial Machinery	3.85	-0.34	0.10	3.00	1.66	0.97	1.20	1.01	0.29
Office Machines	0.55	-0.62	-0.70	11.06	3.71	1.02	1.20	1.62	4.11
Electrical Goods	4.43	-0.56	-0.04	1.51	0.76	1.08	1.07	1.04	2.55
Transport Equipment	8.92	0.24	0.23	0.83	1.38	1.20	1.20	1.24	4.30
Food, Drink & Tobacco	25.79	0.02	0.18	0.25	0.39	0.94	0.72	0.35	0.08
Textiles, Clothing & Footwear	10.63	0.40	0.40	0.22	1.36	0.78	1.06	0.07	0.01
Paper & Printing Products	5.16	-0.03	0.21	0.56	0.67	1.12	1.04	0.61	0.03
Rubber & Plastic Products	3.14	0.17	0.75	0.34	0.61	1.07	1.05	1.18	0.94
Other Manufactures	4.84	0.17	0.13	0.72	0.74	0.66	1.09	2.47	0.11

Source: E3ME Database.

Table 2.9b.   Descriptive sectoral statistics (relative to manufacturing total) – Spain, 1990

E3ME sector	Output share	Balassa (EU)	Balassa (non-EU)	Import penetration	Export intensity	Wage rate	Unit labour costs	Capital intensity	R&D intensity
Ferrous & Non-ferrous Metals	8.24	-0.29	-0.07	1.11	1.11	1.76	0.92	0.01	0.44
Non-metallic Mineral Products	5.79	0.07	0.10	0.49	0.76	0.97	0.86	0.07	0.06
Chemicals	10.81	-0.48	-0.16	1.14	1.07	1.81	0.86	0.04	1.61
Metal Products	6.73	-0.31	0.27	0.67	0.79	0.88	1.22	0.02	0.15
Agricultural & Industrial Machinery	3.70	-0.51	-0.22	2.12	2.14	0.88	1.06	0.03	0.60
Office Machines	1.40	-0.56	-0.86	2.78	3.57	0.97	0.83	12.24	2.88
Electrical Goods	5.43	-0.58	-0.47	1.66	1.06	1.20	1.19	4.34	2.17
Transport Equipment	14.22	-0.06	-0.26	1.04	1.58	1.51	1.35	0.60	4.52
Food, Drink & Tobacco	22.87	-0.26	-0.27	0.48	0.47	0.83	0.66	1.51	0.07
Textiles, Clothing & Footwear	8.18	-0.27	-0.26	1.05	0.98	0.65	1.09	0.31	0.01
Paper & Printing Products	4.78	-0.18	-0.40	0.75	0.66	1.07	1.13	0.29	0.00
Rubber & Plastic Products	3.30	-0.09	0.44	0.53	1.05	1.07	0.97	0.33	0.45
Other Manufactures	4.48	-0.17	-0.31	0.45	0.54	0.61	1.25	1.40	0.23

Source : E3ME Database.

2.4.6.   France (see Tables 2.10a and 2.10b)

Between 1980 and 1987, France's manufacturing industry experienced a deterioration in its competitive position, notably in the Motor Vehicles industry, a key sector in the French economy and a major employer. Buigues *et al.* (1990) identified Motor Vehicles as being more sensitive to the SMP in the case of France than for the EU as a whole. Of course, the single market could have had benefits for potentially vulnerable sectors if these responded by renewing industrial structures. The outlook for the Motor Vehicles industry, therefore, was uncertain because its restructuring efforts (such as updating its range of models) were still incomplete at the end of the 1980s. By contrast, Shipbuilding was expected to be relatively unaffected by the SMP primarily because the sector was highly internationalized (see the import-penetration and export-intensity ratios in Tables 2.10a and 2.10b). The industry was also already operating in accordance with rules discussed at Community level. Another strong Transport Equipment sector was Aerospace Equipment. This highly protected industry was characterized by substantial investment in industrial plant and the role of public procurement had been relatively important. Overall, the Transport Equipment industry experienced an increase in its total share of manufacturing output from 12% to 13% between 1980 and 1990, but on the European market its share fell from 23% to 21%.

Over the same period, Food, Drink and Tobacco maintained the highest share of manufacturing output in the French economy (18%) and in European terms its total share increased slightly from 18% to 19% (see Tables 2.10a, 2.10b, A9a and A9b). Buigues *et al.* (1990) noted that this was a traditional French industry which enjoyed a special position due either to its high brand image (champagne, other wines) or particular know-how (bottled waters and soft drinks). Tables 2.10a and 2.10b show that the sector was also characterized by high intra-industry trade (both within the EU and with the rest of the world). Export intensity and R&D expenditure had increased between 1980 and 1990 but wage rates and unit labour costs also experienced an increase over the same period.

Finally, the Chemicals industry held a strong position in terms of its share in French manufacturing output (a rise from 9% to 10% between 1980 and 1990). On the European market, it retained its share of 17% (see Tables 2.10a, 2.10b, A3a and A3b) but its openness to trade and R&D expenditure both fell relative to the average for French manufacturing. More positively, unit labour costs had decreased and capital investment expenditure had risen (again relative to the French manufacturing average). Buigues *et al.* (1990) pointed out, though, that Pharmaceuticals appeared to be a very sensitive industry. It had experienced difficulties in France due to the insufficient level and results of its R&D efforts (compared with those of Germany, Switzerland or Japan). Furthermore, the industry was also hindered by the size and international linkages of its laboratories. Therefore, if the Pharmaceuticals industry was to be strengthened, it was suggested that a radical response was required. A willingness to adapt, however, was already emerging at the end of the 1980s. In summary, these traditionally strong sectors of France were expected to benefit from the completion of the single market, although in some cases the importance of public procurement represented a threat.

Industries which appeared to be in a weak position included Footwear, Clothing and Electrical Goods. Tables 2.10a and 2.10b show that for these consumer goods, the levels of intra-EU trade were already relatively high, as were import penetration ratios. Indeed, foreign competition was strong and the single market was expected to boost French imports from other EU countries.

The openness of the French economy meant the implementation of the SMP was not expected to have strongly negative effects. France already had a high level of imports in labour-intensive sectors. The economy has a comparative advantage in capital and R&D intensive sectors, thus these sectors were expected to gain slightly.

Table 2.10a. Descriptive sectoral statistics (relative to manufacturing total) – France, 1980

E3ME sector	Output share	Balassa (EU)	Balassa (non-EU)	Import penetration	Export intensity	Wage rate	Unit labour costs	Capital intensity	R&D intensity
Ferrous & Non-ferrous Metals	8.38	-0.12	-0.04	1.58	1.32	1.25	1.04	1.62	0.46
Non-metallic Mineral Products	4.34	-0.07	-0.05	0.62	0.53	1.00	0.86	1.37	0.27
Chemicals	9.54	-0.01	0.26	1.32	1.41	1.34	0.97	1.34	2.19
Metal Products	7.61	-0.18	0.54	0.37	0.44	0.94	0.99	0.83	0.13
Agricultural & Industrial Machinery	7.72	-0.24	0.46	1.37	1.39	1.11	1.09	0.73	0.42
Office Machines	2.58	-0.10	-0.16	2.01	1.69	1.61	1.00	0.74	1.50
Electrical Goods	7.14	-0.23	0.42	1.08	1.07	0.98	1.03	0.84	2.94
Transport Equipment	12.38	0.20	0.56	1.29	1.78	1.12	1.24	1.32	2.87
Food, Drink & Tobacco	18.29	-0.03	0.25	0.51	0.51	0.91	0.74	1.02	0.11
Textiles, Clothing & Footwear	7.77	-0.10	0.03	1.11	0.93	0.72	1.04	0.55	0.07
Paper & Printing Products	6.38	-0.17	-0.29	0.65	0.39	1.12	1.02	0.82	0.04
Rubber & Plastic Products	3.24	-0.05	0.49	0.92	0.96	0.90	1.04	0.98	1.04
Other Manufactures	4.56	-0.33	-0.11	0.93	0.57	0.68	0.92	0.84	0.39

Source: E3ME Database.

Table 2.10b. Descriptive sectoral statistics (relative to manufacturing total) – France, 1990

E3ME sector	Output share	Balassa (EU)	Balassa (non-EU)	Import penetration	Export intensity	Wage rate	Unit labour costs	Capital intensity	R&D intensity
Ferrous & Non-ferrous Metals	6.57	-0.06	-0.08	1.36	1.31	1.37	0.81	1.25	0.42
Non-metallic Mineral Products	3.81	-0.12	0.12	0.56	0.55	1.01	0.91	1.13	0.21
Chemicals	10.73	-0.04	0.21	1.28	1.38	1.40	0.89	1.17	2.08
Metal Products	7.14	-0.21	0.27	0.40	0.37	0.92	1.08	0.99	0.11
Agricultural & Industrial Machinery	7.16	-0.22	0.17	1.22	1.14	0.99	1.08	0.64	0.57
Office Machines	3.46	-0.12	-0.31	1.85	1.55	1.39	0.91	0.80	0.91
Electrical Goods	8.36	-0.18	0.20	1.16	1.10	1.07	1.06	0.71	2.53
Transport Equipment	13.27	0.05	0.38	1.26	1.60	0.95	1.06	1.27	2.89
Food, Drink & Tobacco	18.25	0.03	0.18	0.51	0.59	0.95	0.84	0.90	0.14
Textiles, Clothing & Footwear	6.11	-0.18	-0.15	1.37	1.18	0.75	1.15	0.67	0.06
Paper & Printing Products	7.28	-0.17	-0.32	0.60	0.43	1.04	1.05	1.40	0.04
Rubber & Plastic Products	3.74	-0.10	0.15	0.91	0.87	0.93	1.10	1.07	0.69
Other Manufactures	4.07	-0.17	-0.19	0.98	0.75	0.66	1.01	1.01	0.28

Source: E3ME Database.

2.4.7. Ireland (see Tables 2.11a and 2.11b)

The sectoral impact of the SMP in Ireland was expected to reflect the country's 'dualistic' industrial structure. Foreign-owned, export-orientated companies were regarded as internationally competitive, whereas indigenous companies were usually less export-intensive and had little experience of international competition.

Food, Drink and Tobacco increased its large share of manufacturing output in Ireland over the 1980s. Between 1980 and 1990 its total share rose from 34% to 40% (see Tables 2.11a, 2.11b, A9a and A9b), and the sector is generally characterized by low levels of import penetration and export-intensity. For example, Ireland's international trade in beer and soft drinks was relatively limited and tended to be confined to trade with Northern Ireland and Great Britain, rather than other countries within the EU. Therefore, it seemed unlikely that the abolition of restrictions would boost exports by these sectors. By contrast, Buigues *et al.* (1990) believed that the Dairy Products industry would be affected by the SMP because international trade in this sector had previously been restricted by non-tariff barriers.

Chemicals holds one of the largest shares of manufacturing output in Ireland, although it fell from 12% to 10% between 1980 and 1990 (see Tables 2.11a, 2.11b, A3a and A3b). The same tables show that the industry is also characterized by high levels of intra-industry EU trade and high import-penetration and export-intensity ratios, relative to the Irish manufacturing average.In addition, wage rates and unit labour costs fell and R&D expenditure rose between 1980 and 1990 (relative to Irish manufacturing as a whole). Buigues *et al.* (1990) believed that the Pharmaceuticals industry would see indirect benefits from freer trade, since much of this industry in Ireland specialized in the production of intermediate compounds or ingredients used as inputs for finished drugs produced elsewhere.

Electrical Equipment increased its share of manufacturing output from 4% to 10% between 1980 and 1990 (see Tables 2.11a, 2.11b, A7a and A7b). Again, it has a high level of intra-industry trade within the EU and export-intensity slightly exceeds import penetration. In fact, the industry exported a large share of its output, while Ireland imported the larger electrical goods such as washing machines and refrigerators which were not produced domestically. The industry was regarded as relatively competitive and successful in its own specialized export markets.

On the other hand, the Textiles, Clothing and Footwear industry appeared to be in a vulnerable position despite its orientation towards intra-industry trade within the EU. Tables 2.11a and 2.11b show that import penetration exceeded export-intensity, which is a characterization as an indigenous Irish industry. According to Buigues et al. (1990), the SMP was expected to have an impact upon the Clothing industry following the abolition of border controls and the opening-up of public procurement. Moreover, increased import competition from developing countries was likely once clothing imports into any part of the EU were permitted to be sold freely in Ireland. The Irish Footwear industry appeared to be in a weak position. Again, this industry showed high import penetration relative to export-intensity.

Office Machines increased its share of manufacturing output from 9% to 14% between 1980 and 1990, thus making it an important sector of the Irish economy (see Tables 2.10a, 2.10b, A6a and A6b). However, as can be seen from Tables 2.11a and 2.11b import penetration greatly exceeds export intensity. The same tables show that R&D expenditure had fallen dramatically between 1980 and 1990 (relative to the Irish average for manufacturing). As Buigues et al. (1990) pointed out, many types of machinery are simply not produced in Ireland and hence import penetration tends to be high (for example in Agricultural and Industrial Machinery). In other words, this may not necessarily reflect the weak competitive position of existing firms, but simply that in some subsectors import penetration is 100%.

Ireland faced a comparative disadvantage in industries characterized by scale economies. The country has a weak position in the Textiles and Footwear industries but the economy was in a strong position in high-technology industries due to the presence of multinationals. On balance, the main concern for the Irish economy was foreign-owned multinational companies that might take the opportunity to expand their production elsewhere. Ireland has a number of locational disadvantages of course, but on the whole they were not expected to diminish any positive impact from the SMP.

Table 2.11a.   Descriptive sectoral statistics (relative to manufacturing total) – Ireland, 1980

E3ME sector	Output share	Balassa (EU)	Balassa (non-EU)	Import penetration	Export intensity	Wage rate	Unit labour costs	Capital intensity	R&D intensity
Ferrous & Non-ferrous Metals	1.39	-0.39	-0.17	2.00	2.24	1.03	1.00	0.50	0.18
Non-metallic Mineral Products	9.51	0.00	0.41	0.32	0.46	1.30	1.06	0.70	0.19
Chemicals	12.11	-0.14	0.37	1.02	1.11	1.49	0.67	1.69	2.04
Metal Products	2.77	-0.31	-0.13	1.30	1.11	0.94	1.87	0.85	0.17
Agricultural & Industrial Machinery	2.45	-0.47	-0.39	1.76	1.68	0.86	1.44	1.29	0.61
Office Machines	9.42	0.51	-0.51	1.21	1.55	0.86	0.48	1.20	2.87
Electrical Goods	4.42	-0.22	-0.14	1.47	1.34	0.79	1.02	2.67	4.72
Transport Equipment	4.55	-0.46	-0.67	1.38	1.01	1.22	1.34	0.21	3.20
Food, Drink & Tobacco	34.33	0.48	0.55	0.38	0.88	1.07	0.90	1.03	0.20
Textiles, Clothing & Footwear	7.37	-0.13	-0.20	1.26	1.18	0.61	1.29	0.37	0.09
Paper & Printing Products	5.03	-0.45	-0.68	0.73	0.44	1.20	1.16	0.41	0.05
Rubber & Plastic Products	3.23	-0.23	-0.02	1.32	1.25	1.17	1.10	0.91	0.21
Other Manufactures	3.35	-0.70	-0.78	1.12	0.34	0.79	1.12	0.84	0.71

Source: E3ME Database.

Table 2.11b.   Descriptive sectoral statistics (relative to manufacturing total) – Ireland, 1990

E3ME sector	Output share	Balassa (EU)	Balassa (non-EU)	Import penetration	Export intensity	Wage rate	Unit labour costs	Capital intensity	R&D intensity
Ferrous & Non-ferrous Metals	1.70	-0.32	-0.53	1.27	0.99	1.51	1.25	1.25	0.30
Non-metallic Mineral Products	5.45	0.11	0.29	0.35	0.49	1.30	1.76	0.76	0.14
Chemicals	10.33	0.19	0.42	2.16	1.72	1.21	0.62	1.38	2.07
Metal Products	2.75	-0.13	0.16	1.02	0.93	0.70	1.70	1.28	0.14
Agricultural & Industrial Machinery	2.08	-0.22	-0.23	1.81	1.86	0.60	1.40	0.81	0.35
Office Machines	14.49	0.61	-0.21	2.64	1.76	0.96	0.34	0.24	0.90
Electrical Goods	10.17	0.17	-0.10	1.04	1.09	1.10	0.79	0.49	2.50
Transport Equipment	1.36	-0.52	-0.69	1.62	1.83	1.50	3.06	3.09	2.60
Food, Drink & Tobacco	40.81	0.43	0.66	0.27	0.49	1.23	1.15	1.58	0.27
Textiles, Clothing & Footwear	3.74	-0.18	-0.39	1.59	1.55	0.55	1.56	0.79	0.04
Paper & Printing Products	2.63	-0.45	-0.78	0.92	0.49	0.87	2.37	1.29	0.05
Rubber & Plastic Products	1.19	-0.21	-0.14	2.32	2.80	1.02	2.41	4.01	0.14
Other Manufactures	3.24	-0.55	-0.70	0.93	0.44	0.36	0.88	1.43	0.50

Source: E3ME Database.

2.4.8. Italy (see Tables 2.12a and 2.12b)

Italy is characterized by a high level of integration with other European countries and is noted for the importance of its Textiles, Clothing and Footwear industry. In this sector Italy increased its share of total EUR-12 output from 32% to 38% between 1980 and 1990 (see Tables 2.12a and b, A10a and A10b). However, the industry is also well known for its strong protection against non-EU countries. Tables 2.12a and b highlight the low levels of import penetration relative to the manufacturing average. Therefore, the removal of non-tariff barriers and subsequent import liberalization was expected to have a considerable impact at both the distributive and productive levels. However, the net effects upon the industry remain unclear. For example, it is possible that greater competition would lead to a reduction in prices. On the other hand, rising prices could also be envisaged if quality was further improved. An increasing range of goods was expected to be produced which would have positive effects for the consumer but an adverse effect upon employment if production was transferred abroad. In short, the potential impact of the SMP upon the Textiles, Clothing and Footwear industry remains ambiguous.

Chemicals increased its Italian share of manufacturing output from 8% to 11% between 1980 and 1990 (see Tables 2.12a and b, A3a and b). Within the industry, the expectation was that the SMP would have only negligible effects. Reflecting this, firms within this sector had not adopted any specific policies or strategies in preparation for the programme's completion. The industry had already undergone a radical transformation of both its ownership and organization structure, making it capable of adapting to the demands of global competition. However, the industry was vulnerable in some respects. It was specialized in lower value added products, and its international links tended to be with less-developed countries. Italy was one of the only EU countries to show a trade deficit in the Chemicals sector at the end of the 1980s. Tables 2.12a and b illustrate the fall in the industry's export-intensity ratio between 1980 and 1990 (relative to the Italian average for manufacturing). R&D expenditure had also fallen in relative terms as the industry's highly-specialized and medium-sized firms appeared reluctant to engage in increased scientific and technological research. Furthermore, they were expected to be vulnerable to the effects of the SMP because of their high concentration on the domestic market.

There were some indications of weakness in R&D-intensive activities. Tables 2.12a and b show a relative decline in R&D expenditure by those industries which tend to be characterized by high R&D, namely Chemicals, Office Machines, Electrical Goods and Transport Equipment. Indeed, the Office Machines industry appears weak. Import penetration, for example, remained very high relative to the manufacturing total and had a poor trade position both on the EU and external side.

In summary, it appeared that few Italian industries were in a position to benefit from the SMP. Large firms, however, remained optimistic, believing that the single market would be beneficial to the Italian economy, but small firms were particularly wary of the programme's implications.

Table 2.12a.   Descriptive sectoral statistics (relative to manufacturing total) – Italy, 1980

E3ME sector	Output share	Balassa (EU)	Balassa (non-EU)	Import penetration	Export intensity	Wage rate	Unit labour costs	Capital intensity	R&D intensity
Ferrous & Non-ferrous Metals	8.64	-0.44	-0.33	1.32	0.65	1.61	1.02	1.79	0.80
Non-metallic Mineral Products	6.22	0.22	0.44	0.46	0.72	1.11	0.85	1.29	0.12
Chemicals	8.06	-0.44	0.10	1.54	1.02	1.40	1.09	1.40	4.74
Metal Products	7.63	0.29	0.76	0.33	0.89	0.88	1.00	0.96	0.14
Agricultural & Industrial Machinery	9.57	0.11	0.63	1.44	2.01	1.28	1.02	0.63	0.25
Office Machines	1.30	-0.19	-0.09	3.32	2.40	0.95	0.96	1.49	3.32
Electrical Goods	5.99	-0.22	0.32	1.55	1.18	1.15	1.21	1.02	3.28
Transport Equipment	6.78	-0.31	0.54	1.86	1.75	1.15	1.32	1.05	4.44
Food, Drink & Tobacco	14.02	-0.56	-0.18	0.65	0.26	0.94	0.73	0.81	0.08
Textiles, Clothing & Footwear	16.35	0.52	0.35	0.57	1.20	0.73	0.96	0.77	0.01
Paper & Printing Products	5.42	0.09	-0.48	0.53	0.36	1.24	1.04	0.99	0.04
Rubber & Plastic Products	3.24	-0.03	0.61	0.74	0.91	1.04	1.02	1.35	1.05
Other Manufactures	6.73	0.17	0.30	0.48	0.74	0.58	0.85	1.06	0.13

Source: E3ME Database.

Table 2.12b.   Descriptive sectoral statistics (relative to manufacturing total) – Italy, 1990

E3ME sector	Output share	Balassa (EU)	Balassa (non-EU)	Import penetration	Export intensity	Wage rate	Unit labour costs	Capital intensity	R&D intensity
Ferrous & Non-ferrous Metals	7.45	-0.30	-0.42	1.45	0.84	1.50	1.03	1.17	0.65
Non-metallic Mineral Products	5.82	0.16	0.45	0.45	0.77	0.98	0.99	1.16	0.08
Chemicals	11.55	-0.48	-0.09	1.25	0.69	1.73	1.04	1.07	2.37
Metal Products	8.09	0.34	0.56	0.27	0.67	0.82	0.96	0.83	0.17
Agricultural & Industrial Machinery	8.57	0.18	0.55	1.15	2.15	1.09	1.12	0.78	0.67
Office Machines	2.08	-0.10	-0.28	2.46	1.90	1.34	1.09	0.98	2.69
Electrical Goods	7.19	-0.17	0.03	1.12	1.09	1.36	1.16	1.42	3.02
Transport Equipment	7.15	-0.22	0.34	1.82	1.53	1.28	1.24	1.12	3.66
Food, Drink & Tobacco	12.27	-0.50	-0.10	0.76	0.37	1.03	0.71	0.84	0.09
Textiles, Clothing & Footwear	14.57	0.46	0.34	0.70	1.21	0.65	0.94	0.77	0.01
Paper & Printing Products	5.34	0.01	-0.49	0.63	0.43	1.12	1.01	1.39	0.01
Rubber & Plastic Products	3.95	0.12	0.43	0.71	1.06	1.11	1.09	1.40	0.63
Other Manufactures	5.91	0.30	0.29	0.84	0.94	0.56	0.81	0.69	0.29

Source: E3ME Database.

2.4.9.   Luxembourg (see Tables 2.13a and 2.13b)

Luxembourg is a small, open economy which experienced increased levels of intra-industry trade both with the rest of the EUR-12 and with the rest of the world between 1980 and 1993 (see Tables 2.3 and 2.4). Primarily, Ferrous and Non-ferrous Metals is the dominant industry in the economy. Its share of Luxembourg's total manufacturing output fell, however, from 55% in 1980 to 47% in 1990. Rubber and Plastic Products holds the next largest share (an increase from 12% in 1980 to 15% in 1990). However, these industries are small compared to the EU total.

Tables 2.13a and b show that Non-metallic Mineral Products, Metal Products, Agricultural and Industrial Machinery, Textiles, Clothing and Footwear, and Food, Drink and Tobacco were characterized by high levels of intra-industry trade both within the EU and with the rest of the world. Furthermore, the export-intensity ratios exceeded their import-penetration ratios. However, this export orientation is most notable in the Transport Equipment and Chemicals industries. R&D expenditure is also very high in these sectors relative to the Luxembourg average for manufacturing.

It was believed that Luxembourg's favourable location would attract foreign-based multinational companies. However, like Denmark, it had the disadvantage of the small absolute size of its economy and limited primary resources compared to other countries in the EU.

Table 2.13a.   Descriptive sectoral statistics (relative to manufacturing total) – Luxembourg, 1980

E3ME sector	Output share	Balassa (EU)	Balassa (non-EU)	Import penetration	Export intensity	Wage rate	Unit labour costs	Capital intensity	R&D intensity
Ferrous & Non-ferrous Metals	55.69	0.72	0.41	NA	1.19	1.24	1.15	1.09	0.72
Non-metallic Mineral Products	4.85	-0.21	-0.11	0.54	0.48	0.74	0.81	1.87	0.64
Chemicals	2.40	-0.74	-0.63	1.10	1.98	1.16	0.76	0.97	4.58
Metal Products	4.65	-0.34	0.24	0.65	0.63	0.75	0.85	0.56	0.35
Agricultural & Industrial Machinery	5.55	-0.31	0.10	0.84	0.87	0.87	1.18	0.63	0.87
Office Machines	0.00	0.00	0.00	0.00	0.00	0.00	0.00	0.00	0.00
Electrical Goods	0.00	0.00	0.00	0.00	0.00	0.00	0.00	0.00	0.00
Transport Equipment	0.71	-0.58	-0.72	1.16	4.09	0.62	0.96	1.06	5.80
Food, Drink & Tobacco	8.92	-0.46	-0.29	0.50	0.33	0.56	0.61	0.82	0.21
Textiles, Clothing & Footwear	1.90	-0.12	-0.09	1.54	2.23	0.62	0.57	0.26	0.14
Paper & Printing Products	2.12	-0.61	-0.84	0.51	0.21	0.70	0.98	0.64	0.07
Rubber & Plastic Products	12.57	0.50	0.67	0.39	0.61	1.03	0.89	0.91	2.84
Other Manufactures	0.61	-0.80	-0.80	0.79	0.45	0.40	0.84	1.35	0.80

Source: E3ME Database.

Table 2.13b.   Descriptive sectoral statistics (relative to manufacturing total) – Luxembourg, 1990

E3ME sector	Output share	Balassa (EU)	Balassa (non-EU)	Import penetration	Export intensity	Wage rate	Unit labour costs	Capital intensity	R&D intensity
Ferrous & Non-ferrous Metals	47.11	0.49	0.22	2.76	1.17	1.28	1.08	1.04	0.90
Non-metallic Mineral Products	7.96	0.08	0.25	0.46	0.50	0.72	0.74	0.39	0.46
Chemicals	4.41	-0.46	-0.23	1.47	2.41	1.07	0.61	3.49	4.13
Metal Products	5.62	-0.14	0.24	0.56	0.68	0.71	1.22	0.47	0.27
Agricultural & Industrial Machinery	4.78	-0.24	0.01	0.82	0.90	0.99	1.27	1.11	1.18
Office Machines	0.00	0.00	0.00	0.00	0.00	0.00	0.00	0.00	0.00
Electrical Goods	0.00	0.00	0.00	0.00	0.00	0.00	0.00	0.00	0.00
Transport Equipment	0.72	-0.61	-0.66	1.14	5.07	0.56	1.05	0.62	4.83
Food, Drink & Tobacco	7.62	-0.29	-0.48	0.48	0.32	0.54	0.68	1.00	0.23
Textiles, Clothing & Footwear	3.00	0.01	-0.02	0.98	1.19	0.81	0.71	1.40	0.12
Paper & Printing Products	2.49	-0.59	-0.81	0.42	0.14	0.75	1.15	0.39	0.06
Rubber & Plastic Products	15.57	0.53	0.52	0.65	0.76	1.21	1.06	0.77	1.59
Other Manufactures	0.67	-0.77	-0.80	0.67	0.40	0.41	0.85	1.18	0.59

Source: E3ME Database.

2.4.10. Netherlands (see Tables 2.14a and 2.14b)

The Netherlands appeared, *prima facie*, to be in a strong position to benefit from the SMP. Firstly, the Dutch economy was very open in terms of trading patterns and its level of intra-industry trade within the EU had greatly increased between 1980 and 1993 (see Table 2.3). Secondly, the Netherlands possessed a number of notable advantages including a high-quality labour force, good infrastructure and a favourable geographical location.

Chemicals and Electrical Goods belong to the category of high-tech process or strongly automated industries and were characterized by strong growth potential. For example, Electrical Goods was less likely to be affected by changes in distribution than production and was known to export relatively more than its European competitors (Tables 2.14a and b show its high export-intensity ratio relative to the Dutch average for manufacturing). In 1990, Chemicals' share of Dutch manufacturing output was still fairly high (14%). Food, Drink and Tobacco held the largest share (27%), but only accounted for 6% of EU output. Tables 2.14a and b show that this sector also had high import-penetration ratios relative to export-intensity ratios. In 1990, intra-industry trade within the EU was the largest relative to the Dutch average for manufacturing.

Some industries appeared to be in a stronger position because of their R&D expenditure. Tables 2.14a and b highlight the increases in R&D expenditure experienced by the Office Machines and Transport Equipment industries. Buigues *et al.* (1990) believed that the expansion of intra-EU trade would benefit these strongly-positioned industries provided they adjusted capacity in preparation for the wider market and increased competition which lay ahead.

By contrast, two different groups of industries appeared vulnerable to the impact of the SMP. The first category included those industries which had been protected by non-tariff barriers: Textiles, Clothing and Footwear and Non-metallic Mineral Products. Both were characterized by low levels of import penetration from other EU countries (see Tables 2.14a and b) and the subsequent removal of remaining non-tariff barriers was expected to lead to strong competitive pressures.

The second category of vulnerable industries included Agricultural and Industrial Machinery and Rubber and Plastic Products. They had already been subject to competitive pressures within the EU and since 1980 had experienced increased export-intensity. The dismantling of internal frontiers was likely to intensify intra-EU competition and act as a catalyst to the ongoing specialization process. Shifts in the geographical distribution of these industries in the EU were a possibility. In the long run these industries were expected to suffer unless the technological performance of Dutch firms improved. Indeed, both industries appeared to require increases in R&D expenditure from the low levels seen in Tables 2.14a and b.

In summary, the Netherlands was already a very open economy. It had a comparative advantage in the capital-intensive industries, but it was unclear if this would be exploited with the implementation of the SMP. The comparative disadvantage in labour-intensive industries meant the economy was vulnerable to increased import-penetration in those industries.

Table 2.14a. Descriptive sectoral statistics (relative to manufacturing total) – Netherlands, 1980

E3ME sector	Output share	Balassa (EU)	Balassa (non-EU)	Import penetration	Export intensity	Wage rate	Unit labour costs	Capital intensity	R&D intensity
Ferrous & Non-ferrous Metals	6.36	-0.12	-0.36	1.07	1.06	1.60	0.96	0.00	0.43
Non-metallic Mineral Products	3.66	-0.30	-0.51	0.81	0.66	0.99	0.85	0.00	0.06
Chemicals	15.62	0.07	0.29	1.26	1.50	1.29	0.98	0.00	3.07
Metal Products	7.16	-0.22	0.17	0.54	0.47	0.96	1.07	0.00	0.12
Agricultural & Industrial Machinery	2.80	-0.29	0.08	1.97	2.44	0.97	1.18	0.00	0.47
Office Machines	0.26	0.01	-0.43	2.04	4.35	2.15	0.99	0.00	1.02
Electrical Goods	8.79	-0.06	0.24	1.46	1.59	1.14	1.03	0.00	3.43
Transport Equipment	6.18	-0.30	-0.03	1.38	1.31	0.97	1.17	0.00	0.91
Food, Drink & Tobacco	28.05	0.22	0.07	0.52	0.74	0.82	0.84	0.00	0.43
Textiles, Clothing & Footwear	4.54	-0.38	-0.35	1.44	1.33	0.78	1.18	0.00	0.06
Paper & Printing Products	9.70	-0.06	-0.61	0.44	0.34	1.04	0.97	0.00	0.06
Rubber & Plastic Products	1.97	-0.30	-0.14	1.12	0.94	0.92	1.03	0.00	0.25
Other Manufactures	4.85	-0.65	0.10	0.92	0.30	0.92	1.00	0.00	0.16

Source: E3ME Database.

**Table 2.14b.   Descriptive sectoral statistics (relative to manufacturing total) –
Netherlands, 1990**

E3ME sector	Output share	Balassa (EU)	Balassa (non-EU)	Import penetration	Export intensity	Wage rate	Unit labour costs	Capital intensity	R&D intensity
Ferrous & Non-ferrous Metals	5.08	-0.22	-0.35	1.11	1.01	1.99	1.11	0.00	0.44
Non-metallic Mineral Products	3.81	-0.19	-0.28	0.60	0.50	1.10	0.83	0.00	0.06
Chemicals	14.85	0.08	0.16	1.53	1.61	0.88	0.79	0.00	2.34
Metal Products	7.97	-0.15	0.02	0.51	0.54	0.96	1.06	0.00	0.19
Agricultural & Industrial Machinery	2.96	-0.16	0.05	2.19	2.50	0.92	1.16	0.00	0.39
Office Machines	0.24	-0.11	-0.66	1.74	8.20	2.51	1.17	0.00	3.19
Electrical Goods	9.83	-0.31	0.01	1.00	0.99	1.20	1.03	0.00	3.30
Transport Equipment	6.56	-0.31	-0.40	1.20	1.13	0.97	1.20	0.00	1.20
Food, Drink & Tobacco	27.49	0.31	0.21	0.53	0.78	0.97	0.90	0.00	0.30
Textiles, Clothing & Footwear	3.18	-0.44	-0.62	1.56	1.77	0.68	1.08	0.00	0.13
Paper & Printing Products	10.33	0.00	-0.50	0.45	0.37	1.08	1.06	0.00	0.04
Rubber & Plastic Products	3.29	-0.21	-0.15	0.96	0.80	0.82	0.96	0.00	0.26
Other Manufactures	4.33	-0.71	-0.44	1.07	0.94	0.91	1.22	0.00	0.27

Source: E3ME Database.

2.4.11. Portugal (see Tables 2.15a and 2.15b)

As noted earlier for Spain, it is difficult to isolate the effects of the SMP on the Portuguese economy from the anticipatory and transition effects of accession to the EU in 1986. Following accession, it appears that the competitive position of Portuguese industry deteriorated both within the EU and against the rest of the world, in particular the NICs.

The import-penetration and export-intensity ratios in Tables 2.15a and b show that the most open sectors of the economy are Agricultural and Industrial Machinery, Office Machines, Electrical Goods and Transport Equipment, and these also tend to have relatively high wage rates and investment and R&D ratios.

Traditionally strong sectors such as Textiles and Food were expected to come under increasing pressure from NICs (either from accession or the SMP) as the introduction of a common external tariff reduced the overall level of protection for these industries. However, Community efforts to help these sectors by financing restructuring were likely to disguise some of the more direct effects.

Buigues *et al.* (1990) believed that sectors such as Metal Goods, parts of Textiles and Shipbuilding (part of Transport Equipment) were not likely to have been particularly affected by barriers to imports and so were less vulnerable to the SMP.

The industries which faced the greatest threat from an increase in trade can be categorized as predominantly domestic or international in orientation. Mainly domestic sectors included certain foodstuffs (in Food, Drink and Tobacco) while international sectors included glass and glassware (contained within Non-metallic Mineral Products) and rolling stock (Transport Equipment). There was a range of industries covering only a small part of Portugal's manufacturing sector, typified by high levels of import-penetration ratios (and often export-intensity ratios). These sectors (Office Machines, Agricultural and Industrial Machinery, Electrical Goods) were starting from a relatively small base, and so may have faced difficulties in a more competitive environment.

Many sectors showed a high deficit for European inter-industry trade in 1980, while Food, Drink and Tobacco, Textiles, Clothing and Footwear, and Paper and Printing Products showed a large inter-industry EU trade surplus, highlighting the fact that Portugal had a small number of strong sectors. This feature appeared to be less evident by 1990, and this is confirmed by the data for the Grubel-Lloyd Index in Table 2.3, which shows a rising level of intra-industry trade with other EU Member States. Conversely, trade outside of the EU seems to have become more one-way in nature between 1980 and 1990, particularly in sectors such as Non-metallic Mineral Products, Agricultural and Industrial Machinery, Electrical Goods, and Other Manufactures. Of these, Agricultural and Industrial Machinery and Electrical Goods were relatively the most highly traded and had a relatively high wage rate (a crude indicator of skill level).

In conclusion, some of the least competitive industries showed the greatest potential for growth but were dependent on foreign companies to develop that potential. The continuing opening up of the economy associated with the SMP was expected to further deteriorate the competitive position of the Portuguese economy. The comparative advantage the economy enjoyed in the labour-intensive sectors was not expected to offset losses in R&D intensive sectors.

Table 2.15a. Descriptive sectoral statistics (relative to manufacturing total) – Portugal, 1980

E3ME sector	Output share	Balassa (EU)	Balassa (non-EU)	Import penetration	Export intensity	Wage rate	Unit labour costs	Capital intensity	R&D intensity
Ferrous & Non-ferrous Metals	4.12	-0.67	-0.45	1.62	0.78	1.39	1.18	1.18	0.60
Non-metallic Mineral Products	5.27	-0.05	0.26	0.49	0.66	0.93	0.93	1.41	0.14
Chemicals	8.84	-0.70	-0.14	1.60	0.89	1.46	1.00	3.41	4.99
Metal Products	5.36	-0.30	0.48	0.58	0.77	1.04	1.02	0.88	0.16
Agricultural & Industrial Machinery	2.04	-0.88	-0.42	3.16	2.05	1.35	1.38	0.78	0.26
Office Machines	0.62	-0.36	-0.51	1.76	1.20	0.00	0.00	0.00	NA
Electrical Goods	3.64	-0.29	0.01	2.03	1.79	1.25	1.01	0.85	3.71
Transport Equipment	6.14	-0.42	-0.21	1.78	1.48	1.34	1.00	0.78	4.15
Food, Drink & Tobacco	28.30	0.32	-0.13	0.20	0.34	0.96	0.63	0.53	0.10
Textiles, Clothing & Footwear	21.42	0.57	0.56	0.37	1.36	0.85	1.27	0.74	0.01
Paper & Printing Products	5.24	0.78	0.44	1.00	2.93	1.20	0.88	0.98	0.04
Rubber & Plastic Products	2.31	-0.72	-0.07	1.04	0.45	0.97	0.94	1.19	1.30
Other Manufactures	6.65	-0.35	0.48	0.91	0.77	0.63	1.00	0.87	0.15

Source: E3ME Database.

Table 2.15b.   Descriptive sectoral statistics (relative to manufacturing total) – Portugal, 1990

E3ME sector	Output share	Balassa (EU)	Balassa (non-EU)	Import penetration	Export intensity	Wage rate	Unit labour costs	Capital intensity	R&D intensity
Ferrous & Non-ferrous Metals	4.26	-0.40	-0.49	1.18	0.79	2.31	0.99	0.05	0.32
Non-metallic Mineral Products	5.65	0.18	0.68	0.42	0.81	1.13	1.10	1.60	0.09
Chemicals	10.63	-0.52	-0.10	1.08	0.76	1.82	1.03	0.22	5.01
Metal Products	4.38	-0.18	0.37	0.77	0.84	0.99	1.13	0.38	0.20
Agricultural & Industrial Machinery	1.53	-0.73	-0.70	2.16	1.81	1.03	1.46	1.00	0.85
Office Machines	1.38	-0.77	-0.53	1.17	0.39	0.00	0.00	0.00	NA
Electrical Goods	4.07	-0.27	-0.11	2.02	2.16	1.48	1.25	1.26	3.70
Transport Equipment	5.82	-0.35	-0.37	1.86	1.82	1.45	1.18	0.19	4.74
Food, Drink & Tobacco	26.14	-0.16	-0.28	0.25	0.21	0.84	0.55	1.00	0.13
Textiles, Clothing & Footwear	23.66	0.35	0.54	0.98	1.62	0.81	1.15	1.09	0.01
Paper & Printing Products	5.78	0.49	0.33	1.18	1.87	1.27	0.88	2.48	0.00
Rubber & Plastic Products	2.02	-0.51	-0.14	1.16	0.53	0.98	1.20	1.54	0.75
Other Manufactures	4.63	-0.53	-0.12	0.62	0.30	0.56	1.21	1.04	0.37

Source: E3ME Database.

2.4.12. United Kingdom (see Tables 2.16a and 2.16b)

The UK economy displays a high level of intra-industry trade, ranking fourth in Europe behind France, Belgium and West Germany on the basis of trade with EU Member States, and second (behind Belgium) using the same index on trade with the rest of the world. UK manufacturing suffered from a severe decline in employment and a deterioration in the trade balance over the period 1980 to 1990.

Of the two industries with an EU trade surplus in 1980, only Office Machines remained with an EU trade surplus in 1990. The high-technology industries (with relatively high levels of R&D) were Office Machines, Electrical Goods, and Transport Equipment. High R&D spending needs to be spread over large sales volumes, and firms sought to exploit the economies of scale associated with lower barriers to trade with the rest of the EU. These sectors had relatively high import-penetration ratios, but also high export-intensity ratios.

The Ferrous and Non-ferrous Metals industry has become relatively more globalized as can be seen from Tables 2.16a and b with both the import-penetration and the export-intensity ratios increasing. The industry in the UK had moved from a trade deficit in 1980 to a slight surplus in 1990.

Industries such as Paper and Printing Products, and Rubber and Plastic Products had suffered from increasing competition and mergers in the 1980s. The industries were less internationalized, as indicated by the relatively low import-penetration and export-intensity ratios.

The sectors which could potentially gain from the SMP were mainly those which had a relatively high investment in R&D. Tables 2.16a and b identify these as Chemicals (especially pharmaceuticals), Office Machines (computers, in particular), Electrical Goods and Transport Equipment.

Below average performance was considered likely for industries such as Textiles, Clothing and Footwear, which had become more inter-industry orientated. It is worth noting, however, that the sector was already highly open, perhaps indicating that any structural change within the sector had already occurred. Certain sectors, such as Food, Drink and Tobacco, Metal Products, Non-metallic Mineral Products, Paper and Printing Products, and Rubber and Plastic Products were among the relatively least open to trade in the UK economy, and showed relatively low levels of investment and R&D intensity, all of which suggested vulnerability to greater trade under the SMP.

To conclude, the UK had a comparative advantage in the R&D intensive sectors, and as the UK economy was already very open, it was expected these industries would benefit from the implementation of the SMP. However, this optimistic outlook was offset by a slightly pessimistic outlook for some of the industries that were not open to trade.

Table 2.16a.   Descriptive sectoral statistics (relative to manufacturing total) – United Kingdom, 1980

E3ME sector	Output share	Balassa (EU)	Balassa (non-EU)	Import penetration	Export intensity	Wage rate	Unit labour costs	Capital intensity	R&D intensity
Ferrous & Non-ferrous Metals	6.35	-0.23	-0.21	1.22	0.86	1.36	1.44	1.66	0.45
Non-metallic Mineral Products	4.55	0.33	-0.10	1.20	1.26	0.98	0.89	1.30	0.20
Chemicals	10.15	-0.00	0.40	0.95	1.31	1.11	0.86	1.95	2.24
Metal Products	5.36	-0.07	0.41	0.54	0.68	1.07	1.15	0.87	0.18
Agricultural & Industrial Machinery	11.82	-0.03	0.52	1.09	1.63	1.03	1.14	0.74	0.46
Office Machines	2.14	0.11	-0.22	2.39	2.06	0.71	0.93	0.66	2.12
Electrical Goods	6.80	-0.23	0.33	1.32	1.30	0.87	1.09	0.88	3.89
Transport Equipment	9.57	-0.18	0.22	1.77	1.90	1.14	1.33	1.21	2.74
Food, Drink & Tobacco	21.25	-0.37	-0.06	0.49	0.33	0.98	0.51	0.74	0.16
Textiles, Clothing & Footwear	6.47	-0.15	-0.14	1.11	0.79	0.63	1.15	0.58	0.09
Paper & Printing Products	8.42	-0.09	-0.38	0.53	0.32	1.32	1.07	0.93	0.06
Rubber & Plastic Products	3.09	-0.06	0.45	0.73	0.91	1.06	1.08	1.20	0.18
Other Manufactures	3.96	-0.61	-0.01	1.39	0.78	0.79	1.14	0.99	0.73

Source: E3ME Database.

Table 2.16b. Descriptive sectoral statistics (relative to manufacturing total) – United Kingdom, 1990

E3ME sector	Output share	Balassa (EU)	Balassa (non-EU)	Import penetration	Export intensity	Wage rate	Unit labour costs	Capital intensity	R&D intensity
Ferrous & Non-ferrous Metals	4.43	0.00	-0.12	1.59	1.90	1.37	1.02	1.58	0.34
Non-metallic Mineral Products	3.99	-0.11	-0.23	0.50	0.44	0.89	0.97	1.76	0.14
Chemicals	11.11	-0.06	0.30	1.05	1.30	1.29	0.85	1.62	3.00
Metal Products	4.69	-0.25	0.11	0.63	0.65	1.00	1.17	0.79	0.13
Agricultural & Industrial Machinery	10.39	-0.12	0.22	0.85	1.07	0.90	1.19	0.73	0.37
Office Machines	5.99	0.04	-0.39	1.70	1.74	0.80	1.04	1.08	3.32
Electrical Goods	9.53	-0.14	0.11	1.15	1.29	1.22	1.22	0.72	2.88
Transport Equipment	8.58	-0.27	0.07	1.56	1.64	1.07	1.10	1.44	2.08
Food, Drink & Tobacco	19.30	-0.32	0.04	0.52	0.39	1.04	0.55	0.71	0.12
Textiles, Clothing & Footwear	5.10	-0.35	-0.39	1.57	1.18	0.64	1.17	0.54	0.04
Paper & Printing Products	9.62	-0.19	-0.38	0.53	0.38	1.14	1.08	0.95	0.05
Rubber & Plastic Products	3.87	-0.21	0.06	0.85	0.83	1.09	1.12	1.18	0.15
Other Manufactures	3.31	-0.43	0.01	1.38	1.08	0.64	1.18	0.70	0.41

Source: E3ME Database.

2.5. Implications for the study

The academic literature reviewed thus far points to certain lessons to be learnt before undertaking off-model work and empirical analysis. Maintaining the order of the literature/issues discussed in this chapter, the implications for the remaining part of the study are as follows:

2.5.1. Trade and the labour market

European trade is primarily of the intra-industry type. This is by nature of the convergence in income per capita between members of the European Union which tends to lead to a demand for similar, but differentiated, products, i.e. a demand for variety of choice. The resources available to Member States, particularly in the area of human capital, also lead to trade in similar types of products between the same sector, as opposed to trade between different sectors, i.e. inter-industry trade. Thus, although inter-industry trade does have some relevance with respect to extra-EU flows of goods and services, the small proportion of trade accounted for by this area limits the extent to which predictions from such theories as HOS and factor content analysis can have for the empirical part of the study.

Having identified intra-industry trade as the predominant feature of flows of goods and services within the EU, the question then arises of finding a coherent theory with which to test the data and from which to make predictions for the impact of the SMP on employment and labour costs. The so-called New Trade Theory offers some useful insights into the transmission mechanism between trade and the labour market, but no definitive link exists between trade and employment. This is important because the lack of a theoretical linkage makes it difficult to alter the structure of E3ME, the model used to undertake the empirical side of the analysis. In the absence of a coherent message from the literature on the relationship between trade and employment, alternative methods are used to determine the likely effects. These methods rely more on looking at the market structure of particular manufacturing sectors, with a view to identifying common trends among similar sets of industries; the factors used to group sectors are market type (segmented or fragmented) and product type (homogeneous or differentiated). Generally speaking, those sectors identified as having segmented markets and producing differentiated products are thought most likely to encourage intra-industry trade and thus most likely to benefit from the single market programme, which is essentially turning Europe into a free and open marketplace for producers and consumers alike. Conversely, those sectors within fragmented markets and producing homogeneous products are most likely to suffer as a result of the single market programme. The removal of non-tariff barriers allows lower cost foreign competition the opportunity to completely displace a sector within a Member State. Sectors that fall somewhere in between the two extremes are expected to benefit/suffer the effects according to how far they are along each dimension, i.e. the degree of market fragmentation and homogeneity of product.

However, even using this methodology has its drawbacks. The effects of intra-industry trade are most likely to occur within sectors, e.g. employees moving from one firm to another within the same industry, whereas the effects of inter-industry trade might be to cause more visible structural shifts in the economy. As such, the measured effects from empirical analysis will probably only identify net movements, i.e. the sum of potentially large opposing forces.

2.5.2. Trade and investment

The literature which combines trade and employment often cites technological change as one of the principal causes of the demise of manufacturing employment, with trade cited as another important yet smaller source of labour market developments. However, there is a body of opinion that proposes a causal linkage between trade and technological developments, which if true would mean that trade has a much greater impact on employment and labour costs than previously realized. There are arguments for and against the existence of such a causality. Those positing the link point towards empirical findings that technological progress is concentrated in skilled-labour-intensive industries, raising inequalities between the wages of skilled and unskilled workers. So-called 'defensive innovation' is also used as a reason, whereby firms faced with increased international competition invest in new techniques to stay ahead of their competitors. The evidence is not overwhelming, however. Critics point out that if the technology was available, firms would take advantage of it anyway. They also use Agriculture as an example of a sector that is relatively closed to trade and yet enjoys high levels of labour productivity.

Given the inconclusive nature of the debate on the causality between trade and investment, it seems a suitable strategy would be to test the hypothesis on data to be used in the empirical part of the study. This way a better indication will be given of whether the trade-investment relationship exists in the European sectoral data, and furthermore whether the relationship should be incorporated within the E3ME modelling structure. This analysis is performed in Chapter 3.

2.5.3. Partial equilibrium studies

A review of partial equilibrium studies was undertaken in Section 2.3. This allowed attention to be given to specific countries/sectors where an issue over trade and employment exists, the

UK textiles sector and its associated falling employment is a prime example of this. These studies also allow a direct comparison of the results obtained from the study (which are country and sector specific) with those already existing in the literature. This provides a useful cross-check of robustness of results, as many other studies use either aggregate data or panel data techniques which makes meaningful comparisons less feasible. The partial equilibrium studies reviewed typically come to similar conclusions, namely that trade has an adverse effect on employment prospects in sectors which are more geared towards inter-industry trade, e.g. Textiles and Clothing. What most of the studies fail to capture is the simultaneity of the relationship between trade and employment, i.e. they treat the changes in trade as exogenous. Clearly, this is a shortcoming as it ignores second round price and income effects. The study by Driver *et al.* (1986) highlights the importance of taking account of indirect effects, i.e. intermediate demand, which is important for some sectors (particularly the metal-based ones) whose products are primarily aimed as inputs to other industries rather than for final consumption.

2.5.4. Background sectoral analysis

The existing empirical work mentioned above, which uses the market structure/product type hypothesis, is somewhat limited in its homogeneous treatment of the sectors across countries. With this in mind, it seems best to take a closer look at the characteristics of the manufacturing sectors of each of the Member States being covered by the study. Across Europe, one would expect that same sector would have a different share of output, productivity, import penetration, export intensity, etc., depending on which Member State was being analysed. This is the purpose of investigating more closely the information contained in the sectoral databanks of the E3ME model, i.e. it allows more focus to be added to the broad implications drawn out from the review of literature on trade and employment.

The data analysis reveals important insights to the structure of manufacturing in each Member State. For example, identifying the size of particular sectors is useful in determining where most changes in employment and labour costs are likely to occur. In general, the largest sectors are Ferrous and Non-ferrous Metals, Chemicals, and Food, Drink and Tobacco. Transport Equipment (e.g. cars) and Agricultural and Industrial Machinery are also large sectors in Belgium, West Germany, France, Italy and the UK, while Textiles, Clothing and Footwear tends to be larger in the southern Member States of Greece, Spain, Italy, and Portugal. A great deal of information is provided from the data, and this is used in Chapter 3 to guide the empirical analysis and inform on the likely Member State/sectoral-specific impact of the single market programme.

3. Off-model analysis

3.1. Trends in intra-EU trade

One of the expected impacts of the SMP was to stimulate intra-EU trade by removing barriers to trade. Firms would be more able to source their inputs from the most competitive supplier in the EU, without tariffs or other barriers favouring national suppliers. Economies of scale would then accumulate, reinforcing the trend towards fewer suppliers operating on a larger scale. This anticipated effect has clearly happened in some sectors, for example the growing cross-border purchases of motor components and the designation of factories as sole European source for certain products.

This impact of the SMP would clearly boost cross-border trade at the expense of intra-country trade. It is not immediately obvious, however, whether this stimulus would be expected to increase intra-EU trade at the expense of extra-EU trade (a zero sum game analogy) or whether all trade can increase through the dynamic effects of growth (a positive sum game). Two hypotheses can be imagined. Firstly, the increase in operating efficiency within the EU from the single market enhances the competitiveness of EU supplies relative to extra-EU imports, thus boosting intra-EU trade and reducing extra-EU trade. Secondly, the increasing specialization of plants is applied at a world level, increasing extra-EU exports and imports. The overall stimulus to economic growth within the EU would also stimulate extra-EU imports, and if sufficient to boost the world trading system, also extra-EU exports.

Figure 3.1 shows the trends in total trade for the EU, adopting a standard EUR-12 definition for the whole period 1960–92, with the percentages calculated in current prices. Thus, the rapid increase in the price of crude oil in 1973 (a commodity almost wholly imported from outside the EU at the time) leads to a sharp drop in the proportion of imports to all EUR-12 Member States which are intra-EU imports. The same is true over 1979–81, though at this time the impact appears to have been mitigated by higher EU production (mostly in the UK). A reverse effect would be at work in 1985, when the price of crude oil dropped substantially.

Since 1960, there has been a general upward trend in the proportion of external trade which is conducted with other Member States. This upward trend appeared to slow in the 1970s (though the effect of world oil prices is significant, as above), before resuming upward growth in the 1980s, but there is no conspicuous acceleration since the mid-1980s that could be attributed a priori to the SMP.

The historical levels and subsequent growth of the share of intra-EU trade by country varies markedly, as shown in Figures 3.2 and 3.3. The selection of countries shows that the increases in the overall EU trade (as shown in Figure 3.1) can be accounted for by relatively few and specific cases. For example, the continuous growth in the proportion of the UK's trade which is with the rest of the EU shows up clearly for both imports and exports. For Greece, Spain and Portugal, there is a sharp reversal of previous decline in the proportion of imports, from 1986 onwards. This may be due to the timing of these countries' accession to the EU, which occurred in 1981, 1986 and 1986 respectively. The growing proportion of exports going to the rest of the EU has risen steadily since the 1960s for Spain, but there was a change of trend towards more integration with the EU for Portugal and Greece from the early 1980s onwards.

Figure 3.1.   Intra-EU trade, 1960–92

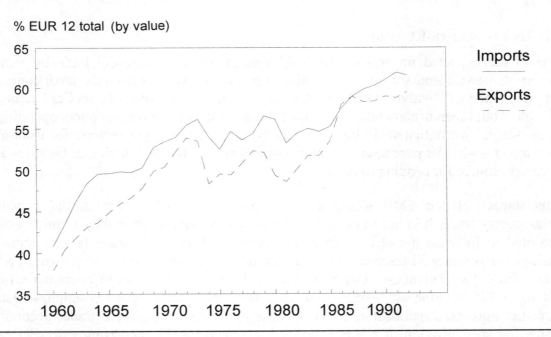

Source: Cheleme database, CCIP, France.

France and Italy show trends in imports very similar to the EU as a whole, while Denmark, the Netherlands, Germany and Belgium/Luxembourg show no strong trends since the 1970s in either imports or exports. Ireland has actually seen a declining proportion of its imports and exports coming from the rest of the EU since the early 1980s. Much of Ireland's trade is conducted with the UK, and an increasing non-EU share may reflect the growth of US-owned electronics production in the country.

In summary, the detailed country data on the direction of trade does not a priori suggest a strong impetus to EU integration at the end of the 1980s. Where strong trends exist, these pre-date the SMP, or are more coincidental with accession to EU membership than the SMP.

The sectoral pattern of intra-EU trade is shown in Figures 3.4 and 3.5. The data has been converted to the E3ME industry classification from CCIP's CHELEME database. The sectors shown are the manufacturing industries and agriculture (i.e. energy and water have been excluded). The clearest trends are for Agriculture, and Food, Drink and Tobacco. The share of trade (both imports and exports) being conducted with the rest of the EU has risen steadily over the whole period shown. The engineering sectors (Metal Products, Agricultural and Industrial Machinery, Office Machines, Electrical Goods, Transport Equipment) show an interesting trend which may reflect the SMP: for imports, there is an apparent increase in the upward trends from about 1985 onwards, while for exports, the previous downwards trends are stabilized over the same period. However, again there is no obvious effect from the SMP on the trends in the direction of trade.

This analysis suggests that although the SMP *may* have increased the proportion of individual Member State's external trade being conducted with other Member States, the impact appears to be limited to a few countries and sectors. There is no evidence of a general effect from the SMP, other than to continue the steady increases already existing since 1960 or 1970 for example.

Figure 3.2.   Intra-EU imports by country

Source: Cheleme database, CCIP, France.

Figure 3.3.   Intra-EU exports by country

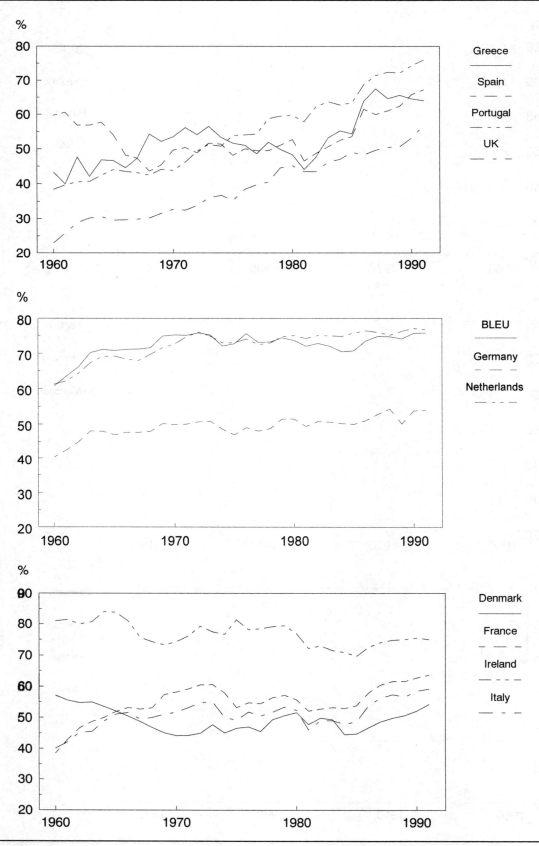

Source: Cheleme database, CCIP, France.

Figure 3.4.   Intra-EU imports by sector

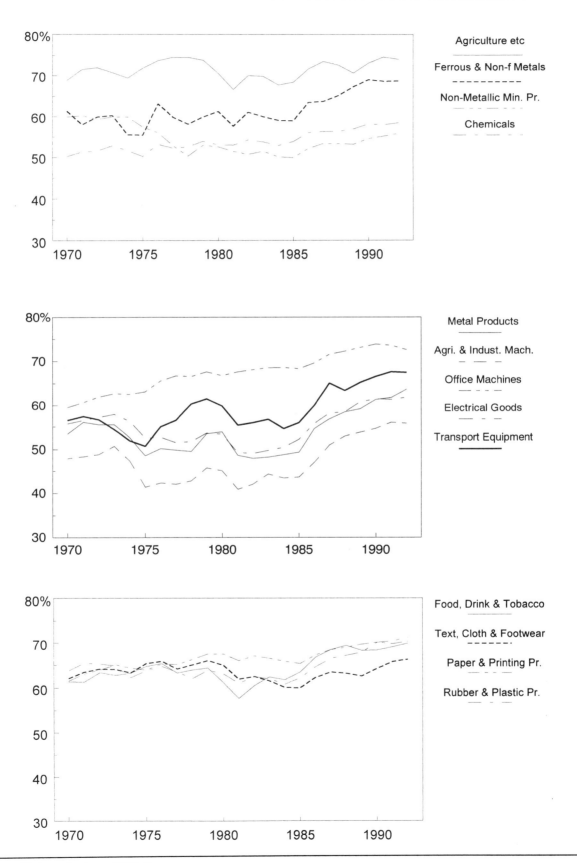

Source: Cheleme database, CCIP, France.

Figure 3.5.   Intra-EU exports by sector

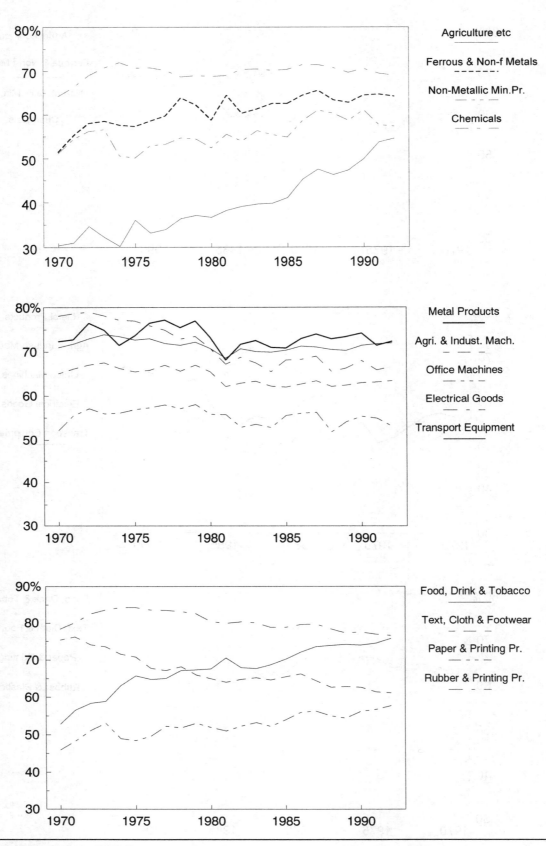

Source: Cheleme database, CCIP, France.

3.2. Trade–investment linkages

In the argument over whether trade or (labour-saving) technology is the main cause for falling relative demand for un-skilled labour there has arisen the question of causality. That higher investment can lead to increased trading possibilities is not at issue, but that the anticipation of increased trade and fiercer competition can provide an incentive to invest more is open to question.

The main argument against the reverse causality is that if producers can lower costs by inducing labour-saving technology, then they will do so anyway, regardless of the extra competition. There seems to be little evidence in support of the hypothesis, i.e. that countries/industries with a higher level of trade intensity have higher levels of labour-saving technological progress. Possibly the best explanation is that some industries operate an aggressive strategy towards international competition (increase investment and R&D), while others are more submissive (decrease investment and R&D).

While the linkage between trade and investment/R&D effects remains unproven, it seems worth testing the hypothesis empirically on the sectoral data used by the E3ME model. Therefore, prior to any possible inclusion of this hypothesis in the model solution stage, a trade intensity variable (calculated by the ratio of total trade to output) was added to the original specification of the investment equations, allowing the data to decide whether they were significant or not. Due to the log–log formulation of the equations, the reported coefficients can be interpreted as elasticities.

The revised specification of the investment equations is as follows:

Long-run trend (see Table 3.1)

$$KR_{rit} = \alpha_0 + \alpha_1 * TXM_{rit} + \alpha_2 * YR_{rit} + \alpha_3 * PKR_{rit}$$
$$+ \alpha_4 * YRWC_{rit} + \alpha_5 * PYRE_{rit} + u_{rit}$$

where

KR = investment volume

TXM = trade intensity

YR = output volume

PKR = price of investment goods

$YRWC$ = real labour costs

$PYRE$ = real energy price

rit = region, industry and time

Short-run dynamics

$$\Delta KR_{rit} = \beta_0 + \beta_1 * \Delta TXM_{rit} + \beta_2 * \Delta YR_{rit} + \beta_3 * \Delta PKR_{rit}$$
$$+ \beta_4 * \Delta YRWC_{rit} + \beta_5 * \Delta PYRE_{rit} + \beta_6 * \Delta RRLR_{it}$$
$$+ \beta_- * \Delta YYN + \beta_8 * \Delta QRX_{rit-1} + \beta_9 * \hat{u}_{rit-1} + v_{rit}$$

where

$RRLR$ = real rate of interest

YYN = ratio of actual to 'normal' output

rit = region, industry, time

The coefficient on TXM in the equations was restricted by sign according to a priori beliefs about the response of certain sectors in certain regions to increased trade intensity. The restrictions were drawn primarily from the analysis performed in Chapter 2. It is quite possible that some sectors may respond positively to increased trade intensity, i.e. so-called 'defensive innovation', while others may react negatively, succumbing more readily to the pressures of competition by reducing levels of investment activity.

There is no evidence of a significant dynamic effect of trade intensity on investment, indicating that it concerns long-run trends and cannot explain variance in investment around this trend very well. In fact, not one significant coefficient was found, so the table for the short-run effects is not included.

For the longer term effects, it would be impossible to go through each of the coefficients in turn and explain why they are positive or negative. Rather, some broad conclusions will be made and one or two specific examples will be explored in the context of the literature review in Chapter 2.

At the EU level, the signs on the coefficients are largely positive, with only two sectors showing an averaged negative trade effect and eleven showing a positive trade effect. There is a built-in correlation between a sector's characteristics, in terms of product and market type, and the effect of trade on investment by virtue of the a priori restrictions. For example, the sectors which fall into the differentiated product/segmented market category, i.e. Chemicals, Transport Equipment and Food, Drink and Tobacco, all have positive coefficient values, indicating an aggressive investment long-term investment response to rising trade intensity.

This is an average EU value, of course, and more information may be gained by examining the results for individual Member States.

It is interesting to note that most of the negative elasticity values are present in Spain and Portugal, while most other Member States display at most only two such results. It is also true that with the exception of Transport Equipment, no sector displays completely consistent signing across all the Member States, demonstrating the diversity in market structures present across the same sector within each EU Member State.

The Food, Drink and Tobacco sector comes close to a consistently positive sign, with only negative effects reported in Italy and Portugal. It is defined as a 'differentiated product/segmented market' type of sector, which indicates a positive sign as a segmented industry will have enough market power and presumably enjoy sufficient economies of scale enough to invest substantially in new production techniques, and increased trade intensity is likely to benefit a differentiated product based on the argument of a larger consumer market promoting increased demand for variety.

Despite the promising results, it was decided not to include the trade intensity term in the investment equations. The reasons for the exclusion were twofold:

(a)   Data limitations. The limited amount of data used to estimate the equations (typically 1970–93) means that an extra variable in a regression has a disproportionate effect on degrees of freedom, reducing the robustness of other coefficient estimates.
(b)   Reduced-form effects. Although having trade terms in the investment equation would supply a direct link between the two variables, it can be argued that trade will affect investment through its impact on output and prices. In other words, trade already has an indirect effect, i.e. through a reduced form of the model, on investment, so there may be no need to add to the causality structure already present.

On balance therefore, it was decided not to include the trade term in the investment equations. Despite this, it is possible to infer what might have happened to results had the trade intensity variable been included. The effect on investment would depend on the coefficient sign. A positive coefficient would compound the positive effect on investment already supplied (indirectly) by the output variable (YR). A rise in trade intensity would thus add a second boost to increased 'defensive' innovation, helping to reduce prices and improve competitiveness and eventually employment prospects. A negative coefficient would work in the other direction, offsetting some of the output effect and thus limiting the effect on employment and labour costs.

Table 3.1.   Results of testing for trade-investment linkages: long-run equation

E3ME sector	BE	DK	DW	EL	ES	FR	IR	IT	LX	NL	PO	UK	EU
Ferrous & Non-ferrous Metals	2.00	0.00	-1.46	0.00	2.00	0.00	0.00	0.00	1.68	0.00	-2.00	0.00	-0.05
Non-metallic mineral Products	1.96	0.98	1.93	0.00	-0.74	-1.38	0.00	0.00	0.00	0.00	-1.14	-0.11	0.43
Chemicals	2.00	0.00	0.63	0.00	0.00	-2.00	0.00	0.00	0.00	0.00	-2.00	0.00	0.02
Metal Products	2.00	0.00	1.17	0.00	-2.00	0.35	0.00	0.00	2.00	0.00	-2.00	0.00	0.57
Agricultural & Industrial Machinery	0.89	0.00	0.60	0.00	-2.00	-0.29	-0.78	0.00	1.49	0.00	-2.00	0.00	0.26
Office Machines	1.00	0.00	0.15	0.00	2.00	1.29	-0.67	0.00	0.00	0.00	0.00	0.00	1.02
Electrical Goods	0.00	0.00	2.00	0.00	1.87	0.00	0.00	-2.00	0.00	0.00	0.00	0.22	0.58
Transport Equipment	2.00	0.00	2.00	0.00	0.13	0.38	0.29	1.01	2.00	0.00	0.00	0.00	1.14
Food, Drink & Tobacco	1.08	0.87	0.00	0.00	1.21	0.67	1.34	-2.00	0.00	0.00	-0.86	0.00	0.16
Textiles, Clothing & Footwear	0.00	-0.58	0.00	0.00	1.97	-0.13	0.00	0.00	0.00	0.00	-1.28	0.00	0.05
Paper & Printing Products	0.00	1.73	0.00	0.00	0.00	0.00	0.00	0.00	-2.00	0.00	-2.00	-0.59	-0.06
Rubber & Plastic Products	1.20	0.00	1.04	0.00	-0.65	0.00	0.00	2.00	0.00	0.00	-1.30	0.00	0.94
Other Manufactures	0.00	-0.89	0.00	0.00	-0.71	0.00	-0.54	0.77	0.00	0.00	1.55	0.73	0.60

Source: E3ME Database.

3.3. Labour market structure

This section looks at the theoretical and applied literature on modelling the labour market. The purpose of the review is to put the E3ME modelling framework in the context of related studies, with a view to amending the structure of equations or simply to inform the reader of alternative methodologies that have been used by other economists and their models. Aside from a general review of how the labour market is treated in E3ME, the topics covered include labour demand, wage determination, product market conditions, structural change and labour supply, i.e. the labour force and population flows.

A discussion of key model parameters (elasticities) is also included. This focuses on employment–output, employment–real wage, employment–productivity and employment–unemployment responsiveness. This comprehensive comparison of the underlying parameters with those from other studies is useful in checking how close the estimates from the E3ME model are to those indigenous models used to look at similar effects for their own countries. Data for France, Germany, Italy and the UK are available for this purpose, while non-EU comparisons are made with Sweden, Japan and the US, among others.

The impact of the SMP on intra-EU trade and the effects of this on employment and labour costs depends partly on:

(a)   the responsiveness of EU labour markets before the SMP was introduced;

(b)   changes in behaviour which would have taken place irrespective of the SMP; and

(c)   changes induced by the dynamic effect of greater product market competition upon labour market behaviour.

Whilst it is not possible to model all these effects and to establish the relative importance of the different contributory factors, we need to be aware of the relevant evidence to date. Model simulation results must be interpreted against this background.

In principle, it would be possible to consider the labour market of each country in turn, but the present context involves the operation of an integrated multi-country model and the emphasis in this off-the-model analysis is on differences in particular responses which are relevant to the impact of the SMP. Thus, consideration of labour supply and migration conditions is much less important.

This section, then, reviews briefly the evidence on structural change, labour market flexibility and occupational patterns in the Member States. No attempt is made to cover fully the differences between countries: this is done regularly in *Employment in Europe*. The main focus is on labour market behaviour which lies on the boundary between what is usually covered in macro-econometric models and what may be important in evaluating the effects of the SMP.

3.3.1. A review of the E3ME labour market methodology

The general modelling structure

The resources provided through the present project need to be concentrated on areas where they are most likely to give rise to credible analyses that identify SMP effects. These are mainly to be found in:

(a)   the treatment of trade and investment flows where SMP impact variables may, in principle, be specified and their effects estimated;

(b)   the general strengthening of the labour market components of the models to ensure that they do actually provide an appropriate transmission mechanism for the expected influences.

Figure 3.6 outlines a revised treatment which might reasonably be attempted for each country (as opposed to what is currently modelled, as portrayed in Figure B2). The latter is based on a study of the following:

(a)   the theoretical literature;

(b)   the general empirical research base;

(c)   the treatments found in operational models of the labour market, i.e. those embodied in aggregate or disaggregated models used regularly for forecasting and/or simulation;

(d)   the need to ensure that key relationships are present through which the effects of the SMP on the labour market may be identified.

It is worth noting that (c) reflects the many compromises required and experience acquired in the construction and running of models which will differ at least to some degree in the purposes for which they have been created. Weaknesses which are acceptable in one analytical context will not necessarily be acceptable in another; strengths in one will not be of great substance in another. It is therefore desirable to review the significance of (c) in the light of the task in hand, i.e. as in (d).

As for (a) and (b), it is the fate of the model-builder to struggle to meet aspirations set by reference to the theoretical literature or to the predominantly *ad hoc* empirical analyses that have been published in the scientific journals. The latter often take advantage of much more time available to spend on the estimation of a few specific relationships. They also often combine the opportunity to exhibit 'best practice' econometric techniques through the availability of particularly good data with greater freedom to select a level of disaggregation which suits the estimation exercise.

So the theoretical literature and the more general research base do provide one set of standards by which to judge the scope for improving on what is contained in existing models. They can also be a source of inconsistency and confusion in the interpretation of a given operational model. This is partly because of difficulties in discriminating between theories on the basis of econometric analysis when different theories can produce essentially the same equation for estimation purposes. It is also because theoretical rationales for different components of the model may be at variance with each other. This means that a given set of parameters and variables which constitute a working model may be interpreted in quite radically different ways. The model-builder will have one, hopefully, coherent view but this is not the only possible interpretation which might be placed upon the model. Note, for example, the ease with which a vintage production function can yield the same labour demand equation as a neo-classical production function when there is difficulty in the measurement of capital stock and crude proxies are introduced to cope with this.

Figure 3.6.   A schematic outline of the labour sector of a model

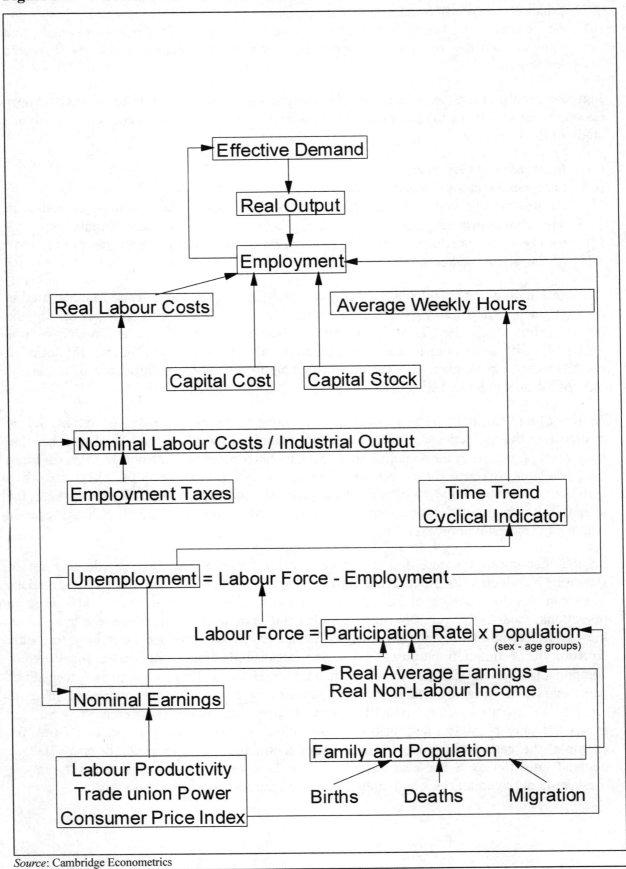

Source: Cambridge Econometrics

Considering very briefly the schema of Figure 3.6 as an agenda for development of E3ME, there are qualifications over its feasibility in the case of certain variables; for example, problems with data on the cost of capital and trade union power mean that not only must these variables be generated exogenously outside the E3ME model but there are also major difficulties in carrying out time-series estimation of the relevant relationships indicated within the labour market sub-model. The pay-off from engaging in much estimation effort in this area is thus likely to be small.

In addition, the levels at which some relationships should be estimated depend on the institutional arrangements in the different countries. This applies particularly to the determination of nominal earnings which may be the subject of national, sectoral or subsectoral bargaining according to the country and industry concerned.

A similar institutional consideration arises in recognizing the role of the public sector, both in industry and services and both as 'consumer of the output produced and as producer/employer'. In those countries and sectors where restructuring, rationalization and/or privatization have been significant phenomena during the last decade (coinciding, therefore, with the build-up of the SMP), there will have been changes in the behaviourial relationships estimated for previous periods. Any SMP effects need to allow for the proper treatment of such sectoral reorganization, otherwise the increases in investment and productivity which are often associated with reorganization will be wrongly attributed to the SMP. However, it must also be recognized that some institutional change can be associated with the SMP because of the rules on public procurement, etc. and so this must also be borne in mind when interpreting the results.

Going beyond the representation of the labour market given in Figure 3.6, there are many possible models for the disaggregation of employment and other variables according to occupational structure, employment status, hours of work patterns and gender. However, the scope for estimating such models is severely limited by the availability of data. In the context of the present modelling exercise, it is not possible to estimate labour demand and supply functions at occupational level and to test for the presence of effects due to the single market programme on the structure of employment within each industry. The data situations for EU Member States were reviewed in Hogarth and Lindley (1994) and do not even allow for the construction of occupational employment models that would distinguish between the secular and cyclical components of change.

The timing of the introduction of the SMP in relation to the cyclical fluctuations of the last decade and the availability of occupational data (which is determined most commonly by the timing of decennial population censuses) rules out the estimation of the simplest of incremental occupational–output models.

However, having said that, it is possible to provide a benchmark for the occupational effects by assuming that proportionate differences between the 'with SMP' and 'without SMP' situations at sectoral employment level will have had the same proportionate effects on occupational employment levels. The same is possible for other disaggregations of aggregate sectoral employment, though these are of less significance for the overall evaluation exercise being undertaken for the Commission, of which this project is just one part (see Appendix C for a detailed treatment of the occupational effects of the SMP).

Two other areas of the model deserve comment at this stage: population and labour supply (see Figure 3.6). Although several features of the SMP could be seen to provide channels by which migration might be stimulated, the empirical evidence suggests a distinctly second- or third-order effect over the time horizon considered here. Effects which work through consumption (as a function of population levels and structure), remission of incomes by migrant workers, and labour supply are therefore not likely to be large enough for any Member State to warrant the major task of introducing a population and migration sub-model into E3ME. Moreover, the countries where this might be deemed most important – including the effects of inter-regional migration within a country (for example, from south to north) – are those for which the data are most problematical.

As regards labour supply effects that come through the completion of the product market and the greater competition this might engender within and between national labour markets, this is also likely to be a very minor aspect of the impact of the SMP programme *per se*. However, with or without some improvement in the treatment of labour supply in the model, it should be possible to trace SMP effects on broad wage and labour cost relativities as derived from the trade and investment effects on employment, unemployment and, thereby, earnings.

Theoretical issues and evidence from the research base

(a)   Labour demand

The specification of the employment functions which were developed in the 1960s and the 1970s relied heavily on the role of output. The desired level of labour demand was a function of output, capital stock and technological progress while the actual level of employment was held to approach the desired level through a simplified mechanism such as the well-known partial adjustment mechanism (Ball and St Cyr, 1966; Killingsworth, 1970). Usually, in the empirical estimations of these labour demand functions capital stock was omitted due to well-known difficulties in measuring it accurately – particularly where the function was to be tested against data for an individual industry – while technological progress was proxied by a time trend which, of course, also captures the effects of increasing capital stock.

In addition to output, capital stock and technical change, a variant of these models related the desired level of labour input to real product wages, too. Empirical estimation of this family of labour demand models, using data for the period prior to 1970, usually produced highly significant coefficients on output and lagged employment – the latter being included to allow for the partial adjustment of actual to desired levels of employment – but insignificant and sometimes wrongly signed coefficients on the real product wage. As a result real product wages tended to be omitted from empirical specifications altogether (see Briscoe and Wilson, 1991).

Yet, during the 1970s and due to the slowdown of employment growth in most of the advanced market economies, the debate on the employment/real wage relationship was reheated. Indeed, in the wake of the first oil-price shock, the evidence that labour demand is negatively related to real labour costs has been substantial: the poor employment record of Europe *vis-à-vis* the United States since the early 1970s is strongly related, *inter alia*, to the differential growth of real wages between the two regions (see OECD (1994a) for a thorough documentation) while the vigorous private sector employment growth and the concomitant protracted recovery of the OECD economies during the late 1980s is also partly assigned to the reversal of earlier unfavourable developments in real wages. This evidence gave new credit to the neoclassical theories of employment determination wherein, on the basis of the profit maximization condition of the firm, i.e. marginal revenues equals marginal costs, the derived demand for labour schedule establishes a negative relationship between employment and the real cost of labour. As a result, during the 1980s, a strong case for a pure form of the neoclassical labour demand function has been established in the literature (Symons and Layard, 1984). In this kind of model there is no room for any aggregate demand variables as the transmission mechanism between variations in aggregate demand and private sector employment is the change in real factor prices. Indeed, in their empirical study of the manufacturing labour demand schedules of six OECD economies (G-7 except Italy), Symons and Layard found a strongly significant effect of real product wages on employment and no or a very weak effect of aggregate demand on employment.

In further work by Layard and Nickell (1985) and Bean *et al.* (1986), demand variables such as fiscal impact and competitiveness have, however, shown up as significant determinants of employment. But, rather than interpreting this as a direct impact of demand on employment, as disequilibrium theory would have it, they argue that it captures the counter-cyclical nature of price mark-ups and its impact on employment. A significant and positive effect of aggregate demand variables on employment – alongside the negative effect of real labour costs – is also shown for a large sample of countries in a study by Bruno (1986).

Finally, empirical labour demand equations indicate that the long run elasticities of private sector demand for labour with respect to real labour costs are similar across the OECD economies while the short run elasticities vary considerably (OECD, 1994b; Turner *et al.*, 1993). This evidence suggests that the OECD economies differ mainly as regards the time required for employment to adjust to changes in the real labour costs. The speed of adjustment is evidently dependent upon the turnover costs of labour, the hiring and firing regulations and the forms of industrial relations that prevail in the domestic production systems (e.g. due to implicit contracts).

As the institutional arrangements of the production process vary across the industrial sectors of a domestic economy one would also expect to observe different labour demand elasticities for different sectors within the boundaries of a domestic economy. Yet, most of the labour demand studies use aggregate data and, as a result, the diversity of labour demand elasticities across different industrial sectors has been hardly examined. One notable exception is Lee *et al.* (1990) who, using disaggregate data from 40 industrial sectors in the UK, estimated labour demand functions for the period 1954–84. The set of regressors includes the real wage rate, a proxy for the expected level of real demand in the sector, a proxy for the demand conditions in the economy and a proxy for technological change. In almost all industrial sectors these variables show up as significant determinants of sectoral employment. The mean and standard deviation (in brackets) for the sectoral long-run wage elasticities are -0.54 (0.58), the mean and standard deviation (in brackets) for the sectoral long-run output elasticities are 0.86 (0.88) and those for the sectoral long-run technological change elasticities are -0.27 (0.72). The standard deviations reveal a considerable variability of long-run estimates across the industries and this, in itself, provides strong support for employing disaggregated analysis rather than aggregate analysis, since the latter cannot capture the sectoral peculiarities that clearly exist. It still remains to be seen whether this direction of research will be pursued more systematically in the future.

(b)   Wage determination

A generic form of the nominal earnings equations that are most commonly incorporated in labour market models today is the expectations-augmented Phillips curve. This version of the Phillips curve relates the rate of change of nominal earnings to a measure of past or expected consumer price inflation, the unemployment rate and a vector of other relevant variables, such as trade union power, the existence of incomes policies and the path of minimum wages, where appropriate. As nominal wage developments play an important role in the determination of private sector employment, the impact of both past inflation and unemployment rates upon them are of critical significance for employment, too. Thus, many theoretical/empirical studies have sought to identify and estimate the size of these impacts. According to them, a slow adjustment of nominal earnings to inflation (nominal wage rigidity) can alleviate the negative consequences of an adverse supply shock, while likewise, a rapid adjustment of nominal earnings to rising unemployment can mitigate the effects of a negative demand shock (Klau and Mittelstadt, 1986). As the evidence indicates, nominal earnings are, indeed, affected by both past inflation and the actual level of unemployment but the size of the impact differs widely across the OECD economies (OECD, 1994b; Coe, 1986).

Finally, bargaining models of wage determination assume that trade unions are mainly concerned with achieving a target real disposable income. As a result, in empirical research, these models relate the growth of real, rather than nominal, earnings to past inflation and unemployment, including also variables such as the rate of productivity growth and the average tax rate on household income (Coe, 1986; Chan-Lee et al., 1987; Lee and Pesaran, 1993).

3.3.2. Product market conditions and labour market reactions

The creation of a much wider product market has not only provoked intensified intra-EU trade but also a great deal of corporate activity involving fundamental re-assessments of business strategies. Foreign acquisitions, licensing, joint ventures, creation of new subsidiaries abroad, and greater investment in existing subsidiaries were all stimulated.

The issue of labour market flexibility is then connected directly to the issue of what kind of socio-economic environment will make one region more attractive than another in a Europe with higher levels of cross-border corporate collaboration, more foreign direct investment and more competitive organizational structures. At the same time, however, European labour market flexibility as a whole was seen as a key factor in competing with other regions of the world economy.

The lowering of barriers in the product market may allow companies to take advantage of differences in labour market conditions of which they may have already been aware, but did not by themselves justify exploiting. Some of these differences may result from different degrees of competition in imperfect product markets and the sharing of monopoly profits between employers and workers in the form of higher pay and voluntary non-wage costs. Completion of the single European market (SEM) should then reduce distortions to competition in the labour market.

An examination of the large differences in labour costs and regulatory conditions suggests that the labour market conditions do vary greatly, but that not all these variations are at the cutting edge of locational decisions. In practice they often appear to be secondary. This is partly for the reason that, compared to other intra-EU variations which bear upon locational decisions, they are not as important, and partly because the variations between EU Member States and other potential non-EU locations are much greater, sometimes by orders of magnitude.[1] Walwei and Werner (1993), for example, using survey results produced by Prognos, show that the environment of economic regulations, market potential and political and social climate were much more important than production costs and even more so than the quality of labour.

Nonetheless, even if labour market conditions do not dominate locational choices by transnational corporations or companies intending more modest degrees of internationalization, they can adversely affect performance. A major concern amongst trade unions especially has been the possibility that national governments with unfavourable mixes of labour market regulation and costs from the employer point of view will seek to alter their positions in the labour market rankings. Competitive imperatives will then be cited as justification for this strategy, whether the focus is on other EU nations or on competition from third countries.

The resulting **'cost-cutting' scenario** is then seen to carry with it the danger that fierce competition via lowering labour standards will drive down the quality of jobs without generating the unit cost levels which can be achieved by the NICs.[2] Moreover, lowering the quality of jobs takes the pressure off management to innovate and to move the company up the value-added spectrum and hence increase its capacity to support high-quality jobs. The cost-cutting scenario is then one in which lowering the quality of jobs leads to lowering the quality of labour (through limiting incentives to train and be trained and generating efficiency wage adjustments on the part of labour supply) and eventually quality of products.

This kind of labour market flexibility would seem to be a poor way of trying to create a virtuous circle of high human capital, high productivity, high wages, profits and investment: the **'quality' scenario**. Whilst curbing monopoly power in product and labour markets is essential, changes in labour market regulation beyond a certain point do not help to achieve this; they merely make weak workers weaker. Achieving the quality scenario, however, was seen to require a fundamental re-assessment of the organization of the economy and its interface with the wider society. In particular, there was a need for greater flexibility in an era of rapid change.

3.3.3. Structural change and labour market flexibility

The deployment of labour: structural change

The notion of flexibility derived from neoclassical price theory is straightforward: it refers to the responsiveness of demand and supply to changes in price and the responsiveness of price to relative movements in demand and supply. The conventional debate about labour market

---

[1]   Hourly industrial (manufacturing and construction) labour cost ratios of roughly 6 to 1, adjusting for differences in industrial structure, can be found within the EU: Germany and Portugal being at the two extremes (Zighera, 1993). The average for the 12 EC countries during the late 1980s was 4½ times that for Portugal, 3 times that for Greece and 50% higher than the Spanish average.

[2]   By contrast with note 1, average manufacturing labour costs in the EC are about three times higher than those of the old NICs (Hong Kong, Korea, Singapore, and Taiwan), but between 10 (Malaysia) and 30 (Indonesia) times higher than new NICs (which also include the Philippines and Thailand) (Petit and Ward, 1995).

flexibility has been dominated by a somewhat narrower underlying perspective: the freedom of demand-side agents (i.e. employers) to vary their purchases in accordance with need and to set prices freely in order to explore the supply response and find an optimal strategy – in other words, the scope for the buyers' prerogatives to be exercised. Suppliers' access to the market and discriminatory pricing are regarded as secondary issues.

There is a tendency to assume that lack of labour market flexibility is caused by regulation, unduly powerful trade unions combined with uncompetitive product markets, poorly designed income maintenance programmes, or badly organized support activities such as education, training, the search for jobs, and the search for employees. However, examining these various elements does not lead to unqualified conclusions about impacts on flexibility and the consequences of this for economic growth and unemployment (Grubb and Wells, 1993).

There is a fairly generally held view that more structural change has taken place since the early 1980s than hitherto and this requires more flexibility in order to minimize its effects on unemployment. Following this up with evidence, however, is not quite as straightforward as it might seem. In particular, in looking for real evidence, it is often forgotten that changes in quantities and prices are not very informative unless the parameters of the demand, supply and wage adjustment functions are known. Moreover, we need to recognize when the lack of flexibility may be due to a **slowness** of response rather than a low degree of adjustment achieved over the long term.

The interpretation of structural change indices, for example, those based on sectoral employment (see Table 3.2), must also allow for the distinction between *ex ante* and *ex post* structural change. Some forms of structural change may themselves reflect failure to compete because of lack of flexibility. Precipitous declines in manufacturing employment caused by inefficiency through restrictive working practices or failure to attract suitably skilled labour because of narrow wage differentials hardly amount to 'coping' with structural change, yet they would usually boost the index. So *ex post* measures of structural change can be interpreted as indicators partly of the task faced by the economy concerned and partly of the amount of redeployment achieved.

Whilst a high level of *ex post* structural change does imply a high level of flexibility (i.e. 'flexible' in Table 3.2) in order to achieve it, a low level may reflect either:

(a)   low flexibility ('inflexible') in the face of high *ex ante* pressure for structural change;
(b)   a medium level of flexibility in the face of a medium level of *ex ante* pressure ('responsive');
(c)   any of the three options if the level of *ex ante* pressure for structural change is too low to offer a test of the capacity for adaptation.

The fact that there do not seem to be great differences in the *ex post* measures of structural change calculated from *employment* data for the 1960s, 1970s and 1980s (Table 3.2) does not necessarily suggest that this has not been a cause of rising unemployment. It may rather imply that these economies could deliver only so much sectoral employment re-allocation because of inflexibilities in labour markets which were common to all three periods. Unemployment then rises because the underlying economic conditions called increasingly for more adjustment.

Table 3.2.   The interpretation of structural change indices

Ex ante:	Extent of *ex post* structural change		
	High	Medium	Low
High	Flexible	Responsive	Inflexible
Medium		Flexible	Responsive
Low			Flexible / Responsive / Inflexible
Growth rate disparities:	**3.0 ≤ G**	**2.0 ≤ G < 3.0**	**1.0 ≤ G < 2.0**
Turbulence measures:	**1.2 ≤ T**	**0.8 ≤ T < 1.2**	**0 ≤ T < 0.8**
1960s: G	DK	D, F	
T	DK	D, F, NL, UK	US
1970s: G	DK	D, F, I, J, NL, UK	
T	DK	B, D, F, I, J, NL, UK, US	
1980s: G		DK, F, I, J, UK	D, NL
T	UK	B, DK, F, I, J, US	D, NL

Source: OECD Jobs Study, Part I, Table 1.7.

Notes: G is the average of the weighted standard deviation of annual growth rates (in logs) of sectoral employment.
T is the average of half the sum of the absolute annual changes in sectoral shares of total employment.

However, similar degrees of sectoral employment re-allocation are recorded in economies as different as France and the United States during each of the three decades (the US was apparently *less* 'flexible' than France in the 1960s). This occurred virtually regardless of the measure used and level of disaggregation and so it is difficult not to conclude that the 'flexibilities' of their labour markets cannot be that different.

On the other hand, the indices do identify Denmark and, to a lesser extent, the UK, associated with particularly low regulation of the labour market, as exhibiting the highest degree of structural change.

Labour force flows

If the scrutiny of employment stock data gives rather limited insight into the extent of flexibility, perhaps the analysis of flow data would be more productive. Of course, a similar problem of assessing the evidence arises because of the distinction between *ex post* and *ex ante* movement, i.e. data relate to actual mobility as opposed to the propensity to be mobile. But if countries show very different scales of movement of the population among activities, it is difficult not to accept that this has implications for the extent of labour market flexibility.

Presumably in the context of labour market flexibility, the key indicators needed are those which reveal **purposeful** movement in which the population shifts from:

(a)   lower to higher value-added activities;
(b)   jobs where there is little scope for training to jobs where individuals will be able to enhance their human capital;

(c)   employment into full-time education or training in order to increase their productive
       power even more substantially (or delays initial entry to employment in order to be better
       prepared);
(d)   unemployment to employment;
(e)   employment in sectors/occupations/geographical areas where the risks of unemployment
       are higher than those where they are lower;
(f)   employment in sectors, etc. with poorer working conditions than those with better
       conditions.

Thus, the mere fact of there being a lot of movement relating to the labour force, employment
and unemployment is a rather poor indicator of flexibility.[3] Fluidity does not necessarily
imply flexibility. Some movement might, indeed, seem to suggest a failure of the labour
market mechanism: very high turnover at organizational level or gross outflows and inflows
at sectoral level may be due more to the inadequate personnel policies of employers as much
as to a well-judged balancing of the relative costs of putting up with high turnover compared
with improving working conditions and rates of retention. The same caveats attach to
attributing low job tenures and low service records with the same industries to dynamic
labour market responses.

We are not sure, therefore, of what to make of the following stylized facts (see European
Labour Force Surveys):

(a)   that mobility in and out of unemployment is over twice as high in Denmark and the UK
       as in Italy and Spain;
(b)   that average job tenure is lower in the Netherlands and the UK (7 or 8 years) and higher
       in France, Germany and Spain (about 10 years);
(c)   that labour turnover (as measured by the percentage of the employed who had been in
       their jobs for less than a year) was highest in Spain and the UK, followed by Denmark
       and France, and lowest in Belgium, Italy and Greece;
(d)   that inter-sectoral mobility is highest in Denmark, France and the UK and lowest in
       Belgium, Germany, Greece and Ireland.

In addition, the occupational stock and flow data for EU countries do not permit satisfactory
analysis either of *ex post* structural change or *ex post* mobility. In the case of spatial labour
market flexibility, the data on geographical mobility are not sufficiently standardized for
differences in the sizes of the spatial areas used to allow us to reach much more than very
broad conclusions, such as that mobility is relatively high in the UK and low in Italy.

Finally, if such data are to be used to indicate differences in labour flexibility between
countries, there would also seem to be a case for using them to indicate how far flexibility
has changed over time for the same country. The results for the UK are especially noteworthy
(Beatson, 1995). Industrial and occupational mobility increased by about 50% during the last
half of the 1980s and then halved by the mid-1990s, more or less bringing it back to the same
levels as a decade ago. Regional migration, which is much less than industrial and

---

[3]   Mobility within the labour force is only one way of achieving a net change in the deployment of labour; differentiated
       rates of initial entry to, and retirement from, the labour force (plus temporary exit and return during working life)
       provide another mechanism. The scope for using the latter mechanism, however, has probably contracted with
       declining youth cohorts, increasing recourse to extended education and training prior to entering the labour market,
       and a marked increase in early retirement stimulated by strategies for coping with high unemployment amongst older
       workers during the 1980s.

occupational mobility, increased more erratically and by a smaller degree but also fell back during the 1990s, actually dropping significantly below the mid-1980s level.

The above results do not suggest a period of success for deregulation followed by one of failure; they reflect more the well-known impact of the cyclical upturn followed by the downturn upon the mobility of labour in an economy. It is clear that more time must elapse before being able to distinguish 'regulation' effects from cyclical effects.

It would be excessively purist not to conclude that the UK is probably the labour market that exhibits the greatest extent of external labour mobility and this probably reflects a greater propensity for mobility. It is the link between this propensity and the institutional and regulatory regime that is not identifiable with present data.

The presentation of key stylized facts about labour force movements in the European Union cannot get round the problem of not having comparable information on the parameters of models that seek to explain those facts. Migration, mobility and turnover data are only now just becoming available in time-series of sufficient comparability and length to support such a modelling exercise, albeit using very simple specifications.

Employment–output responses

Flexibility from the perspective of the organization comes through the ability to vary the use of and payment for labour services in line with the optimal response the organization wishes to make to actual and anticipated changes in the demand for its products. Labour services may be varied by altering hours worked and/or the number of employees; Hart (1987) distinguishes such adjustments as operating at the intensive or extensive margin, respectively. Alternatively, we might note that the former involves action via the internal labour market of the organization, whereas the latter involves the external labour market.

Desired changes in labour services are unlikely to require only the adjustment of the amount of labour input but also some change to the use of different types of labour. This may be handled internally through altering the pattern of hours worked amongst the various occupational groups, or by external market operations via redundancy and recruitment. The internal option will be facilitated by developing a high level of functional flexibility within the existing labour force. This requires appropriate collective agreements on flexibility across different functions of the organization and across tasks and an effective capacity for retraining people as required. The external option depends on there being an efficient external labour market for the skills in question.

It is beyond the scope of this study to review the evidence on internal flexibility. The relevant point in the present context is that different mixes of internal and external flexibility may produce equally effective outcomes. Comparisons of indicators of external flexibility cannot be regarded as conclusive in comparing overall system flexibility. There is, though, another source of evidence which cuts across these detailed considerations and simply focuses on the responsiveness of the labour input to changes in output and wages. However, the division between external and internal market adjustment (i.e. concerning employment and hours, respectively) is settled.

If we consider that the impact of a high propensity for labour mobility should be to improve the capacity of employers to adjust employment levels quickly to what is demanded, we might expect employment to be most responsive to changes in output in the case of deregulated labour markets. However, the E3ME model estimation results (Table 3.3) for the EU using the most comparable data available and aggregating from sector-specific estimates (Barker *et al.*, 1995) suggest that Italy has the highest long-run elasticity of employment with respect to output, and Portugal, Spain and the UK have the lowest. In the middle ground lie Belgium, Denmark, France and Germany with Ireland on the high side (Greece is excluded and the Netherlands' parameters were perverse). These results allow for changes in average weekly hours worked, for which the employment elasticities are particularly high in Belgium, Denmark, France and Spain and low in Germany and Italy, with Ireland, Portugal and the UK in the middle ground.

The above puzzle might be resolved by looking at short-term parameters and, indeed, they show a much smaller response of employment to output in the case of Italy (near zero) than for the UK. However, it is still the case that Portugal and Spain are roughly as responsive as the UK – France, Germany and Ireland are much more so.

A similar check to presuming that regulatory differences might be expected to reveal themselves in a straightforward way via the response of labour demand to economic variables is found when examining the employment elasticity with respect to changes in the real product wage. OECD (1994c) findings show that, whilst technology and industry structure do not differ dramatically and might be expected to yield similar long-run elasticities, the speeds of adjustment vary greatly (Table 3.4). Moreover, it turns out that the UK is slower to respond than France or Germany, having a median lag of four years, twice that of those countries. So, despite its much more deregulated labour market, the key price effect is slower to work. On the other hand, Italy has a long-run elasticity of half that of the UK and a slightly slower speed of adjustment.

Estimates of real product wage parameters (averaged across sectors) are also available from E3ME (see Table 3.5). A more detailed breakdown of the wage–employment elasticities is provided in Table B15 of Appendix B in the Technical Annex. These give rather different results. At the country level, the weighted long-term elasticities are similar to each other, being lower for France, Germany and the UK than the OECD findings, whereas those for Italy are about the same (Table 3.4). These kinds of discrepancy are not unusual in econometric work of this kind in which changes in data, sample period and model specification can lead to quite large changes in parameters and country rankings. Nonetheless, they suffice to provide a further caution against assuming that the degree of regulation of product and labour markets will show up directly in the responsiveness of sectoral employment to variables such as output and wages.

Table 3.3.   Labour demand parameters – E3ME output elasticities

	Long-term	Short-term
BE	0.50	0.11
DK	0.46	0.52
DO	NA	NA
DW	0.58	0.34
EL	NA	NA
ES	0.32	0.16
FR	0.52	0.33
IR	0.77	0.27
IT	0.94	0.02
IS	NA	NA
LX	0.47	0.39
NL	0.08	0.02
PO	0.34	0.13
UK	0.35	0.19
Weighted average	0.70	0.34

Source: E3ME model.

Table 3.4.   Responsiveness of labour demand to real product wages

	Long-term labour demand elasticity	Median lag (years)
France	-1.0	2.0
Germany	-1.0	2.0
Italy	-0.5	5.0
United Kingdom	-1.0	4.0
Sweden	-0.9	7.0
Japan	-0.8	3.0
United States	-1.0	1.0

Source: OECD Jobs Study, Part II, Table 5.1.

Table 3.5.   Labour demand parameters – E3ME real wage elasticities

	Long-term	Short-term
BE	-0.57	-0.11
DK	-0.47	-0.39
DO	NA	NA
DW	-0.59	-0.14
EL	NA	NA
ES	-0.47	-0.01
FR	-0.72	-0.24
IR	-0.34	-0.20
IT	-0.56	-0.13
IS	NA	NA
LX	-0.56	-0.15
NL	-0.24	-0.04
PO	-0.46	-0.07
UK	-0.43	-0.27
Weighted average	-0.56	-0.31

Source: E3ME model.

Wage flexibility

The final macro- or meta-level indicators of labour market flexibility concern the responsiveness of wages to economic conditions. Without this response, the potential for labour demand and supply adjustments which help ultimately to reduce unemployment will not be triggered, however large the relevant price elasticities in the demand and supply functions.

It is possible to examine changes in the wages structure in great detail, comparing the degrees of dispersion in wages of different sectors, occupations, geographical areas; variations among employers within these groups; and variations among individuals.

The main considerations are:

(a)   the institutional arrangements for determining pay and other working conditions;
(b)   the forms of payment adopted;
(c)   the utilization of different forms and their relative importance in the financial and non-financial benefits obtained from work;
(d)   the aggregate profile of nominal and real wages;
(e)   the structure of wages and working conditions;
(f)   changes in wages and working conditions;
(g)   models which seek to explain the principal stylized facts.

Only in the UK has there been significant reduction in the extent of collective bargaining, though in the last half of the 1980s some did occur in the Netherlands. In France and Portugal, the coverage of employees actually rose. By 1990 about 45% of employees were covered in the UK compared with 70% or more in most other EU countries.

The limited data available on modes of payment suggest that no EU country particularly outstrips others in the use of schemes which relate pay more closely to the performance of the organization or the individual employee. If anything, France, Luxembourg and the UK seem to use profit-related pay more than is the case in other Member States (over 25% of the employed are in such schemes), but Belgium, Ireland and the Netherlands are not that far behind (European Commission, 1986). The very partial evidence on trends suggests that performance-related schemes are increasing their coverage.

Examination of earnings dispersion among EU countries is dominated by the marked increase during the 1980s observed for the UK along sectoral, occupational and regional dimensions. Looking at the mid-1970s to mid-1990s, little change was actually observed in sectoral dispersion except for Italy where there was a major reduction from a level that was higher even than Belgium, Germany, Ireland and the UK, towards the low level associated with Denmark (Beatson, 1995). International data for subsequent years are not sufficient to determine what has happened to dispersion, nor is there convincing evidence of an increase in the responsiveness of sectoral wage differentials to sectoral labour market conditions for those countries with relevant data over the 1970s and 1980s.

As regards differentials relating to occupational or educational levels, these are very difficult to construct for reliable comparative analysis. A narrowing of both appears to have occurred in the 1970s followed by some widening in the 1980s, again more noticeable in the UK than elsewhere, with German differentials remaining stable and those for France and the

Netherlands actually narrowing (OECD, 1994a, Chapter 5). It is not known how much the differentials may have changed their degree of responsiveness over the 1980s.

Regional wage dispersion in the EU has been at its greatest in France with little change during the 1970s or 1980s. More generally, regional differentials tended to narrow in the 1970s; during the 1980s, this was reversed to some extent, most of all in the UK but not for Germany. Note that the most striking difference among EU countries is that, whereas for France and the UK regional wages are negatively related to regional unemployment, in Germany and Italy the relationship is positive (OECD, 1990). However, Johnes and Hyclak (1989), estimating regional Phillips curves for three countries, show that regional wage elasticities with respect to unemployment are higher in Germany compared with Italy and the UK.

For regional industrial labour costs, the cost–unemployment relationship is negative for France, Italy, Spain (particularly strong) and the UK, whilst being insignificant for Germany and strongly positive for Portugal. The regional labour cost range is especially high in France and Portugal and low in Germany and the UK.

Finally, turning to the responsiveness of aggregate real wages to labour market conditions, this varies considerably between countries, as shown in Table 3.6. Those with a particularly high level of responsiveness of real wage growth to short-term unemployment also have relatively high, though smaller, responses to long-term unemployment (LTU): Belgium, Germany, Italy and the Netherlands (note also a similar result for Sweden where the impact of LTU is especially marked). For the other countries, not only are both elasticities low but the ratio of long-term to short-term elasticities tends to be smaller: Denmark, France, Greece and, notably, the UK with a zero LTU elasticity. The ratios for Ireland and Spain are somewhat higher, akin to those for the responsive group of countries whereas the LTU elasticity for Portugal is actually positive.

Other things being equal, we might expect LTU to be a higher proportion of total unemployment in those countries where its existence has a lower relative impact on real wage growth. This does not, however, appear to be the case. Belgium, Ireland and Italy recorded LTU at about 60% of total employment in 1992 whereas Denmark, France and the UK with lower ratios of the two elasticities have much lower proportions of LTU.

As with the employment elasticities in the previous subsection, the results of estimating wage models do depend greatly on specification, data, etc. Overall, as Beatson (1995) shows, certain rankings seem to recur. Generally, the UK shows the least flexibility and Germany the most. France is more flexible than the UK but rarely comes above Germany. Italy is erratic, though usually above the UK.

In the present context, it is nonetheless worth looking at some of the results available which pay attention to both dynamics and sectoral disaggregation.

Changes in wage flexibility over time have also been explored by a number of authors. The results are somewhat mixed. Chan-Lee et al. (1987) conclude that less wage inflation took place in the first half of the 1980s than would have been expected; the case is not a strong one even for the UK where it is at its best. Moreover, Poret (1990) finds no change for the UK, though some evidence of a structural break for France, and Anderton et al. (1992) suggest that for the decade as a whole there is no effect for the UK, but observe one for Italy.

3.3.4.   Testing a revised labour market structure

Tables 3.9a and b show the results (in terms of significant coefficient values) of a revision made to the long- and short-run labour demand equations. Namely to test the inclusion of the price of investment goods as an explanatory variable. This follows more closely the recommendations as portrayed in Figure 3.6. A priori, one would expect the effect of a rise in the price of investment goods to have a positive effect on the demand for labour, assuming that capital and labour are substitutes in the production process. These expectations are largely borne out in the estimated coefficients for the price of investment.

This proposed change was not incorporated into the final model specification used to perform the simulations. The two principal reasons were firstly the large proportion of negative or insignificant coefficients (shown as zeros in the tables) and secondly the unsatisfactory method of attempting to approximate the cost of capital (which should include other variables like the rate of interest) by the price of investment goods. Therefore, a direct link between the demand for capital goods and labour is not included in the model specification, although some substitution between labour and capital demand may occur through secondary model effects.

Table 3.6.   Wage flexibility in the EU

	Semi-elasticities of real wage growth with respect to unemployment			Long-term unemployed as % total unemployment[1]
	short-term unemployment	long-term unemployment	total unemployment	
Belgium	-1.32	-0.53	-0.78	59
Denmark	-0.84	-0.21	-0.66	26
France	-0.77	-0.16	-0.58	34
Germany	-1.72	-0.54	-1.26	39
Greece	-0.92	-0.16	-0.62	49
Ireland	-0.62	-0.20	-0.39	60
Italy	-1.21	-0.44	-0.54	58
Netherlands	-1.34	-0.45	-0.99	48
Portugal	-1.09	0.13	-0.62	30
Spain	-0.46	-0.14	-0.34	44
United Kingdom	-0.68	0.00	-0.54	35
Sweden	-1.60	-0.74	-1.59	N.A.

Source(s): OECD (1993), Tables 3.1, 3.2 and 3.5; European Commission (1994), Graph 172.
[1] Unemployment greater than 12 months – 1992 ELFS.

Table 3.7. Responsiveness of real product wages to unemployment

	Long-term real product wage semi-elasticity	Median lag (years)
France	-3.5	1.5
Germany	-3.0	4.0
Italy	-3.5	2.0
United Kingdom	-1.0	1.0
Sweden	-10.0	2.5
Japan	-5.0	3.0
United States	-1.0	1.0

Source: OECD Jobs Study, Part II, Table 5.2.

Table 3.8a. Wage equation parameters – E3ME unemployment rate elasticities

	Long-term	Short-term
BE	-0.02	0.00
DK	-0.01	0.00
DO	NA	NA
DW	-0.02	0.00
EL	NA	NA
ES	-0.09	-0.01
FR	-0.01	0.00
IR	-0.04	-0.03
IT	-0.07	0.00
IS	NA	NA
LX	0.00	0.00
NL	-0.02	-0.04
PO	-0.07	0.00
UK	-0.01	0.00
Weighted average	-0.03	-0.00

Source: E3ME model.

Table 3.8b. Wage equation parameters – E3ME productivity elasticities

	Long-term	Short-term
BE	0.27	0.11
DK	0.51	0.47
DO	NA	NA
DW	0.47	0.40
EL	NA	NA
ES	0.12	0.49
FR	0.42	0.39
IR	0.12	0.22
IT	0.49	0.50
IS	NA	NA
LX	0.25	0.38
NL	0.31	0.44
PO	0.28	0.40
UK	0.38	0.29
Weighted average	0.42	0.45

Source: E3ME model.

3.3.5. Measuring effects by type of labour

As labour supply constraints are most likely to be differentiated according to occupation or skill, they raise further the question of whether or not it is feasible, in the first place, to establish the occupational impacts of the SMP assuming that quantity constraints via labour supply are **not** binding. An outline of a simple method of doing this is given in the note on 'Modifying the Labour Market Treatment Included in E3ME' (Figures 3.6 and B2). Even so, this still falls somewhat short of providing a full treatment of the effects of changes in intra-EU trade upon the skill intensity of national production, EU production and trade flows. This is because the initial trade responses, assuming corporate strategies do not change, will generate a first round of dynamic effects but will then be followed by (or, indeed, accompanied by) anticipatory behaviour through changes in corporate strategies affecting investment in human and physical capital.

A second and further round of dynamic effects will then materialize in which changes in skill intensity do not arise simply through differential trade flow responses but through the movement of the relevant industries up the value-added spectrum, etc. It is these effects which are particularly vulnerable to occupational supply constraints. The intention is to deal with this aspect 'off-the-model' but it may be possible to provide a simple illustration of its potential importance while also explaining clearly why capturing such an effect formally within the model is so difficult to do.

Table 3.9a.   Results of testing revised labour market structure: long-run equation

Long-run equation

E3ME sector	BE	DK	DO	DW	EL	ES	FR	IR	IT	IS	LX	NL	PO	UK	EU
08	1.214	0.000	0.000	0.469	0.000	0.544	0.272	0.164	1.848	0.000	0.000	0.000	0.079	0.000	0.622
09	3.406	0.000	0.000	0.641	0.000	0.509	0.000	0.000	0.496	0.000	0.361	0.000	0.000	0.737	0.537
10	0.516	0.000	0.000	0.002	0.000	0.091	0.000	0.000	0.090	0.000	0.000	0.000	0.000	0.000	0.044
11	0.629	0.000	0.000	1.418	0.000	0.387	0.000	0.128	0.201	0.000	0.000	0.000	0.107	0.000	0.570
12	0.000	0.000	0.000	1.420	0.000	0.216	0.000	0.064	0.000	0.000	1.281	0.000	0.000	0.151	0.603
13	0.911	0.000	0.000	0.000	0.000	2.547	0.418	0.674	0.271	0.000	0.000	0.000	0.000	0.126	0.258
14	0.079	0.000	0.000	0.000	0.000	0.000	1.284	0.053	0.272	0.000	0.000	0.000	0.000	0.000	0.230
15	0.000	0.000	0.000	4.757	0.000	0.607	0.000	0.323	0.000	0.000	0.396	0.000	0.000	0.039	1.721
16	0.515	0.394	0.000	0.021	0.000	0.825	0.000	0.345	0.000	0.000	0.000	0.000	0.000	0.146	0.172
17	2.019	1.413	0.000	0.000	0.000	0.699	0.089	0.000	0.000	0.000	0.000	0.000	0.000	0.144	0.181
18	0.000	0.503	0.000	0.935	0.000	0.976	0.000	0.000	0.557	0.000	0.000	0.000	0.000	0.470	0.553
19	1.449	0.000	0.000	4.946	0.000	1.439	0.000	0.274	0.221	0.000	2.924	0.000	0.000	0.000	1.939
21	0.000	0.650	0.000	1.476	0.000	0.465	0.665	0.000	0.061	0.000	0.342	0.000	0.207	0.201	0.556

Source: E3ME model.

Table 3.9b.   Results of testing revised labour market structure: dynamic equation

Dynamic equation

E3ME sector	BE	DK	DO	DW	EL	ES	FR	IR	IT	IS	LX	NL	PO	UK	EU
08	0.352	0.000	0.000	0.000	0.000	0.358	0.000	0.000	0.000	0.000	0.000	0.000	0.083	0.000	0.023
09	1.487	0.000	0.000	0.235	0.000	0.000	0.000	0.376	0.000	0.000	0.000	0.000	0.247	0.633	0.216
10	0.078	0.000	0.000	0.000	0.000	0.000	0.321	0.000	0.110	0.000	2.574	0.000	0.000	0.053	0.020
11	0.000	0.000	0.000	0.000	0.000	0.000	0.383	0.000	0.103	0.000	0.000	0.000	0.015	0.191	0.100
12	0.000	0.000	0.000	0.306	0.000	0.939	0.000	0.000	0.000	0.000	1.303	0.000	0.218	0.217	0.103
13	0.001	0.000	0.000	0.000	0.000	0.335	0.000	0.021	0.192	0.000	0.000	0.000	0.000	0.000	0.039
14	0.194	0.000	0.000	0.000	0.000	0.267	0.281	0.479	0.073	0.000	0.000	0.000	0.492	0.031	0.086
15	0.235	0.000	0.000	0.000	0.000	0.000	0.181	0.499	0.299	0.000	0.108	0.000	0.000	0.000	0.076
16	0.138	0.000	0.000	0.340	0.000	0.000	0.000	0.137	0.000	0.000	0.287	0.000	0.000	0.969	0.246
17	0.033	0.000	0.000	0.393	0.000	0.137	0.000	0.003	0.000	0.000	0.000	0.000	0.058	0.548	0.161
18	0.000	0.240	0.000	1.816	0.000	0.016	0.000	0.000	0.346	0.000	0.038	0.000	0.008	0.159	0.672
19	0.457	0.000	0.000	0.000	0.000	0.000	0.000	0.000	0.000	0.000	1.128	0.000	0.028	0.164	0.042
21	0.301	0.000	0.000	0.491	0.000	0.000	0.000	0.000	0.000	0.000	0.000	0.000	0.000	0.066	0.129

Source: E3ME model.

3.3.6. Conclusions

Overall, there is little evidence to suggest that underlying behavioural parameters have changed during the last decade. The evidence drawn for the analysis of quantities and prices which reveals the extent of movements in the allocation of labour and in relative wages cannot itself be conclusive. But even a rather loose interpretation of these data, inclined to accept rather than reject the hypothesis of rising labour market flexibility (as defined by increased responsiveness of demand, supply and wage adjustments to economic conditions) produces a weak a priori case.

One possible implication of the above is, of course, that if increased flexibility has not occurred, it cannot have been caused by government and corporate responses to the SMP. Without an effect, we need not look for a cause. However, another interpretation is that the evidence is too partial to allow such a strong conclusion. Limitations relating to both data and model specifications, the possibility of delays in the responses of labour market institutions and actors to more vigorous product market competition, and the difficulty of distinguishing cyclical from longer-term factors in the last half of the 1980s and early 1990s, suggest that there is room for much more research.

Meanwhile, the working assumption, on the basis of this review of labour market evidence, is that it would not be appropriate to impose parameter changes (including constant term adjustments) in the labour market equations on the grounds that European labour markets have become more flexible because of the pressures of product market reform.

3.4. A proxy for the effect of the single market programme

3.4.1. Choice of variable – sensitivity analysis

The choice of a variable to represent the integration effect of the single market programme is far from easy. Initially, using a trade indicator to act as a proxy for the single market (as proposed in Baldwin and Venables (1995)) seemed a way forward in the analysis. On closer inspection, however, a variable such as the proportion of intra-EU to total trade is not a suitable candidate, mainly because trade is an endogenous variable and has to be treated as such – the same goes for most other variables in the model.

After much deliberation the decision was taken to construct a synthetic variable which takes a zero value up to (and including) 1985 and unity in 1992. Between these years, the functional form of the variable is important, as it will dictate the speed over which the implementation of SMP policies is expected to have taken place. In the simplest case, one could imagine a standard dummy variable which shifts immediately from zero to unity in a sharp, one-period, movement. It seems more intuitive, however, that SMP policies and their effects have been transmitted gradually over time, so that a continuous (as opposed to a discrete) variable is preferred for the analysis. Therefore the type of slope employed for the synthetic variable needs to be reviewed.

The alternative assumptions for the functional form of the SMP variable are investigated by means of their resulting estimated impact on intra- and extra-EU exports and imports by sector. Four possible slopes of the synthetic variable are examined: linear, exponential, logistic and deceleration (see Figure 3.7). Each form proposes a different view on the likely phasing of the impact of the SMP. By comparing these different forms, estimation results using empirical evidence can be used to guide selection of the 'best' specification of the SMP variable.

Inspection of the R-squared statistics in Table 3.10 shows that in all cases the variation in imports is better explained by the dynamic equation than the variation in exports. There is little difference among the R-squared statistics for the four different specifications of the SMP variable. Perhaps surprisingly, the best candidate for the import equations is the worst for the export equations. In principle, there could be reasons why imports might respond differently from exports to the SMP, but this interpretation is probably greater than the data will bear.

Table 3.10 presents estimates of the partial effects (i.e. those represented by the coefficient of the SMP variable in the trade equations, with no account taken of the SMP on the other variables in the equations). The table distinguishes short-run impacts of the SMP (i.e. the impact of a change in the SMP variable in the current period) from long-run impacts (the final effect, taking account of lagged adjustments).

A striking result is that the impact of the SMP on imports is positive, under all specifications for the SMP variable. In other words, the SMP has boosted both intra-EU and extra-EU imports, after taking account of the major factor determining imports (the level of economic activity). The effect on intra-EU imports is not surprising: lower trade barriers lead to more trade in both directions. However, the sizeable boost to extra-EU imports is, prima facie, surprising. It should be recalled that an improvement in EU competitiveness resulting from the SMP would be reflected in other variables in the equations. What the coefficient on the SMP variable shows is the impact on imports for given levels of economic activity and price competitiveness. Nevertheless, the positive effect requires some explanation. It may reflect the lower costs to non-EU producers of exporting to a single market.

While altering the functional forms produces some changes in the estimated SMP impact, the quantitative implications of the results are unchanged. Given the inconclusive results obtained by basing the choice on statistical criteria (see Tables 3.10a and b) an exponential shape was deemed the most appropriate form to use as it would imply a slow start-up but then a steady increase in the process in the run-up to 1992 (a well-publicized date by which time the SMP was expected to have been completed).

Table 3.10a. Sensitivity tests on the SMP proxy: R-squared criteria

| | Exports | | Imports | |
	Extra-EU	Intra-EU	Extra-EU	Intra-EU
		Adjusted R-squared dynamic equation		
Linear	0.43	0.27	0.69	0.69
Exponential	0.38	0.24	0.70	0.71
Logistic	0.41	0.25	0.68	0.69
Deceleration	0.44	0.26	0.69	0.69

Source: E3ME model.

Table 3.10b. Sensitivity tests on the SMP proxy: short- and long-run effects (%)

| | Exports | | Imports | |
	Extra-EU	Intra-EU	Extra-EU	Intra-EU
Short-run				
Linear	18	19	36	14
Exponential	25	8	30	15
Logistic	13	14	24	10
Deceleration	8	11	23	16
Long-run				
Linear	7	26	33	30
Exponential	10	25	32	27
Logistic	4	19	25	24
Deceleration	6	22	25	29

Note(s): Weighted averages of the coefficients estimated for the SMP variable across 30 commodities and the EU11 countries, using the value of trade in 1991 as weights. The coefficients are expressed as percentages (i.e. multiplied by 100).
Source: E3ME model.

3.4.2. Location of the SMP proxy

Having decided on the form of the variable designed to proxy the effect of the SMP, the rest of this chapter is devoted to reporting the equation results from introducing this variable into certain areas of the E3ME model. Since there are a limited number of observations and a complicated existing explanation with many explanatory variables, there is a danger of 'over-fitting' and a subsequent loss of enough degrees of freedom if the whole equation structure is allowed to change. For each chosen set of stochastic equations, the SMP variable has therefore been entered as an effect on the constant term only. Including the SMP term on the other explanatory variables one by one would have enabled a more detailed investigation into the effects of the SMP but at the cost of over-complicating an already complex exercise.

Figure 3.7.   Functional forms for the SMP variable

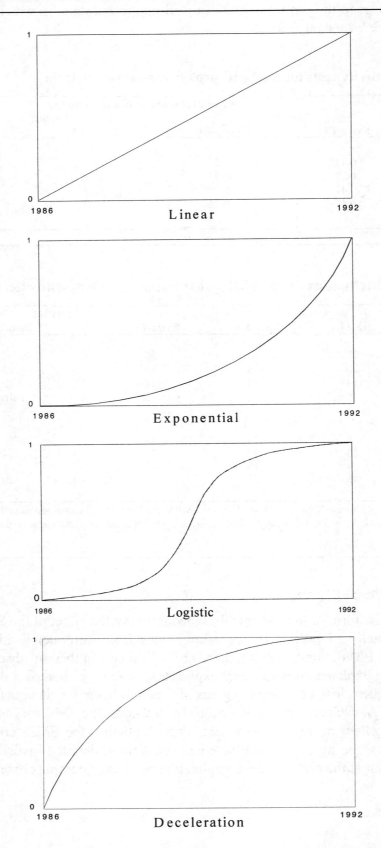

In all cases (including the investment equations analysed above), the equations are estimated using instrumental variables, with the single market variable included as one of the instruments. Equations are modelled using an Engle–Granger two-step methodology which involves first estimating the long-run equation in levels and then estimating the dynamic equation in differences with a lagged dependent variable and the lagged error term from the levels equation. A complete explanation of the derivation and definition of all equation variables can be found in Appendix B of the Technical Annex. Equations are typically of log–log format, with variables measured in ECU and 1985 prices were applicable.

3.4.3. Effects of the SMP proxy

Tables 3.11a and b and 3.12a and b summarize the effects of the SMP on intra-EU and extra-EU exports and imports. These effects were built into the trade volume equations by means of quantitative and qualitative restrictions on the impact of the SMP variable. The literature review in Chapter 2 and the results of testing the single market proxy (see Tables 3.13a to 3.16b) were both used to measure the range of the SMP effect, from +++ (showing strong positive effects) to – – – (showing strong negative effects).

Tables 3.11a and b highlight the preponderance of positive effects of the SMP on intra-EU trade although the impact on imports is more diverse than for exports, where the effects are greatest for Italy, Spain, Luxembourg and Portugal. However, the effects for Spain and Portugal may have been exaggerated due to the difficulty in isolating the impact of the SMP from the effects due to accession to the EU. Various other contemporaneous issues also exist, apart from that of Spanish and Portuguese accession. These are foreign direct investment and structural funds spending. Both of these factors could potentially distort the estimated impact of the SMP, but due to lack of sufficient data it did not prove possible to find an acceptable solution to the problem. These issues are discussed further in Section 4.4. At the national level, though, all Member States are expected to experience increases in trade resulting from the SMP.

Tables 3.11b and 3.12b also show that there seems to be consistency among the sectors and Member States in terms of sign, with most sectors displaying positive elasticities across the Member States. In general, strong positive effects are evident for many countries, and in particular Belgium, Denmark, Germany, Ireland, Luxembourg and the Netherlands.

Exports

Comparing Tables 3.11a and 3.12a shows that the impact of the SMP on intra-EU exports is typically greater than for extra-EU exports. No positive effects for extra-EU exports are recorded for any Italian sector, while a strong positive and evenly distributed impact is recorded for intra-EU exports. The net balance of effects for extra-EU exports (obtained by summing over the negative and positive signs) is negative for all countries except Germany, Luxembourg, the Netherlands and Portugal.

The strongest positive effects for intra-EU exports in Ireland are for Electrical Goods. In Chapter 2 we noted that in 1990 this sector had a high level of intra-industry trade within the EU with export-intensity exceeding import penetration. This industry in Ireland tends to export a large share of its output, while importing the larger electrical goods which are not produced domestically. The industry was identified as being relatively competitive and successful in its own specialized export markets.

Luxembourg enjoys strong positive effects on intra- and extra-EU exports in the Non-metallic Mineral Products and Chemicals industries. In 1990, the former industry was characterized by high levels of intra-industry trade both within the EU and with the rest of the world. Export intensity also exceeded import penetration. However, this export orientation is also particularly notable in the Chemicals industry. R&D expenditure was also very high in both of these sectors relative to the Luxembourg average for manufacturing (see Tables 2.12a and b).

The impact of the SMP on Spanish exports was by far the most positive for intra-EU trade. For example, strong positive effects are evident for intra-EU exports in Non-metallic Mineral Products, Chemicals, Metal Products, Agricultural and Industrial Machinery, and Textiles, Clothing and Footwear. By contrast, none of these sectors shows positive effects for extra-EU exports. A positive impact is evident for both intra- and extra-EU exports in the Spanish Transport Equipment industry. In Chapter 2 we identified this sector as having the greatest potential to benefit from the SMP mainly because of its strong orientation towards intra-EU trade.

In Germany's Office Machines industry, a positive effect is evident for extra-EU exports but a negative effect is apparent for intra-EU exports. This sector in Germany only accounts for a small percentage of total manufacturing output and extra-EU trade displayed a rising deficit between 1980 and 1990. The sector tends to be the most open to trade relative to other sectors within Germany, with a relative import penetration ratio of over 2 in 1990. Office Machines is also classed as a differentiated product/fragmented market, indicating some brand classification but through a large number of firms. Investment and R&D intensities tend to be higher than the manufacturing average, suggesting that the German industry is perhaps more concentrated than the EU average. When compared on an EU basis, relative capital intensity is higher than the average but R&D intensity is one of the lowest. Therefore, the sector was expected to be adversely affected by the SMP, although whether the SMP variable has picked up any effects due to the programme or instead has measured the increasing globalization of the sector, is open to question.

To summarize, the results confirm that the SMP has a greater (and a more positive) impact on intra-EU exports than on extra-EU exports. This was expected because the removal of trade barriers within the EU should lead to more two-way trade between Member States, with only minimal effect on exports to countries outside of the EU.

Imports

The impact of the SMP on intra-EU imports appears to be strongest for Spain, where the SMP has the largest possible impact (i.e. +++ in our tables) in 8 of the 13 manufacturing industries. It is noted in Chapter 2 that in the late 1980s one notable feature of the Spanish economy was the sharp increase in manufacturing imports.

For intra-EU imports in Ireland, the strongest impact of the SMP is on Office Machines. This is an important sector of the Irish economy but in 1990 import penetration greatly exceeded export intensity in 1990 (see Tables 2.9a and b, A6a and b). As noted in Chapter 2, many types of machinery are simply not produced in Ireland and hence import penetration tends to be high. This may not necessarily reflect the weak competitive position of existing firms, but merely that in some subsectors import penetration is 100%.

Some negative effects on extra-EU imports do exist, however, in particular for Germany in Electrical Goods, Transport Equipment, Rubber and Plastic Products, and Paper and Printing Products industries. This is potentially significant as Germany's share of EUR-12 output in these sectors is relatively large. Tables 2.6a and b also show that Transport Equipment in 1990, for example, was a relatively open sector (particularly for exports) and has a relatively high wage rate and above average capital and R&D intensities. Transport Equipment also displays a tendency towards one-way trade flows, although it is classified as producing a differentiated product in a segmented market. As such, the sector appears well placed to benefit from the SMP and the fall in imports to this sector should provide a boost to employment when the results are transferred to the model. It is also interesting to note that trade for this sector is more intra-industry orientated for EU trade than with the rest of the world. This characteristic may have some bearing on the size of the coefficient in the extra-EU imports equation being larger (in the absolute sense) than for intra-EU imports.

In Ireland, the SMP appears to have had a particularly strong positive effect on extra-EU imports. For example, as noted in Chapter 2, the Irish Textiles, Clothing and Footwear industry was expected to experience increased import competition from developing countries once clothing imports into any part of the EU were permitted to be sold freely in Ireland. This feature of the SMP is common for the Textiles sector of most Member States.

In Chapter 2 we identified France's Textiles, Clothing and Footwear industry as being in a weak position prior to the SMP. In particular, foreign competition was strong and the single market was expected to boost French imports from other EU countries. However, Tables 3.11b and 3.12b show that the effect of the SMP on extra-EU imports is stronger than the effect on intra-EU imports. This is also the case for the Netherlands' Textiles, Clothing and Footwear industry. By 1990, this industry was characterized by low levels of import penetration from other EU countries (see Tables 2.12a and b) and the subsequent removal of remaining non-tariff barriers to trade was expected to lead to strong competitive pressures.

Table 3.11a.   Effect of the SMP on intra-EU exports

E3ME sector	BE	DK	DW	EL	ES	FR	IR	IT	LX	NL	PO	UK
Ferrous & Non-ferrous Metals	0	+	+		−	+	+	++	++	+	++	+
Non-metallic Mineral Products	0	+	+		+++	−	+	+	++	−	− −	+
Chemicals	+	+	+		++	+	+	+	+++	0	0	+
Metal Products	+	++	+		++	+	0	+	+++	+	− −	+
Agricultural & Industrial Machinery	0	+	+		++	+	++	−	+	0	0	+
Office Machines	−	0	−		−	−	+	+	0	+	0	−
Electrical Goods	0	0	0		−	+	+++	+	0	0	0	+
Transport Equipment	+	−	+		+	+	−	++	++	+	+	+
Food, Drink & Tobacco	+	+	−		++	+	0	++	+	+	+	+
Textiles, Clothing & Footwear	+	0	++		−	+	0	0	++	0	−	−
Paper & Printing Products	+	−	−		−	−	+	+	−	0	0	−
Rubber & Plastic Products	+	+	0		0	+	−	++	0	++	+	+
Other Manufactures	++	++	+++		0	+++	0	+	++	− −	+++	++

Note: +,++,+++ = strength of positive impact of SMP (+ < ++ < +++)
       0 = no impact of SMP
       −,− −,− − − = strength of negative impact of SMP (− < − − < − − −)
Source: Cambridge Econometrics.

Table 3.11b. Effect of the SMP on intra-EU imports

E3ME sector	BE	DK	DW	EL	ES	FR	IR	IT	LX	NL	PO	UK
Ferrous & Non-ferrous Metals	+	−	+		0	+	0	+	++	+	+	+
Non-metallic Mineral Products	+	+	+		+	+	+	+	+	0	++	++
Chemicals	++	+	++		+	+	+	+	0	0	0	0
Metal Products	+	+	+		+++	+	−	++	+	+	++	++
Agricultural & Industrial Machinery	+	0	0		+++	+	+	−	+	0	+	+
Office Machines	+	++	+		+	−	+++	+	0	−	−	−
Electrical Goods	+	+	+		+++	+	−	+	0	0	−	−
Transport Equipment	−	+	++		+++	−	+	+	+	0	+	+
Food, Drink & Tobacco	+	+	++		+++	+	++	−	0	+	+	+
Textiles, Clothing & Footwear	+	++	+		+++	+	+	+	−	+	+	+
Paper & Printing Products	+	++	+		++	0	++	−	−	+	−	−
Rubber & Plastic Products	+	+	+		+++	−	+	+	++	++	+	+
Other Manufactures	++	+	+		+++	++	+	−	0	0	+	+

Note: +,++,+++ = strength of positive impact of SMP (+ < ++ < +++)
      0 = no impact of SMP
      −,−−,−−− = strength of negative impact of SMP (− < −− < −−−)
Source: Cambridge Econometrics.

Table 3.12a. Effect of the SMP on extra-EU exports

E3ME sector	BE	DK	DW	EL	ES	FR	IR	IT	LX	NL	PO	UK
Ferrous & Non-ferrous Metals	−	0	0		−	−	0	0	+	−	+++	0
Non-metallic Mineral Products	+	−	+		−	−	−	0	+++	−	−	−
Chemicals	+	−	+		0	−	+	−	+++	0	+	+
Metal Products	−	0	0		0	−	0	−	−	0	−	0
Agricultural & Industrial Machinery	0	−	+		0	−	0	−	−	0	+	0
Office Machines	−	−	+		−	−	+	−	0	+	0	0
Electrical Goods	0	−	−		−	−	+	−	0	+	−	−
Transport Equipment	+	0	+		+	−	0	−	++	+	−	0
Food, Drink & Tobacco	+	0	−		−	+	−	−	0	−	−	0
Textiles, Clothing & Footwear	+	−	+		0	+	−	0	+	−	+	0
Paper & Printing Products	−	0	0		−	0	0	0	−	++	+	0
Rubber & Plastic Products	−	+	0		−	−	0	0	−	+	+	−
Other Manufactures	−	+	+		+	−	−	0	0	−	−	−

Note: +,++,+++ = strength of positive impact of SMP (+ < ++ < +++)
      0 = no impact of SMP
      −,−−,−−− = strength of negative impact of SMP (− < −− < −−−)
Source: Cambridge Econometrics.

Table 3.12b. Effect of the SMP on extra-EU imports

E3ME sector	BE	DK	DW	EL	ES	FR	IR	IT	LX	NL	PO	UK
Ferrous & Non-ferrous Metals	0	+	+		+	–	0	0	++	++	+	+
Non-metallic Mineral Products	+++	+	+		0	–	++	++	–	+++	–	–
Chemicals	+++	–	++		0	+	+++	0	++	+	0	0
Metal Products	++	+	+++		+++	+	0	++	+	++	+++	++
Agricultural & Industrial Machinery	+++	+	++		–	0	+	0	++	+	0	+
Office Machines	–	++	++		++	0	+++	+	0	0	0	–
Electrical Goods	+++	+	–		0	+	–	–	0	0	++	0
Transport Equipment	–	+	–		++	–	+++	0	– –	+	–	+
Food, Drink & Tobacco	++	++	0		++	+	+	+	+	+	–	–
Textiles, Clothing & Footwear	+++	++	+++		+++	+++	+++	+	0	+++	+	++
Paper & Printing Products	+	+	–		+++	0	+	+	–	0	++	–
Rubber & Plastic Products	++	0	–		++	0	+++	0	+++	+	++	++
Other Manufactures	+++	++	+++		+	+++	0	+	0	+++	+++	–

Note: +,++,+++ = strength of positive impact of SMP (+ < ++ < +++)

    0 = no impact of SMP

    –,– –,– – – = strength of negative impact of SMP (– < – – < – – –)

Source: Cambridge Econometrics.

Trade volume

The purpose of this study is to channel the effect of the single market through intra-EU trade, so the SMP variable must be imposed here. The tables below report the coefficient of the SMP variable on the following stochastic equations:

(a)   intra-EU imports,
(b)   extra-EU imports,
(c)   intra-EU exports,
(d)   extra-EU exports.

The anticipated effect on trade volumes will depend on how industry reacts to the increased level of competition anticipated by the removal of non-tariff barriers to trade. The literature survey in Chapter 2 has shown how this in turn depends on industry/market structure.

Before discussing some of the results in more detail, it is worth noting an issue which has been addressed in the model results, namely that of ensuring accounting constraints for intra-EU trade flows. Total intra-EU imports (of a particular sector) must equal total intra-EU exports (of the same sector) due to the closed nature of the area. This has been ensured in the SMP scenarios by accepting the effects of the SMP synthetic variable in each Member State's imports from the rest of the EU, then constraining the sum of the estimated effects on their exports to the EU to that of their imports, sector by sector. In other words, the overall scale of the single market effects on trade are assumed to be given by the effect of the policies in widening the choice of feasible imports; these higher imports imply higher exports which are allocated across Member States in accordance with the estimated effects of the policies on their exports.

Another important feature to look for in the results is the effect on intra-EU trade compared to extra-EU trade. Questions such as 'are the intra-EU and extra-EU effects compensating or reinforcing' arise, i.e. are there trade creation versus trade diversion effects?

The revised specifications of the four sets of equations being estimated for the SMP effect are as follows, with the *r*, *i* and *t* subscripts representing region, industry and time, respectively:

(a)   Import volume

Intra-EU imports (see Tables 3.13a and b)

$$QIM_{rit} = \alpha_0 + \alpha_1 * SMP_t + \alpha_2 * QRDDI_{rit} + \alpha_3 * PQRM_{rit} + \alpha_4 * PQRD_{rit} + \alpha_5 * REX_{rt} + \alpha_6 * YRKE_{rit} + u_{rit}$$

$$\Delta QIM_{rit} = \beta_0 + \beta_1 * \Delta SMP_t + \beta_2 * \Delta QRDD_{rit} + \beta_3 * \Delta PQRM_{rit} + \beta_4 * \Delta PQRD_{rit} + \beta_5 * \Delta REX_{rt} + \beta_6 * \Delta YRKE_{rit} + \beta_7 * YYN_{rit} + \beta_8 * \Delta QIM_{rit-1} + \beta_9 * \hat{u}_{rit-1} + v_{rit}$$

where

QIM = intra-EU import volume

SMP = proxy for the effect of the single market programme

$QRDDI$ = sales to the domestic market

$PQRM$ = import price

$PQRD$ = price of sales to the domestic market

REX = exchange rate

$YRKE$ = technological progress

YYN = ratio of actual to 'normal' output

Extra-EU imports (see Tables 3.14a and b)

$$QEM_{rit} = \alpha_0 + \alpha_1 * SMP_t + \alpha_2 * QRDDI_{rit} + \alpha_3 * PQRM_{rit} + \alpha_4 * PQRD_{rit} + \alpha_5 * REX_{rt} + \alpha_6 * YRKE_{rit} + u_{rit}$$

$$\Delta QEM_{rit} = \beta_0 + \beta_1 * \Delta SMP_t + \beta_2 * \Delta QRDD_{rit} + \beta_3 * \Delta PQRM_{rit} + \beta_4 * \Delta PQRD_{rit} + \beta_5 * \Delta REX_{rt} + \beta_6 * \Delta YRKE_{rit} + \beta_7 * YYN_{rit} + \beta_8 * \Delta QEM_{rit-1} + \beta_9 * \hat{u}_{rit-1} + v_{rit}$$

where

QEM = extra-EU import volume

(b)   Export volume

Intra-EU exports (see Tables 3.15a and 3.15b)

$$QIX_{rit} = \alpha_0 + \alpha_1 * SMP_t + \alpha_2 * QXZI_{rit} + \alpha_3 * \frac{PQRX_{rit}}{REX_{rt}}$$

$$+ \alpha_4 * \frac{PQRZ_{rit}}{REX_{rt}} + \alpha_5 * YRKE_{rit} + u_{rit}$$

$$\Delta QIX_{rit} = \beta_0 + \beta_1 * \Delta SMP_t + \beta_2 * \Delta QXZI_{rit} +$$

$$\beta_3 * \Delta \frac{PQRX_{rit}}{REX_{rt}} + \beta_4 * \Delta \frac{PQRZ_{rit}}{REX_{rt}} + \beta_5 * \Delta YRKE_{rit} + \beta_6 * \Delta QIX_{rit-1} + \beta_7 * \hat{u}_{rit-1} + v_{rit}$$

where

QIX = intra-EU export volume

$QXZI$ = other-EU domestic demand

$PQRX/REX$ = export price

$PQRZ/REX$ = other-EU price

Extra-EU exports (see Tables 3.16a and 3.16b)

$$QEX_{rit} = \alpha_0 + \alpha_1 * SMP_t + \alpha_2 * QAXI_{rit} + \alpha_3 * \frac{PQRX_{rit}}{REX_{rt}}$$

$$+ \alpha_4 * \frac{PQRE_{rit}}{REX_{rt}} + \alpha_5 * YRKE_{rit} + u_{rit}$$

$$\Delta QEX_{rit} = \beta_0 + \beta_1 * \Delta SMP_t + \beta_2 * \Delta QAXI_{rit}$$

$$+ \beta_3 * \Delta \frac{PQRX_{rit}}{REX_{rt}} + \beta_4 * \Delta \frac{PQRZ_{rit}}{REX_{rt}} + \beta_5 * \Delta YRKE_{rit} + \beta_6 * \Delta QEX_{rit-1} + \beta_- * \hat{u}_{rit-1} + v_{rit}$$

where

QEX = extra-EU export volume

$QAXI$ = rest of world activity index

$PQRE/REX$ = rest of world price

Trade prices

Prices are also important; this is arguably where the effects of the single market originate. There is a problem, however, due to lack of different price data between extra-EU and intra-EU destinations. However, one may expect to pick up some effect for the following reasons:

(a)   Most countries within the EU (analysed in this study) are quite well integrated (with the possible exception of Spain and Portugal) so that one would expect most intra-EU prices to be lower as a result of the SMP.

(b)   Regarding extra-trade prices, one may see little overall effect given that most SMP policies are aimed at intra-EU. For countries like Spain and Portugal, where accession coincided with the SMP, any Common External Tariffs imposed may reduce overall barriers to the outside world, implying a lowering of import prices.

(c)   In the short term, there may be certain costs associated with the adjustment process to new rules, policies, etc., so the initial impact on prices (from this effect alone) could well be positive.

(d)   Alongside (c) is the argument that increased competition, especially in markets dominated by trade which is both intra-industry and intra-EU, may induce producers to move up-market with their products in terms of quality and price, and so further differentiate themselves from their competitors.

So as with trade volumes, the effects on price could well be mixed. Again, it really depends on the type of sector and product being produced, and because the sectors being studied are themselves aggregates of many smaller industries, there could be many opposing effects taking place at the subsector level that are impossible to quantify, leaving only the aggregate net effect to measure.

The impact of introducing SMP effects in price equations endogenously into the model would be to decrease export and import volumes through reducing price competitiveness. However, if the increased prices are due to improvements in the quality of commodities (i.e. the single market has encouraged producers to specialize in higher priced varieties within the range of varieties in a commodity), it is not appropriate to treat these increases in prices as leading to a reduction in competitiveness: indeed if this were the case, there would be a trade-price effect reducing trade volumes which should be removed from the base scenario. It was therefore decided that an SMP effect should not be introduced into the trade price equations in the base simulations.

Table 3.13a. Results of testing single market proxy – intra-EU import volume: long-run equation

	1 BE	2 DK	3 DO	4 DW	5 EL	6 ES	7 FR	8 IR	9 IT	10 IS	11 LX	12 NL	13 PO	14 UK	15 EU
8 Ferr. & Non-f. Metals	0.149	-0.039	0.000	0.339	0.000	0.000	0.096	0.000	0.207	0.000	0.865	0.161	0.369	0.351	0.210
9 Non-metallic Min.Pr.	0.133	0.173	0.000	0.243	0.000	0.355	0.028	0.106	0.374	0.000	0.173	0.028	1.335	0.855	0.287
10 Chemicals	0.530	0.188	0.000	0.668	0.000	0.090	0.150	0.410	0.180	0.000	0.000	0.000	0.407	0.000	0.281
11 Metal Products	0.315	0.215	0.000	0.333	0.000	1.351	0.324	-0.074	0.534	0.000	0.207	0.240	1.051	0.631	0.479
12 Agri. & Indust. Mach.	0.162	0.000	0.000	0.000	0.000	1.028	0.173	0.097	-0.213	0.000	0.217	0.093	0.008	0.147	0.182
13 Office Machines	0.150	0.554	0.000	0.373	0.000	0.260	-0.339	1.500	0.309	0.000	0.000	-0.500	0.000	-0.500	-0.009
14 Electrical Goods	0.342	0.072	0.000	0.207	0.000	1.500	0.014	-0.264	0.133	0.000	0.000	0.000	0.755	-0.500	0.202
15 Transport Equipment	-0.146	0.071	0.000	1.000	0.000	1.500	-0.226	0.172	0.039	0.000	0.409	0.070	1.287	0.240	0.360
16 Food, Drink & Tobacco	0.488	0.342	0.000	0.550	0.000	1.305	0.190	0.658	-0.108	0.000	0.000	0.360	1.500	0.128	0.367
17 Tex., Cloth. & Footw.	0.193	0.589	0.000	0.226	0.000	1.500	0.303	0.230	0.484	0.000	-0.112	0.131	1.032	0.173	0.378
18 Paper & Printing Pr.	0.162	0.845	0.000	0.303	0.000	0.620	0.000	0.538	-0.500	0.000	-0.130	0.221	1.345	-0.319	0.122
19 Rubber & Plastic Pr.	0.220	0.133	0.000	0.485	0.000	1.340	-0.156	0.487	0.333	0.000	0.989	0.608	0.411	0.295	0.338
21 Other Manufactures	0.584	0.243	0.000	0.315	0.000	1.456	0.618	0.138	-0.500	0.000	0.000	0.000	0.300	0.297	0.339

Source: E3ME model.

Table 3.13b. Results of testing single market proxy – intra-EU import volume: dynamic equation

	1 BE	2 DK	3 DO	4 DW	5 EL	6 ES	7 FR	8 IR	9 IT	10 IS	11 LX	12 NL	13 PO	14 UK	15 EU
8 Ferr. & Non-f. Metals	-0.386	0.032	0.000	0.206	0.000	-0.035	0.000	0.230	0.211	0.000	0.449	0.355	0.589	0.210	0.101
9 Non-metallic Min.Pr.	-0.500	0.170	0.000	0.307	0.000	-0.500	0.060	0.751	0.034	0.000	0.127	0.626	1.435	0.422	0.179
10 Chemicals	0.196	0.000	0.000	0.000	0.000	0.445	0.204	0.072	0.045	0.000	0.519	0.000	0.545	0.442	0.168
11 Metal Products	0.056	0.187	0.000	0.223	0.000	0.325	0.000	0.300	0.178	0.000	0.427	-0.500	0.587	0.812	0.177
12 Agri. & Indust. Mach.	0.230	0.174	0.000	0.000	0.000	0.440	0.539	0.140	0.260	0.000	-0.500	-0.248	0.747	-0.293	0.177
13 Office Machines	0.000	1.261	0.000	0.271	0.000	0.000	0.404	0.000	0.153	0.000	0.000	0.617	0.000	0.000	0.224
14 Electrical Goods	0.043	-0.171	0.000	-0.500	0.000	0.000	0.000	-0.435	0.000	0.000	0.000	0.056	0.634	-0.500	-0.136
15 Transport Equipment	0.726	0.406	0.000	-0.018	0.000	0.678	0.000	0.000	0.000	0.000	0.303	0.245	0.668	1.275	0.362
16 Food, Drink & Tobacco	0.000	0.000	0.000	0.037	0.000	0.430	-0.256	0.005	-0.173	0.000	0.000	0.381	1.372	0.167	0.053
17 Tex., Cloth. & Footw.	0.110	-0.304	0.000	0.476	0.000	0.651	0.000	0.110	0.225	0.000	0.037	0.000	0.598	-0.022	0.217
18 Paper & Printing Pr.	0.000	0.354	0.000	0.219	0.000	0.571	-0.249	0.040	0.070	0.000	0.000	-0.058	0.849	0.000	0.050
19 Rubber & Plastic Pr.	0.101	0.211	0.000	0.350	0.000	0.076	0.000	-0.500	0.318	0.000	0.488	0.000	0.746	0.025	0.136
21 Other Manufactures	0.257	0.238	0.000	0.423	0.000	0.570	0.260	0.000	0.000	0.000	0.071	-0.119	1.080	0.000	0.174

Source: E3ME model.

Table 3.14a. Results of testing single market proxy – extra-EU import volume: long-run equation

		1 BE	2 DK	3 DO	4 DW	5 EL	6 ES	7 FR	8 IR	9 IT	10 IS	11 LX	12 NL	13 PO	14 UK	15 EU
8	Ferr. & Non-f. Metals	0.000	0.335	0.000	0.331	0.000	0.137	-0.287	0.000	0.000	0.000	0.988	0.897	0.331	0.166	0.160
9	Non-metallic Min.Pr.	1.324	0.037	0.000	0.334	0.000	0.000	-0.248	0.813	0.745	0.000	-0.500	1.500	-0.500	-0.500	0.323
10	Chemicals	1.299	-0.291	0.000	0.742	0.000	0.000	0.195	1.500	0.044	0.000	0.652	0.296	0.093	0.000	0.376
11	Metal Products	0.615	0.154	0.000	1.259	0.000	1.045	0.168	0.000	0.665	0.000	0.322	0.754	1.054	0.588	0.841
12	Agri. & Indust. Mach.	1.500	0.089	0.000	0.764	0.000	-0.040	0.000	0.118	0.000	0.000	0.526	0.154	0.117	0.129	0.352
13	Office Machines	-0.359	0.722	0.000	0.785	0.000	0.606	0.000	1.500	0.092	0.000	0.000	0.000	0.000	-0.500	0.221
14	Electrical Goods	1.486	0.072	0.000	-0.500	0.000	0.000	0.133	-0.500	-0.274	0.000	0.000	0.339	0.650	0.000	-0.147
15	Transport Equipment	-0.500	0.365	0.000	-0.500	0.000	0.672	-0.036	1.500	0.000	0.000	-0.785	-0.342	-0.500	0.245	0.005
16	Food, Drink & Tobacco	0.526	0.667	0.000	0.000	0.000	0.516	0.311	0.461	0.156	0.000	0.445	1.243	-0.500	-0.499	0.075
17	Tex., Cloth. & Footw.	1.443	0.861	0.000	1.500	0.000	1.500	1.500	1.139	0.020	0.000	0.000	0.054	0.027	0.593	1.124
18	Paper & Printing Pr.	0.152	0.019	0.000	-0.029	0.000	1.500	0.000	0.376	0.325	0.000	-0.311	0.200	0.704	-0.170	0.081
19	Rubber & Plastic Pr.	0.629	0.082	0.000	-0.500	0.000	0.500	0.000	1.196	0.038	0.000	1.500	1.500	0.559	0.670	0.059
21	Other Manufactures	1.601	0.740	0.000	1.333	0.000	0.126	1.367	0.000	0.116	0.000	0.000	1.500	1.500	-0.500	1.001

Source: E3ME model.

Table 3.14b. Results of testing single market proxy – extra-EU import volume: dynamic equation

		1 BE	2 DK	3 DO	4 DW	5 EL	6 ES	7 FR	8 IR	9 IT	10 IS	11 LX	12 NL	13 PO	14 UK	15 EU
8	Ferr. & Non-f. Metals	-0.056	-0.269	0.000	0.000	0.000	0.000	0.000	1.500	-0.033	0.000	0.000	0.000	0.000	1.500	0.219
9	Non-metallic Min.Pr.	1.500	-0.252	0.000	0.753	0.000	0.000	0.051	0.000	0.688	0.000	0.000	1.500	0.000	0.000	0.565
10	Chemicals	0.628	0.000	0.000	0.000	0.000	-0.500	0.000	1.036	0.294	0.000	0.000	-0.100	0.324	0.000	0.062
11	Metal Products	0.000	1.500	0.000	0.331	0.000	0.137	1.500	0.000	0.516	0.000	0.572	0.000	0.000	1.228	0.636
12	Agri. & Indust. Mach.	1.500	0.000	0.000	0.299	0.000	0.000	1.500	0.000	0.000	0.000	0.000	1.500	0.000	-0.500	0.384
13	Office Machines	1.500	1.500	0.000	0.000	0.000	0.000	0.148	0.000	0.410	0.000	0.000	1.500	1.500	0.000	0.141
14	Electrical Goods	0.254	0.000	0.000	0.442	0.000	1.500	0.000	-0.369	1.500	0.000	0.000	0.207	1.500	0.000	0.446
15	Transport Equipment	0.035	0.000	0.000	0.697	0.000	0.000	0.000	0.778	0.000	0.000	0.000	1.500	1.500	0.000	0.306
16	Food, Drink & Tobacco	0.100	0.258	0.000	0.537	0.000	0.000	0.000	0.480	1.500	0.000	-0.291	0.409	0.000	0.000	0.299
17	Tex., Cloth. & Footw.	0.346	0.636	0.000	0.000	0.000	0.290	0.395	0.913	1.500	0.000	0.000	1.500	-0.061	1.500	0.558
18	Paper & Printing Pr.	0.000	0.087	0.000	0.361	0.000	1.500	0.177	0.463	0.031	0.000	0.000	0.466	1.500	1.500	0.663
19	Rubber & Plastic Pr.	0.530	0.000	0.000	-0.500	0.000	0.000	0.043	0.404	-0.464	0.000	0.000	1.500	0.000	0.000	-0.117
21	Other Manufactures	0.659	0.000	0.000	0.000	0.000	0.129	-0.064	-0.194	0.000	0.000	1.500	1.500	1.500	0.000	0.174

Source: E3ME model.

Table 3.15a. Results of testing single market proxy – intra-EU export volume: long-run equation

		1 BE	2 DK	3 DO	4 DW	5 EL	6 ES	7 FR	8 IR	9 IT	10 IS	11 LX	12 NL	13 PO	14 UK	15 EU
8	Ferr. & Non-f. Metals	0.000	0.311	0.000	0.367	0.000	0.718	0.328	0.143	0.737	0.000	0.586	0.132	1.500	0.793	0.400
9	Non-metallic Min.Pr.	0.000	0.194	0.000	0.239	0.000	1.000	-0.435	0.136	0.351	0.000	0.721	0.228	1.500	-0.500	0.182
10	Chemicals	0.197	0.104	0.000	0.072	0.000	0.566	0.177	0.000	0.498	0.000	1.172	0.000	0.727	0.000	0.133
11	Metal Products	0.308	0.583	0.000	0.094	0.000	0.540	0.229	0.000	0.442	0.000	1.007	0.368	1.500	-0.500	0.228
12	Agri. & Indust. Mach.	0.000	0.071	0.000	0.000	0.000	0.523	0.043	0.707	0.466	0.000	0.418	0.000	1.500	0.000	0.128
13	Office Machines	-0.500	0.000	0.000	-0.140	0.000	0.350	-0.500	0.009	0.162	0.000	0.000	0.000	-0.500	0.000	-0.105
14	Electrical Goods	0.368	0.000	0.000	0.000	0.000	0.559	0.224	1.500	0.271	0.000	0.000	0.000	1.500	0.000	0.194
15	Transport Equipment	0.150	-0.460	0.000	0.150	0.000	0.150	0.150	-0.500	0.562	0.000	-0.184	0.140	1.500	0.150	0.198
16	Food, Drink & Tobacco	0.426	0.038	0.000	0.075	0.000	0.867	0.428	0.402	0.561	0.000	0.467	0.418	0.698	0.149	0.341
17	Tex., Cloth. & Footw.	0.390	0.000	0.000	1.000	0.000	-0.355	0.097	0.000	0.000	0.000	0.597	0.192	0.000	-0.500	0.192
18	Paper & Printing Pr.	0.092	-0.500	0.000	-0.500	0.000	0.611	-0.320	0.336	0.383	0.000	-0.500	0.028	0.718	0.051	-0.074
19	Rubber & Plastic Pr.	0.167	0.443	0.000	0.270	0.000	0.000	0.053	-0.367	0.676	0.000	0.000	0.551	1.500	0.141	0.265
21	Other Manufactures	0.671	0.873	0.000	1.130	0.000	0.000	1.161	0.000	0.126	0.000	0.863	-0.500	0.301	1.477	0.751

Source: E3ME model.

Table 3.15b. Results of testing single market proxy – intra-EU export volume: dynamic equation

		1 BE	2 DK	3 DO	4 DW	5 EL	6 ES	7 FR	8 IR	9 IT	10 IS	11 LX	12 NL	13 PO	14 UK	15 EU
8	Ferr. & Non-f. Metals	0.000	0.000	0.000	0.360	0.000	0.724	0.379	0.000	0.412	0.000	1.085	0.544	0.109	0.719	0.400
9	Non-metallic Min.Pr.	0.000	0.000	0.000	0.284	0.000	0.599	-0.156	0.000	0.459	0.000	0.000	0.254	0.592	0.000	0.221
10	Chemicals	0.251	0.000	0.000	0.000	0.000	0.269	0.061	0.000	0.457	0.000	1.619	0.112	0.000	0.000	0.097
11	Metal Products	0.337	0.181	0.000	0.137	0.000	0.573	0.335	0.000	0.442	0.000	0.702	0.250	1.340	-0.347	0.239
12	Agri. & Indust. Mach.	0.156	0.000	0.000	0.176	0.000	0.382	0.000	0.088	0.416	0.000	0.273	0.000	0.168	0.000	0.165
13	Office Machines	-0.215	0.000	0.000	-0.163	0.000	0.000	0.000	0.000	0.000	0.000	0.000	0.000	-0.500	0.000	-0.046
14	Electrical Goods	0.216	0.176	0.000	0.000	0.000	0.164	0.080	0.270	0.000	0.000	0.000	0.000	0.000	0.000	0.043
15	Transport Equipment	0.145	0.000	0.000	0.000	0.000	0.043	0.148	-0.500	0.570	0.000	-0.500	0.000	0.792	0.000	0.118
16	Food, Drink & Tobacco	0.048	0.000	0.000	0.009	0.000	0.524	0.202	0.000	0.000	0.000	0.000	0.000	0.719	0.000	0.073
17	Tex., Cloth. & Footw.	0.147	0.000	0.000	0.150	0.000	0.153	0.127	0.097	0.092	0.000	0.000	0.170	0.000	0.148	0.115
18	Paper & Printing Pr.	0.176	-0.430	0.000	0.000	0.000	0.166	-0.309	-0.037	0.294	0.000	-0.327	0.122	0.325	0.168	0.046
19	Rubber & Plastic Pr.	0.070	0.094	0.000	0.000	0.000	0.276	0.216	0.000	0.426	0.000	0.507	0.343	0.554	0.097	0.169
21	Other Manufactures	0.000	0.000	0.000	0.078	0.000	-0.179	0.117	0.000	0.000	0.000	0.587	-0.149	0.168	0.319	0.058

Source: E3ME model.

Table 3.16a. Results of testing single market proxy – extra-EU export volume: long-run equation

	1 BE	2 DK	3 DO	4 DW	5 EL	6 ES	7 FR	8 IR	9 IT	10 IS	11 LX	12 NL	13 PO	14 UK	15 EU
8 Ferr. & Non-f. Metals	-0.376	0.000	0.000	0.000	0.000	-0.411	-0.500	0.000	0.000	0.000	0.092	-0.447	1.500	0.000	-0.148
9 Non-metallic Min.Pr.	0.265	-0.185	0.000	0.122	0.000	-0.500	-0.385	-0.500	0.000	0.000	1.220	0.098	0.000	-0.500	-0.055
10 Chemicals	0.000	-0.308	0.000	0.142	0.000	0.000	-0.312	0.000	-0.500	0.000	1.500	0.000	0.318	-0.500	-0.120
11 Metal Products	-0.500	0.000	0.000	0.000	0.000	0.000	-0.500	0.000	-0.500	0.000	-0.337	0.000	-0.500	0.000	-0.153
12 Agri. & Indust. Mach.	0.000	-0.380	0.000	-0.031	0.000	0.000	-0.500	0.012	-0.500	0.000	-0.006	0.000	0.067	0.000	-0.171
13 Office Machines	-0.500	0.370	0.000	0.410	0.000	1.500	-0.257	-0.649	-0.500	0.000	0.000	0.000	0.000	0.000	0.088
14 Electrical Goods	0.191	-0.388	0.000	-0.397	0.000	-0.500	-0.500	0.000	-0.303	0.000	0.000	0.161	1.439	-0.500	-0.353
15 Transport Equipment	0.415	0.000	0.000	-0.500	0.000	0.460	-0.331	0.000	-0.500	0.000	-0.138	0.000	-0.300	0.000	-0.286
16 Food, Drink & Tobacco	-0.500	0.000	0.000	-0.500	0.000	-0.500	-0.500	-0.500	-0.500	0.000	0.000	-0.500	0.054	0.000	-0.365
17 Tex., Cloth. & Footw.	0.028	-0.500	0.000	0.192	0.000	0.000	0.080	-0.500	0.000	0.000	0.146	-0.275	0.158	0.000	0.052
18 Paper & Printing Pr.	-0.397	0.000	0.000	0.000	0.000	-0.500	0.000	0.000	0.000	0.000	-0.500	0.519	0.267	0.000	0.005
19 Rubber & Plastic Pr.	-0.500	0.117	0.000	0.005	0.000	-0.500	-0.500	0.000	0.000	0.000	-0.500	0.279	0.234	-0.414	-0.150
21 Other Manufactures	-0.500	0.414	0.000	0.309	0.000	0.229	-0.500	-0.500	0.000	0.000	0.000	-0.325	-0.500	-0.500	-0.027

Source: E3ME model.

Table 3.16b. Results of testing single market proxy – extra-EU export volume: dynamic equation

	1 BE	2 DK	3 DO	4 DW	5 EL	6 ES	7 FR	8 IR	9 IT	10 IS	11 LX	12 NL	13 PO	14 UK	15 EU
8 Ferr. & Non-f. Metal	0.000	0.000	0.000	1.500	0.000	-0.080	0.000	0.000	0.000	0.000	-0.163	0.000	1.500	0.000	0.538
9 Non-metallic Min.Pr.	1.500	-0.464	0.000	0.000	0.000	0.000	-0.500	-0.597	0.000	0.000	0.000	1.500	0.000	0.000	0.131
10 Chemicals	-0.068	0.000	0.000	-0.087	0.000	0.000	0.000	0.000	1.500	0.000	1.500	-0.036	0.138	0.000	0.081
11 Metal Products	0.000	-0.113	0.000	-0.534	0.000	0.264	1.500	-0.369	-0.271	0.000	0.000	0.000	-0.385	0.000	-0.108
12 Agri. & Indust. Mach.	0.391	0.000	0.000	0.084	0.000	0.000	-0.319	0.000	0.000	0.000	0.113	0.000	1.500	0.000	0.022
13 Office Machines	0.000	0.000	0.000	0.000	0.000	1.500	0.000	0.000	0.000	0.000	0.000	0.000	-0.444	0.000	0.021
14 Electrical Goods	0.343	0.000	0.000	-0.189	0.000	0.000	0.000	1.500	-0.226	0.000	0.000	0.356	1.479	0.000	-0.033
15 Transport Equipment	0.000	0.000	0.000	0.063	0.000	1.500	0.000	0.000	0.000	0.000	0.000	0.000	1.500	0.000	0.067
16 Food, Drink & Tobacco	-0.133	-0.500	0.000	0.000	0.000	0.000	0.048	0.000	0.000	0.000	0.000	0.000	0.000	1.500	0.198
17 Tex., Cloth. & Footw.	1.500	0.000	0.000	0.123	0.000	-0.162	0.000	1.500	0.000	0.000	0.356	0.162	0.000	0.000	0.105
18 Paper & Printing Pr.	0.000	0.000	0.000	-0.002	0.000	-0.061	0.000	0.000	0.608	0.000	0.000	1.500	0.172	-0.178	0.094
19 Rubber & Plastic Pr.	0.000	0.000	0.000	0.043	0.000	-0.500	-0.607	0.000	0.000	0.000	0.000	0.262	0.361	-0.227	-0.101
21 Other Manufactures	-0.500	0.367	0.000	1.500	0.000	0.000	0.000	1.500	0.000	0.000	0.000	1.500	-0.668	0.000	0.522

Source: E3ME model.

Investment demand

It was originally intended that investment demand would be analysed by including the synthetic SMP variable in the investment equation, based on the logic that the SMP may separately have made more opportunities available by removing certain barriers (be they real or psychological). However, due to the constraints of data availability, it did not prove possible to add any more regressors into the investment equation, so this particular effect has not been empirically tested.

3.5. Modelling the Greek economy

From the start of the project, it was clear that incorporating Greece into the E3ME model would not be possible due to problems of data availability and consistency. Therefore, it was decided to attempt to estimate the effects of the SMP on Greece using off-model, but nonetheless quantitative, techniques. The following section describes the methodology used and presents a table of results for Greece which summarizes the effect on sectoral employment of the SMP.

3.5.1. Obtaining the net trade effect due to the SMP by running the E3ME trade equations

A simple generic form of the equations used to estimate trade in the E3ME model is presented in Equations 1 and 2. Equation 1 describes the long-run relationship (in levels) while Equation 2 links the long-run with the short-term dynamics (in first differences). The SMP variable is present in both equations and is described in more detail in Section 3.4.

Stage 1:

$$Y_t = \alpha_0 + \alpha_1 SMP_t + \alpha_2 X_t + u_t \quad (1)$$

Stage 2:

$$\Delta Y_t = \beta_0 + \beta_1 \Delta SMP_t + \beta_2 \Delta X_t + \lambda \Delta Y_{t-1} + \gamma ECM_{t-1} + v_t \quad (2)$$

To calculate the actual effect of the SMP variable on the dependent variable, one needs to combine the two equations. This is done in Equation 3. A further expansion is provided in Equation 4, where the actual impact of SMP_t on Y_t can be determined, as one might expect, from a mix between short- and long-term factors.

$$\Delta Y_t = \beta_0 + \beta_1 \Delta SMP_t + \beta_2 \Delta X_t + \lambda \Delta Y_{t-1}$$
$$+ \gamma(Y_{t-1} - \alpha_0 - \alpha_1 SMP_{t-1} - \alpha_2 X_{t-1}) + v_t \quad (3)$$

Expanding further to give the final simulating equation:

$$Y_t = (\beta_0 - \gamma \alpha_1) + (1 + \lambda + \gamma)Y_{t-1} - \lambda Y_{t-2} + \beta_1 SMP_t$$
$$- (\gamma \alpha_1 + \beta_1)SMP_{t-1} + \beta_2 X_t - (\beta_2 + \gamma \alpha_2)X_{t-1} + v_t \quad (4)$$

The existence of Y_{t-1} on the right-hand side of Equation 4 means that a static simulation is not sufficient to calculate the impact of the SMP proxy on the dependent variable. Note that lack of data meant that intra- and extra-EU trade equations could not be estimated. Instead, estimations for total exports and imports were performed. The following methodological sequence was therefore followed to calculate the SMP effect:

(a)   the equation is dynamically simulated with the SMP variable as described in Section 3.4;
(b)   a second dynamic simulation is performed with the SMP variable set to zero;
(c)   the effect on the dependent variable (exports and imports) of the SMP variable is the difference between paths determined in (a) and (b).

3.5.2.   Calculating the effect on output

From the methodology outlined at the beginning of Section 3.5, a net trade effect could be calculated for each sector. As far as possible, this was estimated in line with a priori expectations made from analysis of the available data (see Table 3.17). The net trade effect was then fed through to a change in output via the (implicit) relationship described by Equation 5.

$$Y_{it} \equiv D_{it} + (X_{it} - M_{it}) \quad (5)$$

where, for sector i and time t,

Y_{it} = total demand

D_{it} = domestic demand

$(X_{it} - M_{it})$ = external demand

This calculation is performed so that proportionally the change in output is smaller than the overall change in trade.

3.5.3.   Applying average elasticities

Having obtained the percentage change of output for each sector due to the SMP effect, a corresponding percentage change in employment can be calculated by applying the average EU11 output elasticity from the existing E3ME employment equations. This elasticity is measured to be 0.56, i.e. a 10% increase in output gives rise to a 5.6% rise in employment, other things being equal. Table 3.18 shows the numbers relating to the three stages mentioned above, ending with the cumulative sectoral employment effects for 1993 (the end of the simulation period).

It was intended that the effect on labour costs would be calculated in a similar manner, i.e. applying the percentage changes in employment to the employment elasticities in the labour cost equations. However, by nature of the way that employment enters into the labour cost equations (via productivity and employment effects) this was deemed too complex for this type of analysis.

For total manufacturing employment the estimated SMP effect is a fall of -0.27% by 1993, equivalent to approximately 2,000 people. Qualitatively speaking, the overall results agree with the somewhat pessimistic effect of the SMP estimated by other studies of Greece, reviewed by the ESRI study *The cases of Greece, Spain, Ireland and Portugal*, European

Commission (1997g). The results are also corroborated by the European Commission Business Survey (1997h), which reports more firms indicating a negative effect on employment prospects than a positive one. This result occurs whether responses are weighted by number of firms, number of employees or turnover.

At a sectoral level labour-intensive sectors such as Textiles, Clothing and Footwear fare quite badly, registering the largest individual employment fall of almost 5%. These results support the argument that highly protected, labour-intensive sectors are not likely to show much benefit from the SMP, and the more highly competitive environment it brings. Most other sectors report a negligible effect of the SMP on employment (a result in line with most other anecdotal evidence). The largest positive employment impact is in Office Machines (+1.12%), a sector which has a low labour intensity relative to Greek manufacturing as a whole. In the long run, it is possible that restructuring of industry will allow an improved competitive position for these types of firms, but in the short run this is likely to be achieved by a reduction in employment.

Table 3.17.   Effect of the SMP proxy on trade in Greece

E3ME sector		Exports	Imports
08	Ferrous & Non-ferrous Metals	+	+
09	Non-metallic Mineral Products	+	0
10	Chemicals	0	+
11	Metal Products	0	0
12	Agricultural & Industrial Machinery	0	+
13	Office Machines	−	+
14	Electrical Goods	0	+
15	Transport Equipment	−	+
16	Food Drink & Tobacco	+	0
17	Textiles, Clothing & Footwear	+	+
18	Paper & Printing Products	−	0
19	Rubber & Plastic Products	0	+
21	Other Manufactures	+	+

Source: E3ME model.

Table 3.18.   Effect of the SMP proxy on employment in Greece

E3ME sector		Net trade (million ECU)	Output % change	Employment % change
08	Ferrous & Non-ferrous Metals	+34.42	+1.3	+0.73
09	Non-metallic Mineral Products	+0.01	0.0	0.0
10	Chemicals	+0.19	0.0	0.0
11	Metal Products	-9.37	-0.4	-0.22
12	Agricultural & Industrial Machinery	0.0	0.0	0.0
13	Office Machines	+8.81	+2.0	+1.12
14	Electrical Goods	0.0	0.0	0.0
15	Transport Equipment	0.0	0.0	0.0
16	Food, Drink & Tobacco	+11.03	+0.2	+0.11
17	Textiles, Clothing & Footwear	-205	-8.8	-4.93
18	Paper & Printing Products	-2.06	-0.1	-0.06
19	Rubber & Plastic Products	0.0	0.0	0.0
21	Other Manufactures	+8.54	+0.6	+0.34
	Manufacturing total	-153.43	-0.48	-0.27

Source: E3ME model.

4. Interpretation of results

This chapter presents the results of the empirical work of the study. Section 4.1 presents the model simulation results, and discusses the estimated impact of the SMP by sector and country. Section 4.2 discusses the implications for employment by type of labour. Section 4.3 presents selected evidence from other studies carried out for the European Commission under the present review of the impact of the SMP, for comparison with this study's results. Section 4.4 outlines some issues that relate to the SMP that have not been directly dealt with in the model. Finally, Section 4.5 examines the sensitivity of the results to larger, imposed changes in exports and imports by sector.

4.1. Model simulations of employment and labour costs

This section explains briefly how the simulations were undertaken and presents the main results for the impact of the SMP on employment and labour costs by country and sector.

4.1.1. The calibrated model

An 'impact' model simulation of the kind examined here has to be compared with a 'base' case. In this study, the 'base case plus SMP' is what actually occurred over the period 1985–93, and the non-SMP *antimonde* 'base' represents an estimate of what would have happened if the SMP had not been implemented. However, it would not be appropriate simply to compare a model result for the *antimonde* with the actual outcome because in this case any failure of the model to explain the actual outcome would be attributed to the SMP. Instead, the model is calibrated to the actual outcome, by including in the solution a set of residuals for each equation so that the model results match the actual data; thus, the calibrated solution gives the actual values in the simulation. These residuals are then kept constant in the simulations, so that the difference between an *antimonde* and the 'base case plus SMP' reflects only the model's response to the impact under examination.

The model was simulated for the period 1985 to 1993 and calibrated for the following stochastic variables, all of which are disaggregated by sector and region except aggregate consumption and residual income which are only by region:

employment – YRE
hours worked – YRH
participation rate – LRP
industrial prices – PYH
export volume – QRX
import volume – QRM
export prices – PQX
import prices – PQM
industrial average earnings – YRW
aggregate consumption – RSC
industrial investment – KR
residual income – RRI

Selected non-stochastic endogenous variables have been calibrated including output and the price of consumers' expenditure. The following other stochastic variables are determined relative to other variables in the model:

investment in dwellings – RDW
disaggregate consumption – CR
energy demand – FR0, FRC, FRO, FRG, FRE

The following variables are exogenous to the model: rest of the world assumptions covering exchange rates, short- and long-term interest rates, industrial production, wholesale prices and commodity prices; EU demographic assumptions, such as population and migration by gender; policy assumptions by region (country), such as current and investment spending by government on areas such as defence, education and health, direct and indirect tax rates; and various energy and environmental assumptions.

4.1.2. Simulations

Two main sets of simulations were performed, testing for the effects of the SMP via intra-EU trade only, and via both intra-EU and extra-EU trade.

(a)   The first set of simulations used a version of the model having the SMP variable included in the intra-EU trade equations only. Both the import and export equations (by sector) have the synthetic SMP variable included, as discussed in Chapter 3. The extra-EU trade equations were estimated without the SMP effect included. The *antimonde* simulation was performed by 'switching off' the synthetic SMP variable, i.e. setting its value to zero over the whole period of the simulation. This causes intra-EU imports and intra-EU exports to change, by the amount of the estimated SMP effect. This then feeds through to output, employment and the other variables in the model including, via indirect influences such as relative labour costs, extra-EU trade. We describe this as model simulation 1.

(b)   The second simulation was with a version of the model having the SMP variable included in both the intra-EU trade equations and the extra-EU trade equations. The SMP variable was then 'switched off' for both the intra- and extra-EU trade equations. This is model simulation 2 with the results giving us the direct effects of the SMP on both intra- and extra-EU trade.

In all cases the synthetic SMP variable starts in 1986, and so this is the year in which the differences from base start to occur. In the first year or so, the short-term dynamic effects will be dominant, but later on the longer term effects come into play.

4.1.3. Notes on interpretation of results

In interpreting the model results the following points should be noted:

The reliability of the model reflects the reliability of the data. In creating a model of Europe with this level of disaggregation there are problems in obtaining consistent data for all the countries of the EU. In constructing the model, extensive collecting, updating and validation of the data was done by institutes in the Member States but for some sectors for some countries the data are not completely reliable. The suspect sectors are typically service sectors, as detailed historical data are often hard to find for many countries. These data deficiencies can cause problems in the model as there are numerous feedback loops from one sector to another, particularly in the wage equations where the wages in one sector depend on wages in other sectors and other regions.

Several procedures have been followed to limit such problems. First, in estimating the equations, extreme values of coefficients have been restricted to upper and lower limits. Second, in some cases, simpler specifications, for example shares of demand, have been adopted. And third, in a few cases the specific values have been fixed at actual levels. The model should be seen as providing reasonably consistent and comprehensive results at the European level.

As stated in more detail in Appendix B, the model is an econometric model with a complete specification of the long-term solution in the form of an estimated equation which has long-term restrictions imposed on its parameters. Economic theory, for example the recent theories of endogenous growth, informs the specification of the long-term equations and hence the properties of the model. The dynamic relationships are specified in terms of error correction models (ECM). Due to the large number of equations estimated, it is not possible to take account of effects which are specific to particular regions and sectors. The implication is that the SMP variable might occasionally reflect changes in sectoral responses that are not the result of the SMP. The comparison of the results with those expected in the literature as well as across sectors and Member States helps to guard against this danger.

The model results show what effect the SMP has had on trade and hence on employment. The estimated coefficients on the SMP have not been restricted to be positive or negative. The coefficients have been restricted as follows: first to remove a few extreme values and second to remove all non-significant negative effects (i.e. those with a t-ratio of less than 1).

Where a significant coefficient has been estimated for the synthetic SMP variable this may reflect other influences not captured in the other variables in the equations. While this is a feature of any econometric modelling exercise, there are reasons to be particularly cautious for those equations where a substantial proportion of the variance of the dependent variable remains unexplained. One way in which this study has attempted to avoid spurious attribution of impact has been to distinguish intra-EU and extra-EU trade. This avoids confusion of two quite different potential effects of the SMP: increasing trade within the EU, and changing the competitiveness of EU producers *vis-à-vis* the rest of the world.

In principle, the impact of the SMP could be tested in a similar way on a variety of other variables in the model. For example, there could be a direct SMP effect on import or export price equations, or a specific effect of trade on investment. As explained in Section 3.2, the absence of a direct link between investment and trade intensity means that any effects on investment generated by the model will be transmitted largely through output, and to a lesser extent through the impact of labour costs in the equation for investment demand. It was also planned to update the labour market equations, however, the confines of the present study did not allow these additional effects to be explored.

The results are presented as percentage difference from 'base' where 'base' is defined as the non-SMP *antimonde*. These differences are therefore defined as:

$$100 * ((\text{with SMP} - \text{without SMP}) / \text{without SMP})$$

Tables 4.1 to 4.10 are the percentage difference from base for exports, imports, employment, industrial unit labour costs and labour costs per employee, averaged across regions (countries) and sectors. The results are presented for the manufacturing sectors only, although the service sectors are included in the simulation, and the results from these sectors will have knock-on effects on the manufacturing sectors. The full results by region (country) and sector (including services) for employment and labour costs are presented in Appendix D.

4.1.4. Intra-EU trade simulation: model simulation 1

Model simulation 1 is limited to the impact of the SMP on intra-EU trade only, i.e. no additional variables were introduced into the extra-EU trade equations.

Pan-European results

Overall, the EU-wide effect of the SMP is to boost trade. Tables 4.1a and 4.2a show that by 1993 total exports (i.e. intra-EU and extra-EU) are 5.5% higher than the non-SMP base, compared to an import effect of 4.9%. At the sectoral level (see Tables 4.1a and 4.2a), the outlying effects occur in Metal Products, with double-digit percentage differences for both exports and imports, and Office Machines and Paper and Printing Products, both of which have negative effects on imports and exports. From Tables 4.1b and 4.2b, the countries which stand out most are Spain, Luxembourg and Portugal, all of which record large export differences while the last two Member States display similar patterns for imports. The profile of the effect is closely linked to the exponential shape of the SMP proxy: by 1989, trade is only 0.6% above the non-SMP base.

The effect on employment is small, but positive. In Table 4.3a, a 0.5% difference from base is recorded by 1993, implying that the positive impact on exports has slightly outweighed the negative effect of a rise in imports. The Ferrous and Non-ferrous Metals sector stands out from the other industries with a large difference from base, as do Metal Products and Food, Drink and Tobacco, both of which record a fall in employment. Across the Member States (see Table 4.3b), Luxembourg, Germany and, to a lesser extent, France, show falls in overall employment, while Ireland and Belgium show the largest percentage increase in employment. As with trade, most of the employment effect occurs in the latter half of the simulation period.

The results for unit labour costs (Tables 4.4a and b), initially show negative effects, i.e. an improvement in competitiveness, but then turn sharply positive from 1992 onwards indicating a loss in competitiveness. However, this result reflects an implausible large impact on unit labour costs in one sector (Agricultural and Industrial Machinery). Discounting this outlier, most sectors show a slight improvement in competitiveness. The results for labour costs per employee (Tables 4.5a and b) show little change for the first four years with only a small decrease for the Manufacturing sector by 1993. Ireland shows the largest decrease of labour costs per employee of around 2% by 1993. The sectors experiencing the largest decreases were Ferrous and Non-ferrous Metals and Metal Products, suggesting these sectors experienced competitive effects (thus forcing down labour costs) due to the SMP.

Results by Member State

Belgium

Relative to the other Member States, Belgian exports and imports follow the EU average profile quite closely. The effect on employment is positive and quite large (2.6%), suggesting that in levels terms there is a gain in net trade. Table 4.4b shows how the difference from base for unit labour costs rises until 1990 with a 1.0% difference and then tails off until 1993, when a small overall gain in competitiveness is recorded. For total labour costs per employee the gain in competitiveness rises to 1.0% difference from base by 1993.

At the sectoral level, the results show that the movement in exports comes largely from Ferrous and Non-ferrous Metals (negative), Electrical Goods (positive), Transport Equipment (positive) and Textiles, Clothing and Footwear (positive). For imports the dominant sectors, in terms of difference from base, are Metal Products and Electrical Goods. This translates into a negative impact on employment in the Metal Goods and Electrical Goods sectors, while Transport Equipment, Textiles, Clothing and Footwear, and Other Manufacturing more than offset this. As well as experiencing a negative impact on employment, the Electrical Goods sector shows a decrease in labour costs per employee of nearly 5 by 1993, indicating that this sector increased in competitiveness due to the SMP.

These results are broadly consistent with the expectations set out in Chapter 2, which suggest that, given the small and open nature of the Belgian economy, the SMP would have a positive overall impact. Industries such as Motor Vehicles and Textiles were expected to benefit from the removal of remaining barriers to trade, while sectors such as Household Electrical Appliances, Shipbuilding, and Clothing and Footwear were not expected to perform as well.

While these expectations were largely confirmed by the model results, they do raise the important issue of aggregation. For example, the Motor Vehicles sector was expected to benefit from the SMP whereas the Shipbuilding sector was not. A similar analysis can be made when comparing the expected reaction of the Textiles sector against that of Clothing and Footwear. Because these sectors are aggregated in the model (appearing as Transport Equipment and Textiles, Clothing and Footwear), there may well be offsetting effects within the broad sectors.

Denmark

For the Danish economy, the effect of the SMP on exports is negligible until the final simulation year, when the difference from base rises to 1.4%. Imports have a higher percentage difference from base of 5.1% by 1993, and the consequent effect on employment is a negative comparison until 1992 when the jump in export occurs and employment shows a positive impact of 0.3%. Unit labour costs have a smooth profile by comparison, and show a positive difference from base of approximately 1%, implying an overall reduction in competitiveness for the manufacturing sector, while labour costs per employee decreased by 0.9% from base.

From the background analysis in Chapter 2 and the qualitative indications in Tables 3.11a to 3.12b, the overall assessment of the Danish economy in relation to the SMP effect was reasonably favourable. Most Danish industries were already highly integrated into the EU economy, however, so any effect of restructuring to cope with the enlarged market would occur over a longer time scale than provided for by the present analysis. The problem of aggregation arose here too, since industries such as Textiles, Clothing and Footwear faced conflicting prospects.

West Germany

The SMP appears to have had a negative impact on the German trade position. In fact, while the impact of the SMP on exports is broadly in line with the EU (about 5%), the effect on imports is nearly double the EU average (9% as opposed to 5%) and larger than the impact on exports. The strongly negative impact of the SMP on the competitiveness of the German manufacturing industries (unit labour costs were 1.2% higher than base) could explain the shift from domestic production to imports, which is reflected in a negative effect on employment (-0.6%, higher than in the other European countries apart from Luxembourg). The results for labour costs per employee, which are 0.5% below base by 1993, show that there has been some adverse productivity effect, i.e. output decreases by more than employment.

At the sectoral level the strongest impact on exports was in Ferrous and Non-ferrous Metals, Metal Products and Textiles, Clothing and Footwear, while Non-metallic Mineral Products, Transport Equipment and Other Manufactures saw the strongest positive impact on imports. As a result of this differentiated impact, the German trade position deteriorated particularly in Transport Equipment, Non-metallic Mineral Products and Other Manufactures. Modest improvements were only detected for Electrical Goods, Textiles, Clothing and Footwear and Ferrous and Non-ferrous Metals.

Some of these results can be interpreted in the light of the discussion in Chapter 2. The low pre-SMP import penetration in industries such as Transport Equipment could explain the boost of imports after the SMP, and the pre-SMP comparative advantage in Electrical Goods could explain the improved performance of German exports. The strong negative impact on unit labour costs and labour costs per employee in Textiles, Clothing and Footwear could help explain the improved trade performance of the industry post-SMP, despite the relative a priori 'weaknesses'. The different impact on unit labour costs in Office Machines (3.5%) and Electrical Goods (only 0.3%) explains the different behaviour of these industries in the period following the SMP, despite the fact that Germany appeared to have an a priori comparative advantage in both sectors.

The impact on employment is negative almost everywhere (apart from Chemicals and Rubber and Plastic Products), particularly in industries such as Metal Products, Non-metallic Mineral Products and Other Manufactures that saw their trade position deteriorate, but also in Textiles, Clothing and Footwear and Electrical Goods.

Spain

From the start it should be noted that it is difficult to disentangle the impact of the SMP on the Spanish economy from the impact of joining the EU; both events go back to 1986–87 and the SMP synthetic variable is introduced starting from 1986. Taking the results at face value, they show a strong positive effect of the SMP on Spanish exports, more than three times the EU average, and a much weaker effect on imports (3.2% higher than in the base simulation), slightly below the EU average. As a result, manufacturing employment is increased by 1.5% higher than the non-SMP base (against 0.5% for the EU as a whole). While the effect on unit labour costs appears to be negligible (-0.05%), the effect on labour costs per employee is positive.

These results contradict the deterioration of the trade balance experienced by this economy in the late 1980s as reported in Chapter 2. In the early 1990s the Spanish economy therefore seems to have reversed the initial negative impact of liberalization (due both to the SMP and to the joining of the EU) on its trade position.

The strongest effects on exports were on Ferrous and Non-ferrous Metals (this industry also shows the only relevant negative effect on imports), Food, Drink and Tobacco and Metal Products. Imports increased particularly in Non-metallic Mineral Products, Food, Drink and Tobacco and Textiles, Clothing and Footwear. As a result the trade position appears to have deteriorated only in Non-metallic Mineral Products and Textiles, Clothing and Footwear, while it appears to have improved particularly in Metal Products, Chemicals, and Electrical Goods. The SMP had a similar impact in Transport Equipment, where imports decreased by 1.3% and exports increased by 13% with respect to the non-SMP base.

The overall positive impact on employment appears mainly to be due to Metal Products, Chemicals, and Textiles, Clothing and Footwear.

The strongest negative effects (i.e. a gain in competitiveness) on unit labour costs were in Ferrous and Non-ferrous Metals while labour costs per employee show an increase. The strongest positive effect on unit labour costs was in Textiles, Clothing and Footwear, consistent with the results for import and export volumes.

France

The impact of the SMP in model simulation 1 was to increase imports more than the increase in exports, but there was no significant effect on employment. The biggest loss was in the Metal Products sector. Offsetting this there was an increase in employment in the Office Machines industry. The other sector to have a significant increase in employment was Ferrous and Non-ferrous Metals.

Ireland

Even though manufacturing imports have risen by more than manufacturing exports due to the SMP, there is a positive effect on employment, indicating that it is the employment-intensive industries that have gained.

Unit labour costs decreased while labour costs per employee show the biggest decrease of 1.8% by 1993. One of the main sectors to benefit is Metal Products, with a large increase in both exports and employment. As stated in Chapter 2, Textiles, Clothing and Footwear appeared to be vulnerable to the SMP, but these results do not confirm this expectation, with exports and employment all higher than in the non-SMP base. The results also indicate that the Office Machines sector suffered from the impact of the SMP. This could be because many types of machinery are simply not produced in Ireland.

Italy

As reported in Chapter 2, the Italian economy did not appear to be in a strong position to face the SMP. Italy is specialized in traditional industries like Textiles, Clothing and Footwear which are historically heavily protected and also under-represented in R&D-intensive industries like Chemicals, Office Machines and Electrical Goods. The economy appeared to be vulnerable to competition from other European countries. However, the simulation exercise suggests a positive impact of the SMP on Italian exports (7.0% with respect to the base simulation, 2% more than the EU average), higher than that for Italian imports (3.2%, below the EU average of 4.8%). The improved trade performance appears to be due to a reduction in unit labour costs by -0.4%, while Germany and the UK saw higher unit labour costs and France a smaller reduction. The SMP appears also to have had a positive impact on employment (0.5%), in line with the EU average.

At the sectoral level, the trade balance improved particularly in Non-metallic Mineral Products, Paper and Printing and Rubber and Plastic Products. Industries with high R&D intensity generally saw an improvement in their trade balance relative to the non-SMP base, as is evident in the results for Office Machines, Chemicals, and Electrical Goods. As expected, the trade balance deteriorated in industries heavily protected before the SMP, such as Textiles, Clothing and Footwear, but not in Transport Equipment.

Increased unit labour costs were experienced only in Textiles, Clothing and Footwear and Food, Drink and Tobacco, with the highest negative effects on Non-metallic Mineral Products and Metal Products. There is only slight fluctuation in labour costs per employee.

Generally speaking, employment increased in industries that saw their trade balance improve such as Food, Drink and Tobacco and Paper and Printing.

Luxembourg

In this simulation, the effect of the SMP has been to increase trade by over 25% compared to base. It is likely that the majority of this trade increase reflects entrepôt trade, as both imports and exports increased by similar proportions. The effect on the economy is generally to cause employment to fall.

Netherlands

In this simulation the SMP has a negligible effect on employment and only a slight increase in imports and exports. This economy was expected to gain from the SMP due to its high quality labour force, good infrastructure and favourable geographical location, but the economy was already very open. One sector that the results indicate has become more traded due to the SMP is Metal Products, where both exports and imports rose considerably relative to base. The resulting net effect is a decrease in employment. Unit labour costs register extremely large increases which balloon upwards to 36% above base by the end of the simulation period. Clearly these results do not seem very satisfactory, and are due entirely to extraordinarily large increases in unit labour costs for Agricultural and Industrial Machinery, and Electrical Goods, while most of the other sectors show a slight decrease in unit labour costs. The fact that employment and labour costs per employee alter only slightly implies huge output losses and corresponding falls in productivity. However, the fact that the aggregate effect is dominated by a couple of outlying sectors suggests the problem may be due to re-exporting, e.g. entrepôt trade, which can cause distortions to trade measurements – the quality of which is a cornerstone to the robustness of the results obtained by the model simulations. This hypothesis is corroborated by the fact that the results for Luxembourg seem to be suffering from a similar experience.

Portugal

Ignoring the difficulties in distinguishing the effects of the SMP from the accession effects, the model indicates that the effect of the SMP on intra-EU trade has been an expansion. Although manufacturing imports have increased more than manufacturing exports, it would appear that the effect of the SMP has been to increase employment by nearly 18,000. The results indicate that the effect of the SMP on Agricultural and Industrial Machinery, Electrical Goods, and Transport Equipment is to increase employment. This is as expected, as these sectors were the most open in the Portuguese economy, with high levels of investment and R&D. These sectors also had relatively high unit labour costs, compared to other sectors, and these appear to have decreased considerably due to the SMP coupled with productivity gains.

As expected, Food, Drink and Tobacco is adversely affected by the SMP. Although exports increase, imports increase by more, resulting in employment losses of around 10,000. It appears that the main beneficiaries from the SMP are those sectors already well positioned in the Portuguese economy, with secondary knock-on effects for the rest of the economy.

United Kingdom

This country appears to have benefited from the SMP. Manufacturing imports are slightly down, while manufacturing exports are up. Taking into account the service sectors, the overall effect on trade is insignificant. Employment increases from base, indicating that the severe decline in employment experienced in the UK would have been worse had the SMP not taken place. The increase in employment is partly explained by the decrease in labour costs per employee of 1.5% from base by 1993. As expected, the high-technology industries of Office Machines, Electrical Goods and Transport Equipment all show increased exports, with employment increasing in particular in Office Machines. The relaxation of trade barriers enabled the Food, Drink and Tobacco industry to increase its exports, but not without associated employment losses and decreases in unit labour costs. Most sectors show a decrease in unit labour costs (increase in competitiveness) due to the SMP.

4.1.5. Intra- and extra-EU trade simulation: model simulation 2

This analysis describes the results of model simulation 2, with a version of the model in which both the intra- and extra-EU trade equations have the SMP variable included. The key feature of simulation 2 compared to simulation 1 is that a strong positive SMP effect on extra-EU imports and a weak positive effect on extra-EU exports are included in the simulation. However, the effect on extra-EU manufactured exports is negative, perhaps due to diversion of EU exports away from extra-EU markets.

Pan-European results

The effect of the SMP variable on total EU manufacturing exports is to produce an overall decrease of 1.62% from base in 1993. There are wide variations among countries. The poorer Member States, Spain, Portugal and Ireland, show an increase in trade, whereas the richer four

EU countries – Germany, France, Italy, and the UK – have experienced decreases in total manufacturing trade. This decrease is almost completely explained by a decrease in extra-EU manufacturing exports.

Luxembourg stands out as showing a large increase in exports due to the SMP, but this probably reflects substantial entrepôt trade which distorts the data: the model gives a large increase in manufacturing imports due to the effects of the SMP. These results are consistent with the results reported in Chapter 3 where the coefficients of the SMP variable for the manufacturing extra-EU export volume equations are generally negative (see Table 3.12b). However, when service sectors are included the overall EU results are higher than in the base case.

With imports the story is slightly different. Again, the poorer Member States experienced increases in manufacturing imports, and the UK and France show a decrease, but Germany shows a large increase in imports. Overall, there is an increase in manufacturing imports for Europe of 5.54% from base. This increase corresponds to the coefficients on the SMP variable for both the intra- and extra-EU import volume equations, which were generally positive. The largest increases in imports are at the intra-EU level, but there are also increases in extra-EU imports.

The overall effect on manufacturing employment in this scenario is a small decrease of 0.53% by 1993 for the EU as a whole. The poorer countries, Spain, Ireland, and Portugal, are the main countries to see a boost to employment. The UK also shows a slight increase in manufacturing employment. The other countries show a decrease in manufacturing employment, with Germany suffering some of the greatest employment losses, consistent with the negative trade effect noted above.

Overall, including the service sectors, there has been a slight increase in employment, indicating that the service sectors appear to have benefited the most from the SMP. Taking into account the service sectors, Germany now shows an increase in total employment by 1993 compared to base, as the gains in the service sectors outweighed the losses in manufacturing. Appendix D contains detailed results for employment (and other key variables) by sector, including services. In the case of Germany, Table D1c shows how the positive employment effect in Other Market Services (over 3% above base by 1993) more than offsets falls in employment in the Inland Transport sector. In general for simulation 1, Inland Transport seems to be the most adversely affected sector, with only the Netherlands and Portugal showing positive SMP gains by the end of the simulation period. The results for simulation 2 for services, i.e. incorporating the effects of extra-EU trade, can generally be seen as providing more optimistic results for employment, particularly for Other Market Services.

The effects on unit labour costs are mixed. Initially, there was a slight increase in unit labour costs for the manufacturing sectors, but by 1993 various countries show a decrease in unit labour costs for the manufacturing sectors. On average across the EU, there has been a small increase in unit labour costs of 0.05% in 1993 compared to if the SMP had not occurred. Luxembourg stands out as the country that shows the biggest decrease in unit labour costs, although the problem over trade data referred to earlier suggests that this estimate should be treated with caution.

The simulation results show that, on average across the EU, the largest falls in labour costs per employee were experienced by the Electrical Goods, Transport Equipment and Metal Products industries, while Other Manufactures experienced the largest increases (see Table 4.10a). The UK and Ireland have the largest falls in labour costs per employee (around 2%) over the 1985–93 period (see Table 4.10b). In contrast, the largest increases in labour costs per employee were experienced by the Netherlands and Luxembourg, although again the latter country's estimate should not be taken at face value. Overall, though, the impact of the SMP on labour costs per employee appears to have been minimal.

Results by Member State

Following is a brief summary of the results for each country.

Belgium

This country was very open to trade before 1986, and was expected to benefit from the SMP due to its central location in Europe, and its relatively small market. The results indicate that trade has increased due to the SMP. Manufacturing imports increased by more than manufacturing exports, but the effect on employment is positive, compared to if the SMP had not happened. This suggests that the labour-intensive sectors have benefited, indicating that Belgium has benefited from the SMP. At the sectoral level, as noted in Chapter 2, Electrical Goods was a vulnerable sector, as it was previously highly protected. This view is supported by the simulation which shows a large increase in imports compared to base by 1993, with a resulting decrease in employment. By 1993, Electrical Goods was the only manufacturing sector in Belgium which had experienced a fall in labour costs per employee due to the SMP. For Other Manufactures exports have increased more than imports due to the SMP, enabling the sector to expand but by 1993 labour costs per employee had also risen sharply.

Denmark

This country shows a deterioration in manufacturing exports, and a rise in manufacturing imports due to the SMP. This resulted in a decrease in employment of 8,460 in 1992 from base, but in 1993 there is a slight recovery with employment only 3,960 jobs less than base. This indicates that the full benefits of the SMP might not have been realized by 1993. These changes in employment from base are small, less than 1% by 1993, and there is also a small increase in unit labour costs. These small effects are consistent with what was expected because Denmark was already a very open economy and thus less vulnerable to any negative effects from the SMP. At the sectoral level there are some surprises. Office Machines was considered susceptible to detrimental effects from the SMP due to its strong import penetration, and relatively high wage rates, but the results show a sharp decrease in unit labour costs from base by 1993, increasing the competitiveness of the industry, thus allowing exports to increase with subsequent employment gains. However, the simulation results also reveal this sector had experienced the largest increases in labour costs per employee over the 1985–93 period. Agricultural and Industrial Machines show a decrease in exports and imports, thus becoming a less traded sector, but exports have decreased more than imports, with consequent employment losses. In general, the effects of the SMP on Denmark seem negligible in this simulation.

Germany

The simulations show a decrease in German manufacturing employment and a fall in labour costs per employee, with only Chemicals and Metal Products showing an increase in employment. The German Chemicals industry is characterized by high levels of R&D expenditure, which could explain its improved performance following the SMP. The Office Machines sector was identified as having a better competitive position on the EU market, but the results indicate that it has lost market share with an increase in imports and only a modest increase in exports. Labour costs per employee have fallen most sharply in Transport Equipment but have shown large increases in Rubber and Plastic Products. If services are included, the results indicate the Other Market Services sector has benefited, giving overall an increase in employment by 1993, compared with the base. In conclusion, it appears that the manufacturing sectors have generally not benefited from the SMP.

Spain

As noted in Chapter 2, there are major difficulties involved in separating the impact of the SMP on the Spanish economy from the anticipatory and transition effects induced by accession to the EU. The SMP will have increased the openness of the Spanish economy, and it was expected that some negative effects would be evident, reflecting the high level of protectionism that existed previously. However, the results indicate an improvement in the trade balance, as exports have increased more than imports for the manufacturing sectors. The effect on employment is positive amounting to over 50,000 jobs, apparent across the manufacturing sectors generally, although labour costs per employee had increased due to the SMP. Buigues *et al.* (1990) expected industries such as Textiles, Clothing and Footwear and Food, Drink and Tobacco to benefit from the SMP, and the results support this. For example, by 1993, labour costs per employee had fallen in both these sectors. The Chemicals industry also gained, with the results showing a large increase in exports of over 15% by 1993 compared to the base, and only a small difference in imports.

France

France initially shows a positive increase in manufacturing imports, but by 1992 there is a negative impact due to the SMP. For manufacturing exports there is an increasingly negative impact throughout the simulation period. Nevertheless, there is virtually no impact on employment, despite an increase in unit labour costs and a decrease in labour costs per employee and in wages and salaries. These results are mirrored across the sectors. For example, the Chemicals sector shows an increase in imports, a decrease in exports, and an increase in unit labour costs. This is consistent with the expectation noted in Chapter 2, reflecting the weaker record of R&D (compared, say, to Germany).

Ireland

There is substantial trade between the Irish and UK economies. In this simulation the SMP seems to have had a large impact on two Irish sectors: Metal Products, and Textiles, Clothing and Footwear. Both these sectors showed large increases in trade due to the SMP, probably due to large increases in imports for these sectors in the UK. The increase in trade resulted in large increases in employment which dominate the overall increase in employment for the economy. Metal Products had also experienced very large falls in labour costs per employee due to the SMP. The manufacturing employment increase in 1993, at over 5%, is the largest percentage increase in Europe, although some sectors show a decline as some high productivity sectors suffer a large increase in imports. Electrical Goods and Transport Equipment both increased employment; both these industries were regarded as relatively competitive and likely to benefit from the SMP. Overall, it appears that manufacturing sectors have gained from the SMP, at least in employment terms, whilst the service sectors have not benefited from the SMP.

Italy

The results indicate that manufacturing in Italy has not benefited from the SMP; imports and labour costs per employee are up, exports and employment are down. This decrease in competitiveness could partly be explained by an increase in unit labour costs of more than the EU average. Italy was already characterized by a high level of integration with other European countries, before the SMP, which could explain why there has not been a large change in trade. With the inclusion of services, there is a slight increase in overall employment. As noted in Chapter 2, few Italian industries were considered likely to benefit from the SMP, which is borne out by this simulation. As expected, the R&D-intensive sectors suffered, in particular Chemicals, Office Machines, Electrical Goods and Transport Equipment.

Luxembourg

This country showed a very large increase in trade in both directions, but the large extent of entrepôt trade suggests that the results should be treated with caution. It may be that, with the dropping of barriers to trade, exports through Luxembourg have increased substantially. It was expected that multinational companies might relocate in Luxembourg, due to its favourable location, but the limited impact supports the idea that entrepôt trade has been the main beneficiary. The results also indicate that the effect of the SMP has been to reduce employment although labour costs per employee have not been affected. Non-metallic Mineral Products, characterized by high levels of intra-industry trade, is one of the two manufacturing industries to have gained employment.

Netherlands

Overall, the Netherlands showed a rise in labour costs per employee and a subsequent fall in employment, both in total, and for manufacturing. Manufacturing exports show an increase of 2.1% in 1993, but manufacturing imports have increased by 6.6%. Trade at the intra-EU level has increased, but there was a slight decrease at the extra-EU level.

This country was expected to gain from the SMP in the high-tech industry of Electrical Goods due to the high-quality labour force the Netherlands possesses. The model bears out this expectation, with Electrical Goods being one of the few sectors to have gained in employment by 1993. Non-metallic Mineral Products has also gained in employment. This industry had been protected by non-tariff barriers, and had low levels of import penetration. The effect of the SMP has been to increase trade, with imports increasing but a much bigger increase in exports. The Transport Equipment industry was expected to be able to compete with the opening up of the industry due to the high level of R&D expenditure in the industry. In this simulation, both exports and imports have increased, but there has been a reduction in employment and unit labour costs (perhaps to remain competitive).

Portugal

The results should be interpreted with care, as it is difficult to isolate the effects of the SMP on the Portuguese economy from the anticipatory and transition effects of accession to the EU in 1986. Taking the results at face value, the simulation indicates that trade has increased for the manufacturing sectors as a result of the SMP. The increase in trade started from relatively low values compared to the other EU countries, thus showing high percentage increases. Manufacturing imports have increased slightly more than manufacturing exports. It appears that Portugal has been forced to reduce unit labour costs, due to the increased openness of trade, in order to remain competitive. However, labour costs per employee had also increased over the simulation period. Manufacturing employment shows an increase of 1.5% in 1993 compared to the base.

Virtually all manufacturing sectors in Portugal have experienced an increase in imports, with the exception of Non-metallic Mineral Products. Similarly, Agricultural and Industrial Machinery, and Office Machines are the only sectors to show a decrease in exports. As Portugal has become a more open economy it has been exposed to increased competitive pressures, and wages and unit labour costs are lower than in the base. The decrease in wages in one sector has 'knock-on' decreases in other sectors, leading to a decrease in unit labour costs across most of the sectors. The Chemicals industry is the only industry to show a substantial increase in unit labour costs, and is the only industry to show a decrease in manufacturing employment by 1993.

The Transport Equipment sector in Portugal in 1985 was already fairly open, in that it had high import and export penetration ratios. The model results indicate that the effect of the SMP on the Transport Equipment sector in Portugal is to increase trade even further. Exports have increased more than the increase in imports, enabling the sector to expand and increase employment. Unit labour costs have also decreased for the sector. Surprisingly, Textiles, Clothing and Footwear in Portugal has suffered, according to the model results. Imports have increased slightly while exports have decreased, resulting in a decrease in employment. The Metal Products sector had a low import and export penetration ratio in 1980. The effect of the SMP has evidently been to open up the sector to increased imports, but this has been offset by a substantial increase in exports, with only a small net effect on employment. This could be explained by manufacturers switching from domestic sales to exports.

United Kingdom

The results of this simulation exhibit a relatively large fall in manufacturing exports of 7.1% by 1993 relative to the non-SMP base. The decrease in exports is due mostly to a decrease in extra-EU exports. Manufacturing imports also decreased, but by less. Overall employment has decreased, along with a rise in unit labour costs but a fall in labour costs per employee.

The Transport Equipment sector in the UK is a relatively highly traded sector as noted in Chapter 2. A priori, it would be expected that this sector would benefit from the SMP in the UK and this is borne out by the model results: there is a slightly negative impact on imports, and a positive impact on exports, giving an increase in employment. The Office Machines industry, which is R&D-intensive, has gained from the SMP according to this simulation, as expected. Exports have increased with a corresponding increase in employment. The Food, Drink and Tobacco industry appears to have experienced a large increase in exports as a result of the SMP. This is partly because it was one of the sectors that was relatively least open to trade in the UK economy. There has been a slight decrease in imports, but employment is also lower.

The Textiles, Clothing and Footwear industry in the UK was expected to suffer from the competitive pressures associated with the SMP. The SMP variable shows a slightly positive impact on exports, and a large positive impact on imports. Overall, the UK has not benefited from the SMP in this simulation, mainly due to the decrease in extra-EU manufacturing exports.

Table 4.1a. **Model simulation 1: impact of the SMP on EU manufacturing exports by sector (%)**

		1985	1986	1987	1988	1989	1990	1991	1992	1993
8	Ferrous & Non-f. Metals	0.00	0.00	0.01	0.10	0.17	0.41	0.92	1.58	3.09
9	Non-Metallic Min.Pr.	0.00	0.02	0.16	0.48	1.15	2.34	5.14	7.32	9.09
10	Chemicals	0.00	0.00	0.03	0.13	0.35	0.80	1.52	2.59	3.64
11	Metal Products	0.00	0.04	0.36	1.41	3.47	7.22	13.95	23.72	30.41
12	Agri. & Indust. Mach.	0.00	0.00	0.01	0.05	0.12	0.22	0.42	0.82	2.34
13	Office Machines	0.00	0.00	-0.02	-0.10	-0.30	-0.69	-1.32	-2.62	-3.65
14	Electrical Goods	0.00	0.02	0.18	0.65	1.74	3.42	5.50	8.47	8.91
15	Transport Equipment	0.00	0.01	0.06	0.19	0.53	1.32	3.05	5.21	7.35
16	Food, Drink & Tobacco	0.00	0.00	0.04	0.17	0.44	0.87	1.87	2.74	3.55
17	Tex., Cloth. & Footw.	0.00	0.00	0.03	0.13	0.39	0.98	2.11	4.13	6.62
18	Paper & Printing Pr.	0.00	0.00	0.00	-0.04	-0.12	-0.19	-0.30	-0.58	-1.25
19	Rubber & Plastic Pr.	0.00	0.00	0.04	0.17	0.46	0.95	2.00	3.70	5.38
21	Other Manufactures	0.00	0.02	0.15	0.54	1.33	2.77	5.07	8.50	10.65
	Total	0.00	0.01	0.06	0.24	0.60	1.29	2.51	4.09	5.53

Source: Cambridge Econometrics Forecast C51F1Y-C51F1R/da1/F1RA.

Table 4.1b. **Model simulation 1: impact of the SMP on EU manufacturing exports by country (%)**

	1985	1986	1987	1988	1989	1990	1991	1992	1993
Belgium	0.00	0.01	0.07	0.22	0.55	1.11	2.24	3.58	5.02
Denmark	0.00	0.00	0.01	0.05	0.14	0.19	0.35	0.23	1.38
Germany	0.00	0.01	0.06	0.21	0.56	1.18	2.27	3.86	5.28
Spain	0.00	0.02	0.19	0.71	1.82	4.11	7.91	12.90	16.79
France	0.00	0.01	0.08	0.31	0.78	1.65	3.07	4.78	5.91
Ireland	0.00	0.01	0.06	0.21	0.49	1.05	2.16	3.32	4.01
Italy	0.00	0.01	0.07	0.27	0.71	1.49	2.86	4.84	7.05
Luxembourg	0.00	0.03	0.31	1.24	3.11	6.83	13.31	22.61	29.11
Netherlands	0.00	0.00	0.02	0.09	0.21	0.45	1.14	2.00	3.14
Portugal	0.00	0.01	0.12	0.46	1.20	2.88	6.08	10.64	15.69
United Kingdom	0.00	0.00	0.04	0.12	0.27	0.54	1.08	1.64	2.15
Total (Manuf)	0.00	0.01	0.06	0.24	0.60	1.29	2.51	4.09	5.53

Source: Cambridge Econometrics Forecast C51F1Y-C51F1R/da1/F1RA.

Table 4.2a.   Model simulation 1: impact of the SMP on EU manufacturing imports by sector (%)

		1985	1986	1987	1988	1989	1990	1991	1992	1993
8	Ferrous & Non-f. Metals	0.00	0.00	0.00	0.03	0.03	0.08	0.27	0.50	1.65
9	Non-Metallic Min.Pr.	0.00	0.02	0.17	0.50	1.23	2.51	5.36	7.32	9.46
10	Chemicals	0.00	0.00	0.04	0.15	0.38	0.84	1.63	2.83	3.86
11	Metal Products	0.00	0.05	0.47	1.64	3.98	7.94	14.13	22.99	29.16
12	Agri. & Indust. Mach.	-0.00	0.00	0.02	0.08	0.18	0.35	0.71	1.37	3.72
13	Office Machines	0.00	0.00	-0.02	-0.08	-0.23	-0.50	-0.91	-1.63	-2.03
14	Electrical Goods	0.00	0.02	0.17	0.59	1.57	2.93	4.64	6.71	6.39
15	Transport Equipment	0.00	0.01	0.07	0.24	0.65	1.58	3.60	5.99	8.02
16	Food, Drink & Tobacco	0.00	0.00	0.04	0.16	0.41	0.72	1.66	2.37	2.94
17	Tex., Cloth. & Footw.	0.00	0.00	0.03	0.12	0.34	0.80	1.60	2.97	4.45
18	Paper & Printing Pr.	0.00	0.00	0.00	-0.03	-0.09	-0.25	-0.26	-0.57	-1.05
19	Rubber & Plastic Pr.	0.00	0.01	0.05	0.18	0.46	1.03	2.00	3.47	4.88
21	Other Manufactures	0.00	0.01	0.11	0.38	0.96	2.01	3.67	6.16	7.76
	Total	0.00	0.01	0.06	0.23	0.58	1.20	2.33	3.71	4.89

Source: Cambridge Econometrics Forecast C51F1Y-C51F1R/da1/F1RA.

Table 4.2b.   Model simulation 1: impact of the SMP on EU manufacturing imports by country (%)

	1985	1986	1987	1988	1989	1990	1991	1992	1993
Belgium	0.00	0.01	0.10	0.36	0.99	1.70	2.88	4.65	6.10
Denmark	0.00	0.01	0.07	0.28	0.70	1.27	2.40	4.01	5.11
Germany	0.00	0.01	0.11	0.39	1.01	2.19	4.17	6.86	8.94
Spain	0.00	0.00	0.01	0.08	0.24	0.59	1.30	2.21	3.25
France	0.00	0.01	0.05	0.18	0.44	0.92	1.61	2.37	2.65
Ireland	0.00	0.01	0.10	0.37	0.87	1.86	3.37	5.74	7.39
Italy	0.00	0.00	0.04	0.15	0.39	0.81	1.49	2.29	3.18
Luxembourg	0.00	0.03	0.28	1.13	2.60	5.99	12.01	20.74	25.50
Netherlands	0.00	0.01	0.08	0.33	0.83	1.55	2.99	4.73	5.70
Portugal	0.00	0.01	0.09	0.36	1.01	2.42	4.98	9.52	18.82
United Kingdom	0.00	0.00	0.01	0.03	0.02	-0.05	0.04	-0.18	-0.41
Total (Manuf)	0.00	0.01	0.06	0.23	0.58	1.20	2.33	3.71	4.89

Source: Cambridge Econometrics Forecast C51F1Y-C51F1R/da1/F1RA.

Table 4.3a.   Model simulation 1: impact of the SMP on EU manufacturing employment by sector (%)

		1985	1986	1987	1988	1989	1990	1991	1992	1993
8	Ferrous & Non-f. Metals	0.00	0.00	-0.01	-0.01	0.08	0.42	1.23	3.25	7.52
9	Non-Metallic Min.Pr.	0.00	0.00	0.01	0.02	0.05	0.10	0.17	0.25	0.42
10	Chemicals	0.00	0.00	0.01	0.04	0.10	0.18	0.35	0.50	0.06
11	Metal Products	0.00	0.00	0.01	0.05	0.12	0.25	0.23	-0.03	-0.46
12	Agri. & Indust. Mach.	0.00	0.00	0.00	-0.01	-0.01	-0.03	0.09	0.03	0.07
13	Office Machines	0.00	0.00	0.01	0.02	0.06	0.13	0.41	0.78	0.64
14	Electrical Goods	0.00	0.00	-0.01	0.00	0.04	0.59	0.48	0.32	0.30
15	Transport Equipment	0.00	0.00	0.00	0.00	0.00	0.00	0.05	0.07	0.09
16	Food, Drink & Tobacco	0.00	0.00	0.00	0.00	0.01	-0.02	0.11	0.04	-0.05
17	Tex., Cloth. & Footw.	0.00	0.00	0.01	0.02	0.07	0.19	0.46	0.77	1.01
18	Paper & Printing Pr.	0.00	0.00	0.00	0.00	0.01	0.07	0.11	0.31	0.50
19	Rubber & Plastic Pr.	0.00	0.00	0.00	0.02	0.06	0.30	0.62	0.96	1.06
21	Other Manufactures	0.00	0.00	0.01	0.02	0.05	0.11	0.17	0.36	0.30
	Total	0.00	0.00	0.00	0.01	0.04	0.16	0.29	0.40	0.49

Source: Cambridge Econometrics Forecast C51F1Y-C51F1R/da1/F1RA.

Table 4.3b.   Model simulation 1: impact of the SMP on EU manufacturing employment by country (%)

	1985	1986	1987	1988	1989	1990	1991	1992	1993
Belgium	0.00	0.00	0.02	0.08	0.21	0.56	1.05	2.04	2.65
Denmark	0.00	0.00	-0.01	-0.03	-0.03	-0.01	-0.10	-0.11	0.33
Germany	0.00	0.00	0.00	0.01	0.01	0.18	0.13	-0.16	-0.56
Spain	0.00	0.00	0.02	0.06	0.17	0.41	0.89	1.40	1.48
France	0.00	0.00	0.00	0.01	0.02	0.05	0.06	0.03	-0.01
Ireland	0.00	0.00	0.04	0.18	0.45	1.17	2.59	4.54	5.92
Italy	0.00	0.00	0.01	0.03	0.08	0.16	0.33	0.54	0.51
Luxembourg	0.00	0.00	-0.01	-0.08	-0.13	0.28	-0.63	-1.19	-2.10
Netherlands	0.00	0.00	-0.02	-0.08	-0.16	-0.18	-0.10	-0.10	0.00
Portugal	0.00	0.00	0.01	0.02	0.07	0.17	0.31	0.61	1.94
United Kingdom	0.00	0.00	0.00	0.00	0.02	0.09	0.26	0.58	1.12
Total (Manuf)	0.00	0.00	0.00	0.01	0.04	0.16	0.29	0.40	0.49

Source: Cambridge Econometrics Forecast C51F1Y-C51F1R/da1/F1RA.

Table 4.4a.   Model simulation 1: impact of the SMP on EU manufacturing unit labour costs by sector (%)

		1985	1986	1987	1988	1989	1990	1991	1992	1993
8	Ferrous & Non-f. Metals	0.00	0.00	-0.02	-0.17	-0.68	-1.36	-2.34	-3.57	-3.31
9	Non-Metallic Min.Pr.	0.00	0.00	-0.01	-0.01	-0.01	0.01	-0.27	-0.42	-0.69
10	Chemicals	0.00	0.02	0.06	0.40	0.35	0.75	1.34	2.68	2.39
11	Metal Products	0.00	0.00	-0.02	-0.02	0.04	-0.04	-0.41	-0.60	0.11
12	Agri. & Indust. Mach.	0.00	0.00	0.00	0.02	0.00	-0.11	1.99	15.01	30.06
13	Office Machines	0.00	-0.01	0.00	0.07	0.15	0.42	0.62	0.16	-1.39
14	Electrical Goods	0.00	0.01	0.08	0.29	1.01	1.26	1.39	1.32	0.19
15	Transport Equipment	0.00	0.01	0.14	0.08	-0.03	-0.12	-0.74	-1.29	-3.18
16	Food, Drink & Tobacco	0.00	0.00	0.00	0.01	0.03	0.01	0.05	0.01	0.10
17	Tex., Cloth. & Footw.	0.00	-0.03	-0.36	-0.88	-1.22	-1.69	-4.46	-7.86	-5.70
18	Paper & Printing Pr.	0.00	0.00	-0.01	-0.03	-0.12	-0.34	-0.59	-1.25	-1.85
19	Rubber & Plastic Pr.	0.00	0.00	0.01	0.01	0.06	0.50	0.24	-0.04	-0.69
21	Other Manufactures	0.00	0.00	0.00	-0.03	-0.05	-0.02	-0.06	-0.04	-0.33
	Total	0.00	0.00	-0.01	-0.03	-0.03	-0.06	-0.09	2.10	5.45

Source: Cambridge Econometrics Forecast C51F1Y-C51F1R/da1/F1RA.

Table 4.4b.   Model simulation 1: impact of the SMP on EU manufacturing unit labour costs by country (%)

	1985	1986	1987	1988	1989	1990	1991	1992	1993
Belgium	0.00	0.01	0.09	0.31	0.97	1.03	0.88	0.53	-0.86
Denmark	0.00	0.01	0.02	0.11	0.23	0.53	0.95	1.14	0.80
Germany	0.00	0.00	0.01	0.03	0.11	0.65	0.67	0.92	1.19
Spain	0.00	0.00	-0.02	-0.06	-0.10	-0.14	0.08	0.17	-0.05
France	0.00	0.00	0.00	0.01	0.04	0.04	-0.14	-0.37	-0.24
Ireland	0.00	0.00	-0.05	-0.19	-0.66	-1.80	-3.76	-6.33	-3.21
Italy	0.00	0.00	0.00	-0.01	-0.02	-0.04	-0.05	-0.15	-0.41
Luxembourg	0.00	-0.03	-0.24	-0.86	-1.68	-2.69	-7.87	-13.69	-15.22
Netherlands	0.00	-0.01	0.03	0.17	0.53	1.70	5.84	22.69	36.40
Portugal	0.00	0.00	-0.02	-0.09	-0.27	-0.65	-1.46	-2.61	-5.33
United Kingdom	0.00	0.00	0.00	-0.01	0.05	0.18	0.37	0.86	1.72
Total (Manuf)	0.00	0.00	-0.01	-0.03	-0.03	-0.06	-0.09	2.10	5.45

Source: Cambridge Econometrics Forecast C51F1Y-C51F1R/da1/F1RA.

Table 4.5a. **Model simulation 1: impact of the SMP on EU manufacturing labour costs per employee by sector (%)**

		1985	1986	1987	1988	1989	1990	1991	1992	1993
8	Ferrous & Non-f. Metals	0.00	0.00	-0.01	-0.03	-0.03	-0.07	-0.06	-0.70	-1.82
9	Non-Metallic Min.Pr.	0.00	0.00	-0.01	-0.01	-0.02	-0.01	0.01	-0.04	0.30
10	Chemicals	0.00	0.01	0.00	-0.01	-0.01	-0.01	0.03	-0.01	0.02
11	Metal Products	0.00	0.00	-0.02	-0.05	-0.15	-0.32	-0.60	-1.04	-1.10
12	Agri. & Indust. Mach.	0.00	0.00	0.01	0.00	0.01	0.08	0.02	0.05	0.55
13	Office Machines	0.00	0.00	-0.03	0.01	-0.01	0.01	-0.03	0.05	0.13
14	Electrical Goods	0.00	0.00	0.02	0.03	0.16	-0.07	-0.36	-0.73	-0.51
15	Transport Equipment	0.00	0.00	-0.01	-0.02	-0.07	-0.06	-0.28	-0.59	-0.39
16	Food, Drink & Tobacco	0.00	0.00	-0.01	0.00	-0.03	-0.12	-0.24	-0.42	-0.43
17	Tex., Cloth. & Footw.	0.00	0.00	0.01	0.01	0.04	0.08	0.13	0.26	0.64
18	Paper & Printing Pr.	0.00	0.00	-0.01	-0.04	-0.08	-0.20	-0.27	-0.49	-0.24
19	Rubber & Plastic Pr.	0.00	0.00	0.00	-0.01	-0.02	-0.06	-0.19	-0.34	-0.27
21	Other Manufactures	0.00	0.00	0.00	-0.03	-0.05	-0.07	-0.23	-0.17	-0.20
	Total	0.00	0.00	0.00	-0.01	-0.02	-0.06	-0.15	-0.33	-0.29

Source: Cambridge Econometrics Forecast C51F1V-C51F1W/da1/F1WA.

Table 4.5b. **Model simulation 1: impact of the SMP on EU manufacturing labour costs per employee by country (%)**

	1985	1986	1987	1988	1989	1990	1991	1992	1993
Belgium	0.00	0.00	0.01	-0.02	-0.01	-0.32	-0.57	-0.83	-1.04
Denmark	0.00	0.00	-0.02	-0.06	-0.15	-0.29	-0.43	-0.64	-0.88
Germany	0.00	0.00	-0.01	-0.05	-0.09	-0.03	-0.45	-0.71	-0.47
Spain	0.00	0.00	0.01	0.04	0.12	0.29	0.53	0.80	1.08
France	0.00	0.00	0.00	0.02	0.08	0.17	0.26	0.32	0.76
Ireland	0.00	-0.01	-0.02	-0.05	-0.11	-0.21	-0.28	-1.10	-1.83
Italy	0.00	0.00	0.01	0.01	0.01	0.00	0.00	-0.06	0.13
Luxembourg	0.00	0.01	0.00	-0.01	0.01	-0.01	0.02	-0.01	0.00
Netherlands	0.00	0.00	-0.02	0.01	0.02	-0.01	-0.07	-0.12	-0.13
Portugal	0.00	0.01	0.01	0.03	0.05	0.09	0.04	0.05	1.34
United Kingdom	0.00	0.00	0.00	-0.04	-0.10	-0.25	-0.50	-1.00	-1.70
Total (Manuf)	0.00	0.00	0.00	-0.01	-0.02	-0.06	-0.15	-0.33	-0.29

Source: Cambridge Econometrics Forecast C51F1V-C51F1W/da1/F1WA.

Table 4.6a.   Model simulation 2: impact of the SMP on EU manufacturing exports by sector (%)

		1985	1986	1987	1988	1989	1990	1991	1992	1993
8	Ferrous & Non-f. Metals	0.00	-0.01	-0.09	-0.25	-0.64	-1.26	-2.08	-3.52	-3.39
9	Non-metallic Min.Pr.	0.00	0.01	0.06	0.08	0.13	0.12	1.19	0.55	0.29
10	Chemicals	0.00	-0.01	-0.08	-0.27	-0.62	-1.07	-1.89	-3.75	-3.85
11	Metal Products	0.00	0.03	0.29	1.14	2.76	5.66	10.76	17.46	21.63
12	Agri. & Indust. Mach.	0.00	-0.01	-0.10	-0.35	-0.89	-1.92	-3.52	-5.69	-5.92
13	Office Machines	0.00	-0.01	-0.06	-0.24	-0.63	-1.36	-2.32	-4.83	-6.37
14	Electrical Goods	0.00	0.01	0.08	0.29	0.76	-0.25	1.43	0.29	-2.87
15	Transport Equipment	0.00	-0.03	-0.21	-0.66	-1.44	-2.57	-4.10	-6.08	-4.98
16	Food, Drink & Tobacco	0.00	0.00	-0.01	-0.01	-0.06	0.30	-2.78	-0.74	-2.24
17	Tex., Cloth. & Footw.	0.00	0.00	0.03	0.12	0.39	0.97	2.06	4.23	7.03
18	Paper & Printing Pr.	0.00	0.00	0.00	-0.02	-0.09	-1.12	0.93	-1.76	-0.98
19	Rubber & Plastic Pr.	0.00	0.00	-0.03	-0.07	-0.19	0.16	-0.06	-0.59	-0.68
21	Other Manufactures	0.00	0.01	0.13	0.53	1.25	2.54	5.56	8.37	10.56
	Total	0.00	-0.01	-0.04	-0.11	-0.26	-0.59	-0.96	-1.63	-1.62

Source: Cambridge Econometrics Forecast C51F1V-C51F1W/da1/F1WA.

Table 4.6b.   Model simulation 2: impact of the SMP on manufacturing exports by country(%)

	1985	1986	1987	1988	1989	1990	1991	1992	1993
Belgium	0.00	0.00	0.04	0.13	0.34	0.66	1.08	2.42	3.06
Denmark	0.00	-0.01	-0.06	-0.15	-0.39	-0.76	-1.61	-1.40	-0.69
Germany	0.00	-0.01	-0.06	-0.13	-0.30	-0.77	-1.06	-1.55	-0.65
Spain	0.00	0.01	0.12	0.51	1.31	3.08	5.60	8.84	11.11
France	0.00	-0.01	-0.08	-0.26	-0.67	-1.51	-2.79	-4.81	-6.47
Ireland	0.00	0.00	0.03	0.08	0.21	-0.07	0.81	1.32	1.74
Italy	0.00	-0.01	-0.05	-0.16	-0.37	-0.69	-1.17	-2.33	-2.60
Luxembourg	0.00	0.04	0.35	1.39	3.49	7.60	14.77	24.49	30.95
Netherlands	0.00	0.00	0.01	0.06	0.11	0.29	0.26	1.43	2.05
Portugal	0.00	0.02	0.16	0.59	1.46	3.27	7.34	11.72	16.76
United Kingdom	0.00	-0.01	-0.12	-0.40	-0.93	-1.91	-3.43	-6.21	-7.12
Total (Manuf)	0.00	-0.01	-0.04	-0.11	-0.26	-0.59	-0.96	-1.63	-1.62

Source: Cambridge Econometrics Forecast C51F1V-C51F1W/da1/F1WA.

Table 4.7a. Model simulation 2: impact of the SMP on EU manufacturing imports by sector (%)

		1985	1986	1987	1988	1989	1990	1991	1992	1993
8	Ferrous & Non-f. Metals	0.00	0.00	0.00	0.02	0.00	-0.05	0.27	0.49	1.51
9	Non-metallic Min.Pr.	0.00	0.04	0.40	1.34	3.24	5.65	11.70	14.86	15.17
10	Chemicals	0.00	0.00	0.05	0.21	0.53	1.20	2.42	4.09	5.72
11	Metal Products	0.00	0.06	0.57	2.08	5.20	10.52	19.53	33.07	41.92
12	Agri. & Indust. Mach.	0.00	0.00	0.04	0.13	0.31	0.59	1.52	2.86	5.29
13	Office Machines	0.00	0.00	-0.04	-0.21	-0.54	-1.39	-1.88	-3.55	-2.99
14	Electrical Goods	0.00	0.02	0.16	0.55	1.44	1.05	3.22	3.91	2.93
15	Transport Equipment	0.00	0.01	0.06	0.21	0.59	1.59	2.58	4.24	5.38
16	Food, Drink & Tobacco	0.00	0.00	0.06	0.24	0.64	1.77	0.08	4.03	4.32
17	Tex., Cloth. & Footw.	0.00	0.01	0.10	0.38	1.04	2.24	4.25	7.49	10.87
18	Paper & Printing Pr.	0.00	0.00	-0.03	-0.08	-0.17	-0.15	-0.14	0.10	-0.49
19	Rubber & Plastic Pr.	0.00	0.00	0.00	0.00	-0.05	-1.29	-1.21	-2.08	-2.54
21	Other Manufactures	0.00	0.01	0.14	0.51	1.32	3.12	5.57	9.51	12.62
	Total	0.00	0.01	0.08	0.28	0.74	1.36	2.48	4.27	5.54

Source: Cambridge Econometrics Forecast C51F1V-C51F1W/da1/F1WA.

Table 4.7b. Model simulation 2: impact of the SMP on EU manufacturing imports by country (%)

	1985	1986	1987	1988	1989	1990	1991	1992	1993
Belgium	0.00	0.02	0.16	0.67	1.79	3.04	5.20	8.51	11.43
Denmark	0.00	0.01	0.06	0.23	0.60	1.29	2.24	4.31	5.54
Germany	0.00	0.02	0.15	0.52	1.40	2.50	4.17	8.70	10.91
Spain	0.00	0.01	0.08	0.39	1.12	2.30	4.56	7.21	9.71
France	0.00	0.00	0.02	0.05	0.07	0.08	0.09	-0.29	-1.17
Ireland	0.00	0.02	0.22	0.80	1.91	4.05	8.22	13.05	16.32
Italy	0.00	0.01	0.07	0.23	0.58	1.16	2.16	3.14	3.87
Luxembourg	0.00	0.04	0.44	1.94	3.86	8.24	15.87	28.47	32.09
Netherlands	0.00	0.01	0.10	0.40	0.98	1.85	3.25	5.36	6.66
Portugal	0.00	0.01	0.12	0.46	1.26	2.84	5.63	10.07	20.22
United Kingdom	0.00	0.00	-0.02	-0.07	-0.24	-0.60	-1.10	-2.44	-2.96
Total (Manuf)	0.00	0.01	0.08	0.28	0.74	1.36	2.48	4.27	5.54

Source: Cambridge Econometrics Forecast C51F1V-C51F1W/da1/F1WA.

Table 4.8a.   Model simulation 2: impact of the SMP on EU manufacturing employment by sector (%)

		1985	1986	1987	1988	1989	1990	1991	1992	1993
8	Ferrous & Non-f. Metals	0.00	0.00	-0.02	-0.06	-0.06	0.13	0.72	2.61	6.72
9	Non-metallic Min.Pr.	0.00	0.00	0.00	0.01	-0.03	-0.06	-0.07	-0.22	-0.48
10	Chemicals	0.00	0.00	0.00	-0.01	-0.02	-0.12	-0.14	0.04	-0.64
11	Metal Products	0.00	0.00	0.02	0.07	0.20	0.58	0.86	1.26	1.84
12	Agri. & Indust. Mach.	0.00	0.00	-0.03	-0.11	-0.28	-0.60	-3.03	-3.45	-3.24
13	Office Machines	0.00	0.00	0.00	0.02	0.06	0.25	-0.52	0.23	0.64
14	Electrical Goods	0.00	0.00	-0.01	-0.03	-0.04	-4.37	-2.84	-1.30	-1.01
15	Transport Equipment	0.00	0.00	-0.01	-0.03	-0.07	-0.16	-0.31	-0.57	-0.86
16	Food, Drink & Tobacco	0.00	0.00	0.00	0.00	0.01	0.26	-0.96	-0.35	-0.61
17	Tex., Cloth. & Footw.	0.00	0.00	0.00	0.00	0.02	0.10	0.42	0.62	0.78
18	Paper & Printing Pr.	0.00	0.00	0.00	-0.01	-0.04	-0.15	0.39	0.30	-0.63
19	Rubber & Plastic Pr.	0.00	0.00	0.00	0.00	-0.01	-1.23	-2.42	-3.07	-4.16
21	Other Manufactures	0.00	0.00	0.00	-0.01	-0.04	-0.10	-0.33	-0.62	-0.91
	Total	0.00	0.00	-0.01	-0.01	-0.03	-0.55	-0.77	-0.52	-0.53

Source: Cambridge Econometrics Forecast C51F1V-C51F1W/da1/F1WA.

Table 4.8b.   Model simulation 2: impact of the SMP on EU manufacturing employment by country (%)

	1985	1986	1987	1988	1989	1990	1991	1992	1993
Belgium	0.00	0.00	0.01	0.05	0.05	0.18	0.25	0.68	0.76
Denmark	0.00	-0.01	-0.06	-0.18	-0.43	-0.84	-1.24	-1.71	-0.77
Germany	0.00	0.00	-0.01	-0.03	-0.07	-1.97	-2.73	-2.27	-2.51
Spain	0.00	0.00	0.02	0.08	0.21	0.50	1.08	1.57	1.52
France	0.00	0.00	0.00	0.00	-0.01	0.03	0.00	0.01	0.02
Ireland	0.00	0.00	0.03	0.18	0.38	0.98	2.33	4.13	5.11
Italy	0.00	0.00	-0.01	-0.04	-0.08	-0.14	-0.33	-0.61	-1.04
Luxembourg	0.00	0.00	-0.02	-0.13	-0.22	0.12	-1.17	-1.85	-2.67
Netherlands	0.00	0.00	-0.03	-0.11	-0.26	-0.27	-0.54	-1.11	-1.89
Portugal	0.00	0.00	0.01	0.03	0.08	0.14	0.31	0.65	1.54
United Kingdom	0.00	0.00	0.00	-0.01	-0.02	0.05	0.08	0.53	0.76
Total (Manuf)	0.00	0.00	-0.01	-0.01	-0.03	-0.55	-0.77	-0.52	-0.53

Source: Cambridge Econometrics Forecast C51F1V-C51F1W/da1/F1WA.

Table 4.9a. **Model simulation 2: impact of the SMP on EU manufacturing unit labour costs by sector (%)**

		1985	1986	1987	1988	1989	1990	1991	1992	1993
8	Ferrous & Non-f. Metals	0.00	0.00	0.04	0.11	-0.11	-0.16	0.34	0.75	1.53
9	Non-metallic Min.Pr.	0.00	0.01	0.10	0.44	0.80	1.13	1.59	1.78	0.54
10	Chemicals	0.00	0.02	0.14	0.72	0.73	1.55	2.44	5.95	5.31
11	Metal Products	0.00	0.00	0.03	0.12	0.39	0.28	0.83	1.84	3.45
12	Agri. & Indust. Mach.	0.00	-0.01	0.00	-0.02	-0.11	-1.50	-2.74	-4.34	-1.85
13	Office Machines	0.00	-0.01	0.02	0.12	0.27	0.26	0.89	1.69	-0.09
14	Electrical Goods	0.00	0.02	0.13	0.47	1.40	-1.37	0.27	1.31	1.35
15	Transport Equipment	0.00	0.01	0.21	0.14	0.13	-0.83	-0.60	-1.15	-3.70
16	Food, Drink & Tobacco	0.00	0.00	0.01	0.08	0.25	0.58	0.55	1.13	2.00
17	Tex., Cloth. & Footw.	0.00	-0.03	-0.38	-0.77	-0.89	-1.01	-4.04	-7.38	-3.29
18	Paper & Printing Pr.	0.00	0.00	0.03	0.13	0.29	1.05	0.47	1.17	-0.04
19	Rubber & Plastic Pr.	0.00	0.00	0.02	0.08	0.22	-0.54	-0.44	-0.32	-0.47
21	Other Manufactures	0.00	0.00	0.02	0.04	0.08	-0.17	0.34	0.76	0.39
	Total	0.00	0.00	0.03	0.11	0.27	-0.14	-0.20	-0.19	0.05

Source: Cambridge Econometrics Forecast C51F1V-C51F1W/da1/F1WA.

Table 4.9b. **Model simulation 2: impact of the SMP on EU manufacturing unit labour costs by country (%)**

	1985	1986	1987	1988	1989	1990	1991	1992	1993
Belgium	0.00	0.02	0.20	0.78	1.96	2.40	2.95	3.25	1.22
Denmark	0.00	0.00	0.02	0.07	0.17	0.43	0.48	0.90	0.78
Germany	0.00	0.00	0.04	0.08	0.12	-4.48	-2.10	-1.61	-2.42
Spain	0.00	0.00	0.02	0.10	0.33	0.70	1.65	2.68	2.43
France	0.00	0.00	0.04	0.16	0.41	0.83	1.55	2.75	3.95
Ireland	0.00	0.01	0.03	0.11	-0.12	-0.41	-1.45	-2.92	-0.28
Italy	0.00	0.00	0.03	0.08	0.19	0.40	0.45	1.14	0.78
Luxembourg	0.00	-0.02	-0.19	-0.64	-1.33	-2.14	-8.36	-13.02	-14.80
Netherlands	0.00	0.00	0.06	0.23	0.61	0.89	0.84	0.59	-0.25
Portugal	0.00	0.00	-0.01	-0.02	-0.11	-0.30	-1.34	-1.87	-3.70
United Kingdom	0.00	0.00	0.03	0.14	0.44	1.08	2.00	3.85	5.44
Total (Manuf)	0.00	0.00	0.03	0.11	0.27	-0.14	-0.20	-0.19	0.05

Source: Cambridge Econometrics Forecast C51F1V-C51F1W/da1/F1WA.

Table 4.10a.   Model simulation 2: impact of the SMP on EU manufacturing labour costs per employee by sector (%)

		1985	1986	1987	1988	1989	1990	1991	1992	1993
8	Ferrous & Non-f. Metals	0.00	-0.01	-0.01	0.01	0.07	0.27	-0.04	0.13	-0.86
9	Non-metallic Min.Pr.	0.00	0.00	-0.02	-0.06	-0.08	-0.32	-0.31	-0.42	0.10
10	Chemicals	0.00	0.00	0.00	0.00	-0.02	0.17	0.16	0.06	-0.11
11	Metal Products	0.00	-0.01	-0.02	-0.06	-0.19	-0.82	-1.07	-1.33	-1.40
12	Agri. & Indust. Mach.	0.00	-0.01	0.00	-0.01	-0.03	-0.56	0.05	-0.06	0.42
13	Office Machines	0.00	-0.01	0.00	0.04	0.05	0.19	0.57	0.89	1.03
14	Electrical Goods	0.00	0.01	0.02	0.02	0.15	-1.45	-0.99	-1.38	-1.26
15	Transport Equipment	0.00	0.00	-0.02	-0.05	-0.15	-1.37	-0.61	-1.09	-1.26
16	Food, Drink & Tobacco	0.00	0.00	-0.01	0.05	0.10	0.16	0.19	0.52	0.94
17	Tex., Cloth. & Footw.	0.00	0.00	-0.01	0.00	-0.01	-0.19	-0.20	0.12	0.45
18	Paper & Printing Pr.	0.00	0.00	0.00	0.00	0.02	0.47	-0.16	0.46	0.59
19	Rubber & Plastic Pr.	0.00	0.00	-0.01	0.02	0.03	-0.30	0.18	0.51	0.84
21	Other Manufactures	0.00	0.00	0.01	0.06	0.16	-0.15	0.60	1.33	1.64
	Total	0.00	0.00	0.00	0.00	0.01	-0.28	-0.12	-0.04	0.01

Source: Cambridge Econometrics Forecast C51F1V-C51F1W/da1/F1WA.

Table 4.10b.   Model simulation 2: impact of the SMP on EU manufacturing labour costs per employee by country (%)

	1985	1986	1987	1988	1989	1990	1991	1992	1993
Belgium	0.00	0.00	0.03	0.13	0.32	0.09	0.33	1.01	1.04
Denmark	0.00	-0.01	-0.01	0.00	-0.01	0.18	0.11	0.76	0.21
Germany	0.00	-0.01	-0.02	-0.06	-0.16	-2.48	-0.35	-0.59	-0.29
Spain	0.00	0.00	0.01	0.03	0.07	0.16	0.26	0.48	0.65
France	0.00	-0.01	-0.02	-0.04	-0.08	-0.34	-0.42	-0.73	-0.63
Ireland	0.00	0.00	-0.03	-0.10	-0.13	-0.30	-0.79	-1.54	-2.17
Italy	0.00	0.00	0.01	0.02	0.03	0.07	-0.10	0.29	0.43
Luxembourg	0.00	0.01	-0.01	0.00	-0.02	0.00	0.01	0.00	0.01
Netherlands	0.00	0.00	0.01	0.05	0.12	0.19	0.30	0.86	1.76
Portugal	0.00	0.00	0.02	0.06	0.13	0.23	0.26	0.49	1.44
United Kingdom	0.00	0.00	-0.02	-0.07	-0.18	-0.41	-0.81	-1.45	-2.22
Total (Manuf)	0.00	0.00	0.00	0.00	0.01	-0.28	-0.12	-0.04	0.01

Source: Cambridge Econometrics Forecast C51F1V-C51F1W/da1/F1WA.

4.2. Effects by type of labour

4.2.1. Impact of the SMP on occupational employment

The method of assessing the occupational implications of the effects of the SMP on intra-EU trade has been outlined in Chapter 3. The results give only a very rough indication of the occupational effects. In particular, the dynamic effects of introducing a higher level of skill intensity in order to capitalize on the opportunities presented by the completion of the single market are not allowed for and there remain significant obstacles to the analysis of skill intensity in relation to value added per person in an internationally comparative context. The results do, however, provide a platform for further analysis incorporating qualitative as well as other quantitative material which cannot be integrated into any model, including E3ME, or into an occupational sub-model attached to it.

4.2.2. Occupational structure for 1991

The occupational structures for each sector are shown in Appendix C, Tables C1–C13 where each row of the table gives the occupational breakdown for each country. Obviously, these embody sampling errors and classification errors, as well as substantive differences between countries. It is important to bear in mind that the occupational data are derived from the European Labour Force Surveys for each country; for manufacturing sectors the underlying samples are relatively small compared with those for the service sectors. In addition, while considerable progress has been made in the construction of a harmonized occupational classification for use in European comparisons, this is still nonetheless in its infancy. Given the lack of previous occupational data at this level in most countries, it is not possible to make an assessment of the relative importance in practice of sampling errors compared with classification errors and actual changes observed.

No detailed commentary will be provided on the tables of occupational structure. Overall, however, we can draw very broad conclusions from the changes in manufacturing occupational employment in the European Union that have taken place during the 1982–91 decade and these are given below.

Though employment in manufacturing has fallen in relative terms between 1982 and 1991, a trend which has continued to date and is expected to do so in the future, the sector continues to employ a very large number of people – about 30 million in EUR-10 depending on the data source. Craft and Related Trades Workers (ISCO 7) was the single largest occupational group in 1991 and accounted for 25% of the work-force. Between 1982 and 1991 this occupation has become less important in relative terms, although it is still dominant. Legislators, Senior Officials and Managers (ISCO 1), Professionals (ISCO 2) and Technicians and Associate Professionals (ISCO 3) have increasingly accounted for a relatively greater share of employment over the historical period and this trend has been projected to continue (Hogarth and Lindley, 1994) albeit at a more modest pace. Occupational change corresponds to the competitive pressures which the sector has faced during the 1980s and the consequent drive to improve productivity. This has resulted in a shift away from labour intensive tasks and the employment of skilled manual workers towards the greater employment of higher level skills associated with professional and technical staff. It should be borne in mind, however, that occupational change in the sector has been relatively modest during the 1980s considering the competitive pressures faced by the sector.

It will be clear from the more detailed data on occupational percentages for sectors and countries (Tables 4.11–4.18) that there are some very striking variations in occupational structures between countries. Moreover, it is not straightforward to relate these to the value-added performance of the industry concerned. So, for example, industries with high proportions of more highly skilled workers do not necessarily coincide with industries which record the highest value-added per person ratios.

4.2.3. Simulation results for 1993

Below we present results from the two alternative simulation exercises. Each one of them provides estimations for the levels of manufacturing employment by sector and country **with** and **without** the implementation of the single market programme (SMP) for the period 1985–93.

Simulation 1 compares the with and without SMP **intra-EU trade effects** on manufacturing employment in Europe, where the impact of the SMP on the extra-EU trade is neutralized; simulation 2 also takes into account the SMP impact on extra-EU trade and compares the with and without SMP **intra-EU and extra-EU trade effects** on manufacturing employment in Europe.

The **sectoral** effects of the SMP in simulations 1 and 2 are shown in Tables 4.11 and 4.15 respectively (EUR-10 excludes Greece and Luxembourg). In simulation 1, the impact of the SMP on manufacturing employment during 1993 varies between a 0.5% decline in metal products and a 7.9% increase in Ferrous and Non-ferrous Metal Products, while the change in employment for the rest of the sectors lies somewhere in between these two extremes. In simulation 2, where the impact of the SMP on the trade of the European Union with the rest of the world is also taken into consideration, the impact of the SMP on manufacturing employment varies between a 4.2% decline in Rubber and Plastic Products and a 7.1% increase in Ferrous and Non-ferrous Metal Products.

The **national** distribution of the impact of the single market programme on manufacturing employment for the two simulation exercises is presented in Tables 4.12 and 4.16 respectively. By 1993, in simulation 1, where the impact of the SMP on extra-EU trade is neutralized, all European Union members, with the exception of Germany, appear to have registered considerable employment gains. Ireland displays the highest proportional gains (5.9% increase in manufacturing employment) while Germany displays a proportional decline of the order of 0.6%. In simulation 2, where the impact of the SMP on extra-EU trade is also taken into account, the picture is quite different. Here, in 1993, the implementation of the SMP appears to have caused a decline in manufacturing employment in Denmark, Germany, Italy and the Netherlands while Belgium, Ireland, Portugal, Spain and the UK register quite small rises.

Overall, in 1993, in the case where the SMP effect on extra-EU trade is not taken into account, the impact of the implementation of the SMP on manufacturing employment appears to be positive. European manufacturing employment appears to have 140,000 more jobs than there would otherwise be, i.e. European manufacturing appears to have 0.5% more jobs than it would have without the implementation of the SMP. On the contrary, in the case where the SMP effect on extra-EU trade is also taken into account, the impact of the implementation of the SMP on manufacturing employment appears to be negative. Here, manufacturing in Europe appears to have 153,000 jobs less than it would have without the implementation of the SMP or, in other words, 0.5% fewer jobs than otherwise. Apparently, the positive impact of the SMP on intra-EU trade is outweighed by its negative impact on extra-EU trade.

If we apply the 1991 occupational structure to the set of results obtained by the two simulation exercises, it is possible to obtain two comparative tables of 'with and without SMP' occupational structures for all manufacturing sectors in each EUR-10 Member State. Table 4.13 gives the aggregate results for EUR-10 that have been calculated on the basis of the results of simulation 1 where the impact of the SMP on extra-EU trade is not taken into account. Setting aside the skilled agricultural workers and the armed forces, whose absolute magnitude is anyway very small, the results show that the already recorded increase in manufacturing employment is spread across all occupations. Yet, the largest **proportional** increase is registered by Plant and Machinery Operators (0.58%) as well as Legislators, Senior Officials and Managers (0.56%) while Craft and Related Trades Workers and Elementary Occupations lag slightly behind (0.53%). Table 4.17 gives the occupational structure projections that are based on results obtained by simulation 2, where the impact of the SMP on extra-EU trade is also taken into account. Here, the recorded fall in manufacturing employment is evenly distributed among the different occupations. The largest proportional decline, however, is registered by Technicians and Associate Professionals (0.97%), Clerks (0.82%) and Professionals (0.79%). It is clear that the occupational consequences, at least at this level of aggregation, exhibit a much more limited range of variation than those found for manufacturing sectors and countries.

The occupational effects by country do introduce more variation because of the way in which the impact of the SMP on intra-EU and/or extra-EU trade is distributed among Member States. The percentage changes in the country–occupational employment distribution are shown in Tables 4.14 and 4.18. Table 4.14 illustrates the country–occupational-structure matrix that has been calculated on the basis of the results of simulation 1 while Table 4.18 illustrates the same matrix on the basis of the results of simulation 2. A positive (negative) number implies an increase (decrease) in employment in that particular cell of the matrix due to the implementation of the SMP.

Here, the greatest variations in occupational impacts between countries are found among Craft and Related Trades Workers. It is down by 2.2% in Germany and up by 11% in Italy, when the impact of the SMP on extra-EU trade is not taken into account, and down by almost 1% in Germany and up by 11% in Italy, when the impact of the SMP on extra-EU trade is taken into consideration.

Table 4.11.   Model simulation 1: impact of the SMP on EU manufacturing employment by sector

Sector		Without SMP A	With SMP B	Differences B-A	(%) (B-A)100/A
8	Ferrous & Non-f. Metals	763647	824195	60548	7.9
9	Non-metallic Min.Pr.	1467919	1474329	6410	0.4
10	Chemicals	1863571	1864630	1059	0.1
11	Metal Products	2764251	2751293	-12958	-0.5
12	Agri. & Indust. Mach.	3091735	3093978	2243	0.1
13	Office Machines	984711	991006	6295	0.6
14	Electrical Goods	2978913	2988046	9133	0.3
15	Transport Equipment	2716866	2719282	2416	0.1
16	Food, Drink & Tobacco	3150816	3148864	-1952	-0.1
17	Tex., Cloth. & Footw.	3199078	3233597	34519	1.1
18	Paper & Printing Pr.	2309707	2321465	11758	0.5
19	Rubber & Plastic Pr.	1296365	1310779	14414	1.1
21	Other Manufactures	2196520	2203318	6798	0.3
	Manufacturing EUR-10	28784098	28924782	140684	0.5

Source: E3ME model.

Table 4.12.   Model simulation 1: impact of the SMP on EU manufacturing employment by country

Country	Without SMP A	With SMP B	Differences B-A	(%) (B-A)100/A
Belgium	711662	730538	18876	2.7
Denmark	511550	513249	1699	0.3
Germany	7596673	7554194	-42479	-0.6
Spain	3491716	3543493	51777	1.5
France	4387442	4387193	-249	-0.0
Ireland	237266	251308	14042	5.9
Italy	5161912	5188370	26458	0.5
Netherlands	1041940	1041892	-48	-0.0
Portugal	916816	934585	17769	1.9
United Kingdom	4727121	4779960	52839	1.1
Manufacturing EUR-10	28784098	28924782	140684	0.5

Source: E3ME model.

Table 4.13.   Model simulation 1: impact of the SMP on occupational employment in EU manufacturing

Occupation		Without SMP A	With SMP B	Differences B-A	(%) (B-A)100/A
1	Legislators, Senior Officials & Managers	1547173	1555792	8619	0.56
2	Professionals	1750247	1755612	5365	0.31
3	Technicians & Associate Professionals	2052646	2057336	4690	0.23
4	Clerks	3436275	3449770	13495	0.39
5	Service Workers etc.	850668	854431	3763	0.44
6	Skilled Agricultural Workers etc.	47551	47909	358	0.75
7	Craft and Related Trades Workers	10311356	10366158	54802	0.53
8	Plant & Machine Operators & Assemblers	6241965	6277981	36016	0.58
9	Elementary Occupations	2544719	2558296	13577	0.53
10	Armed Forces	1497	1499	2	0.13
	Manufacturing EUR-10	28784098	28924782	140684	0.49

Source: E3ME model.

Table 4.14.   Model simulation 1: impact of the SMP on occupational employment by country (%)

	1 Legislators, Senior Officials and Managers	2 Professionals	3 Technicians and Associate Professional	4 Clerks	5 Service Workers etc.	6 Skilled Agricultural Workers etc.	7 Craft and Related Trades Workers	8 Plant and Machine Operators & Assemblers	9 Elementary Occupations	0 Armed Forces	Total
Belgium	3.4	0.3	2.0	2.0	1.6	5.4	3.4	3.1	1.6	15.5	2.7
Denmark	0.1	0.0	-0.2	-0.0	0.2	-0.7	0.5	0.5	0.6	-1.2	0.3
Germany	-0.5	-0.5	-0.5	-0.5	-0.3	-0.7	-0.8	-0.4	-0.4	-0.1	-0.6
Spain	1.7	0.6	1.0	1.2	1.3	1.0	2.0	1.1	1.3	0.0	1.5
France	-0.2	0.3	0.3	0.1	-0.2	0.1	-0.2	-0.1	-0.1	0.0	-0.0
Ireland	4.3	1.8	3.0	4.4	3.3	0.0	10.8	4.9	3.5	0.0	5.9
Italy	0.6	0.8	0.8	0.8	0.6	1.7	0.3	1.1	0.9	0.0	0.5
Netherlands	-0.3	0.0	0.0	0.1	0.6	3.7	-0.1	-0.1	-0.0	0.0	-0.0
Portugal	1.7	2.8	2.3	2.1	1.2	-0.4	2.3	1.3	1.5	0.0	1.9
UK	0.8	1.0	0.7	1.0	0.4	0.0	1.1	1.4	1.8	0.0	1.1
EUR-10 total	0.6	0.3	0.2	0.4	0.4	0.8	0.5	0.6	0.5	0.1	0.5

Note: Percentage increase (+) or decrease (−) in employment resulting from the introduction of the SMP.
Source: E3ME model.

Table 4.15.   Model simulation 2: impact of the SMP on EU manufacturing employment by sector

Sector		Without SMP A	With SMP B	Differences B-A	(%) (B-A)100/A
8	Ferrous & Non-f. Metals	767027	821332	54305	7.1
9	Non-metallic Min.Pr.	1474662	1467325	-7338	-0.5
10	Chemicals	1868989	1856828	-12162	-0.7
11	Metal Products	2754889	2806275	51386	1.9
12	Agri. & Indust. Mach.	3186469	3082966	-103503	-3.2
13	Office Machines	990146	996493	6347	0.6
14	Electrical Goods	3038374	3007555	-30819	-1.0
15	Transport Equipment	2731636	2707969	-23667	-0.9
16	Food, Drink & Tobacco	3194488	3174181	-20307	-0.6
17	Tex., Cloth. & Footw.	3247654	3274747	27093	0.8
18	Paper & Printing Pr.	2352294	2337089	-15204	-0.6
19	Rubber & Plastic Pr.	1398023	1339134	-58889	-4.2
21	Other Manufactures	2206034	2185509	-20526	-0.9
	Manufacturing EUR-10	29210686	29057403	-153283	-0.5

Source: E3ME model.

Table 4.16.   Model simulation 2: impact of the SMP on EU manufacturing employment by country

Country	Without SMP A	With SMP B	Differences B-A	(%) (B-A)100/A
Belgium	726575	732081	5506	0.8
Denmark	514522	510556	-3966	-0.8
Germany	7831161	7634281	-196880	-2.5
Spain	3499148	3552163	53015	1.5
France	4410035	4410915	880	0.0
Ireland	231701	243539	11838	5.1
Italy	5209079	5154943	-54136	-1.0
Netherlands	1062788	1042665	-20123	-1.9
Portugal	920768	934923	14155	1.5
United Kingdom	4804910	4841336	36426	0.8
Manufacturing EUR-10	29210686	29057403	-153283	-0.5

Source: E3ME model.

Table 4.17.   Model simulation 2: impact of the SMP on occupational employment in EU manufacturing

Occupation		Without SMP A	With MP B	Differences B-A	(%) (B-A)100/A
1	Legislators, Senior Officials & Managers	1574323	1564400	-9923	-0.63
2	Professionals	1776818	1762734	-14084	-0.79
3	Technicians & Associate Professionals	2087220	2066888	-20332	-0.97
4	Clerks	3495963	3467282	-28681	-0.82
5	Service Workers etc.	560288	859936	-352	-0.04
6	Skilled Agricultural Workers etc.	47766	47376	-390	-0.82
7	Craft and Related Trades Workers	10440012	10401885	-38127	-0.37
8	Plant & Machine Operators & Assemblers	6337593	6312963	-24630	-0.39
9	Elementary Occupations	2589148	2572432	-16716	-0.65
10	Armed Forces	1554	1508	-46	-2.99
	Manufacturing EUR-10	29210686	29057403	-153283	-0.52

Source: E3ME model.

Table 4.18. Model simulation 2: impact of the SMP on occupational employment by country (%)

	1 Legislators, Senior Officials and Managers	2 Professionals	3 Technicians and Associate Professional	4 Clerks	5 Service Workers etc.	6 Skilled Agricultural Workers etc.	7 Craft and Related Trades Workers	8 Plant and Machine Operators & Assemblers	9 Elementary Occupations	0 Armed Forces	Total
Belgium	1.1	-1.2	0.4	0.1	0.7	3.3	1.2	1.4	-0.2	10.3	0.8
Denmark	-0.7	-1.2	-1.3	-0.9	-0.6	-0.4	-0.9	-0.7	-0.1	-2.8	-0.8
Germany	-2.7	-2.3	-2.7	-2.5	-2.8	-1.9	-2.2	-2.8	-2.8	-4.1	-2.5
Spain	1.7	0.3	0.4	1.1	2.1	0.8	1.9	1.5	1.1	0.0	1.5
France	-0.7	0.3	0.1	0.1	-0.2	-0.7	-0.1	0.1	-0.0	0.0	0.0
Ireland	3.0	-0.4	2.2	3.3	3.3	0.0	11.0	3.6	2.5	0.0	5.1
Italy	-1.1	-1.8	-1.7	-1.2	-0.3	-0.9	-0.9	-1.1	-1.0	0.0	-1.0
Netherlands	-1.8	-1.5	-1.7	-1.7	-2.0	-1.7	-2.0	-2.0	-2.0	0.0	-1.9
Portugal	1.4	1.9	1.7	1.5	1.3	0.8	1.7	1.3	1.4	0.0	1.5
UK	0.3	0.9	0.6	0.8	0.1	0.0	0.5	1.2	1.7	0.0	0.8
EUR-10 total	-0.6	-0.8	-1.0	-0.8	-0.0	-0.8	-0.4	-0.4	-0.6	-3.0	-0.5

Note: Percentage increase (+) or decrease (-) in employment resulting from the introduction of the SMP.
Source: E3ME model.

4.3. Parallel studies and business survey evidence

This section analyses and compares the results of this study with those obtained by related studies of the EU manufacturing sectors and the impact that the single market programme has on them.

4.3.1. Parallel studies

Four parallel studies are covered which summarize the impact of the SMP on the following manufacturing sectors:

(a)   Processed Foodstuffs,
(b)   Textiles and Clothing,
(c)   Food, Drink and Tobacco Processing Machinery,
(d)   Chemicals.

European Commission (1997d)): Processed Foodstuffs (compiled by BER)

Firstly, it is important to note that this analysis only encompasses seven product categories of the Food and Drink manufacturing industry. They are the following: Pasta and similar products, Industrial Baking, Cocoa, Chocolate and Sugar Confectionary, Other Food Products, Alcohol and Spirits, Products of Brewing and Malting, and Soft Drinks/Mineral Water. In 1994, they accounted for 46% of total employment in the Food and Drink manufacturing industry.

For most of the above product sectors, BER's study revealed that intra-EU trade had been enhanced by the SMP. Other Foods and Soft Drinks/Mineral Water in particular both experienced trade expansion and an increasing share of trade within the EU. Furthermore, 96 out of 108 companies in BER's industry survey claim that the SMP has either 'significantly' or 'to some extent' eliminated trade barriers. In other words, market access had been greatly eased. BER's time-series analysis, however, did not support these claims.

The impact of the SMP on cross-border sales and marketing was difficult to assess. Of course, as BER points out, differences in cross-border sales and marketing development were bound to vary according to subsector and Member State. Their industrial survey produced mixed results. On the one hand, most companies felt that the SMP increased intra-EU trade, but on the other, only a minority believed market prices had been reduced.

The survey also revealed that over half of the large firms interviewed had experienced changes in the size and number of plants. BER's analysis of structural breaks provides evidence that the SMP had influenced the scale and scope of operations.

The industrial survey also confirms the existence of major cross-border merger and takeover activity in the Processed Food and Drink sector in the early 1990s. It would appear that the SMP has only consolidated such activity rather than propelled it. On the other hand, there is evidence that the SMP has resulted in companies shifting their production locations abroad.

Market concentration has also increased over the last decade and this partly reflects companies' greater thinking on a pan-European basis. BER reports that over half of the companies interviewed believe the SMP has facilitated market entry. Only firms located in the southern Member States disagreed with this general consensus.

According to BER's structural break analysis, positive trends in productivity and competitiveness emerged for a few sectors after 1986. The industry survey results confirm that changes in R&D and product development strategies have been undertaken by half the companies interviewed in the survey.

The structural breaks analysis and industry survey both suggest that the long-term decline in employment in the Food and Drink sector will persist. However, changes in the patterns of employment have not been uniform across all Member States. For example, German re-unification and the accession of Portugal and Spain to the EU in 1986 have both had an impact on employment in the respective countries.

Finally, BER finds no correlation between the impact of the SMP and changes in either price levels or the extent of price dispersion.

European Commission (1997b): Textiles and Clothing (compiled by CEGOS)

According to the CEGOS study, employment in the Textiles and Clothing industry in the EU has been falling since 1978. However, between 1983 and 1988 this decline appeared to decelerate. Portugal and Greece, and to a lesser extent Spain, have shown the smallest rate of employment fall. Indeed, in the Clothing sector, Greece and Portugal have seen increases in employment of 50% and 18% respectively between 1984 and 1992.

Between 1992 and 1994, employment in the Textiles and Clothing industry has continued to fall. In this industry 131,000 jobs were lost between 1992 and 1993. Moreover, this tendency has affected all sectors across the European Union. The IMF forecasts a loss of 465,000 jobs between 1992 and 2002 in Clothing (34% of employment in this sector). The Netherlands is expected to be the least affected (a fall of 22%), while Portugal, Spain and Germany are likely to be the most affected, experiencing falls in employment of 49%, 40% and 39% respectively. Sewers, cutters and finishers appear to be the most vulnerable workers.

Competition from the Third World is identified as one factor which has had a negative impact on the Textiles and Clothing industry in the EU. In these countries, lower labour costs put firms in the EU at a disadvantage. They have tried to respond by increasing productivity and production has become automated and also relocated. However, employment has had to be sacrificed and a number of companies have even disappeared. The recession and the fall in consumption of the industry's products have bolstered these closures.

The SMP has not reversed this trend. However, even if its impact is only marginal, certain factors lead us to think that the SMP has actually slowed down the tendency. For example, we have previously seen its contribution to the development of competitiveness in the Textiles and Clothing industry.

European Commission (1997a): Food, Drink and Tobacco Processing Machinery (compiled by DRI)

In summary, the SMP appears to have had no impact on sectoral development which includes intra-EU trade, productivity, employment, industrial structure, and end-user prices. However, as DRI points out, their dataset is from 1980 to 1992 or 1993, during which most sector-specific measures directed towards technical harmonization had not yet been completed. The reduction of general barriers to cross-border trade and measures in sectors serving the Food, Drink and Tobacco Processing Machinery industry have provided the main impact. In the former category, we include the elimination of customs formalities, while the liberalization of transport services is an example of a measure employed by an interdependent sector.

As DRI points out, developments in Food and Allied Products have a visible impact on the Food, Drink and Tobacco Processing Machinery industry. For example, changes in the demand for equipment over the last decade have had a stronger influence on machinery manufacturers than sector-specific single market legislation. The Food and Allied industries have undergone major changes over the period which have influenced the size and form of equipment demand. However, it is the trend toward mergers and acquisitions that has radically transformed the profile of equipment demand. Large equipment suppliers have responded by pursuing strategies of horizontal integration. As DRI points out, though, because the concentration of the market is very small, these mergers have had little impact on the overall degree of competition in the Food, Drink and Tobacco Processing Machinery industry.

DRI also highlights the cost savings made possible by the SMP. In the short term, there are savings in development costs due to the reduction of machinery variants with different safety requirements. Research and development costs in the sector account for about 5% of sales. DRI estimates potential savings of 5% of sales. Savings are also attained by lowering the cost of equipment sold in markets where safety regulations were more stringent than those stipulated in EU legislation. Savings of up to 5% of machine value per unit are possible.

European Commission (1997c): Chemicals (compiled by KPMG)

The European Chemicals sector (including man-made fibres but excluding pharmaceuticals) experienced a 30% fall in employment levels between 1978 and 1993. When compared with the large increases in turnover over the same period, this fall in employment is likely to be due to strong productivity gains. According to KPMG, all EUR-12 countries, except Ireland, have experienced a reduction in employee numbers over the last 10 to 15 years with the rate of decline accelerating. The UK, Italy, Ireland and Spain, however, have experienced a smaller rate of reduction in employment levels from 1987 to 1992. Since 1989, though, this general tendency of reduced employment has been more marked in the EU than in the US or Japan.

Annual reports also show that the number of employees is decreasing, as expected. Of the 13 companies with five years of data for 1990 to 1994, nine show a reduction of the number of employees in the range of 14–44%, two remained the same, and two companies reported an increase in number of employees although one of these was explained by an acquisition.

Nearly half the companies interviewed in KPMG's face-to-face survey also confirm that they have greatly reduced their employment levels over the last five to ten years. This survey focused on larger companies, though, and it is probable that these firms implemented relatively larger employment cuts due to the substantial mergers and acquisitions and restructuring activity that occurred in the sector.

Overall, company annual reports suggest two main reasons for reductions in the work-force. Firstly, business growth has been insufficient to sustain high levels of employment, and productivity gains have been required. Secondly, there is an increasing supply of qualified contractors and specialists enabling companies to contract out discrete pieces of work. Therefore, during a period of considerable restructuring in the chemicals industry, any attribution of observed employment changes to the SMP may be erroneous and misleading.

The extent to which the SMP has influenced internal EU mobility was also considered by the KPMG survey. For example, EU legislation was expected to facilitate cross-border mobility in regions such as northern Italy and Rhone-Alpes in France. Nearly half the companies questioned in the survey felt that the SMP had stimulated at least some internal EU job mobility. However, they suggested that the absence of harmonization or ease of transferability of pension schemes had hindered the process.

Productivity and competitiveness were expected to be affected indirectly by the SMP. For example, increased competitive pressures could create the need for greater efficiency, and cost savings in source and trade-related costs were also a possibility. KPMG's other survey evidence supported this hypothesis, with more than 40% of the companies questioned suggesting that the SMP had contributed to the productivity improvements attained over the last five to ten years. However, the survey revealed that the SMP has provided little stimulus to sales efforts to non-EU countries. This may reflect the considerable increase in competition at the global level, the indirect way in which the SMP would help, and the increase, on average, in short-term production costs, which companies felt the SMP to be accountable for.

Overall, though, the net effect of the SMP on the EU chemicals industry appears to be inconclusive. For example, the industry's restructuring trends make it even more difficult to isolate the SMP effects from the general industry employment effects. Therefore, it may only be feasible to identify and assess possible qualitative effects.

4.3.2. Eurostat business survey results

This section reviews briefly the results of the business survey undertaken as part of the current evaluation of the SMP (European Commission (1997h)). The review is divided into subsections examining the survey's results with respect to employment, pricing, liberalization of capital movements, sales to non-EU countries, the overall success of the SMP for industry and the possible need for additional measures to eliminate obstacles to EU trade.

Employment

The business survey's country-by-country comparison indicates that the SMP has had a small impact on employment. Indeed, for the EU states, on average, the firms questioned give a weight of almost 70% to the 'no change in employment' category. However, this conceals some interesting results. For example, positive effects on employment are reported in Italy. Irrespective of whether the results are weighted by number of enterprises, employees or turnover, over 20% of the Italian firms in the survey suggest that the SMP has had a beneficial impact on employment. The survey also suggests that employment has been bolstered in Denmark, the Netherlands and Ireland. On the other hand, negative effects on employment were reported in Germany, Spain and France.

The survey's sector-by-sector comparison suggests that the largest gains in employment have been enjoyed by Machinery and Equipment, with one-fifth of firms reporting positive effects on employment (12.8% of the firms in terms of employees and turnover). However, negative effects have outweighed positive effects in Tobacco Products.

The survey also provides information for different groupings of firms on the basis of firm size. For firms reporting a positive impact, there appears to be a negative relationship between firm size and employment gain. The smaller the firm, the greater the impact of the SMP on employment creation. For negative effects, there was no difference between firms of different sizes.

Pricing

The SMP was expected to lead to lower prices due to enhanced competition. However, the business survey indicates that the SMP has had little importance in influencing pricing decisions. For the EU states, on average, firms attach a weight of over 40% to the 'pricing of no importance' category. In Germany, for example, over 65% of firms believe that the SMP has had no effect on their pricing strategy. However, there was a greater indication in Belgium that the SMP had been 'quite important' in influencing pricing decisions.

In the sector-by-sector results, Food Products and Beverages and Machinery and Equipment are the largest sectors. In these two industries, most firms report that the SMP is 'not important' in determining their pricing strategies. This feeling tends to be shared across all sectors.

The survey results also reveal that there is no apparent relationship between firm size and the impact of the SMP on pricing. Again, most firms, irrespective of size, attach a large weighting to the 'pricing is not important' category.

Liberalization of capital movements

The complete liberalization of capital movements was achieved quickly in the EU and the final restrictions concerning short-term capital movements have been removed. Most firms questioned in the survey, however, believe that the SMP itself has had little impact upon the process. For the EU states, on average, firms attach a weight of over 50% to the 'no change' category. However, firms in Spain and Italy felt that the SMP had played an important role.

The sector-by-sector results show little difference in attitude across different industries, but larger firms tended to view the SMP as having had a more beneficial impact on capital movements.

Sales to non-EU countries

The survey's country-by-country comparison suggests that firms believe that the SMP has had no effect on firms' sales to non-EU countries. For the EU states, on average, firms attach a weight of over 75% to the 'no change' category.

The capacity to respond strategically to the sectoral impact of the SMP might be expected to be greater for those firms already having a strong international orientation mainly in manufacturing. However, the survey shows that in manufacturing industry most firms claim the SMP has had little impact upon sales to non-EU countries. Furthermore, the smallest firms report the most positive response, whereas the greatest gains might have been expected for firms able to benefit from economies of scale in an integrated European market.

'The single market programme has been a success for your sector in the EU'

Overall, the survey suggests that the SMP has had no effect on the success of various sectors in the EU. For the EU states, on average, firms attach a weighting of over 40% to the 'no opinion' category. However, positive responses were reported in Germany, Spain, France and, in particular, Italy. Indeed, for Italy, 60% of firms claim the SMP has been beneficial for their sectors in the EU.

The different sectors in manufacturing can be grouped into three categories: high-technology public-procurement sectors, traditional or regulated markets and sectors with moderate non-tariff barriers. Buigues *et al.* (1990) identified Office Machines as belonging to the first category, and firms in this sector believe the SMP has been a success for the sector. The SMP was also expected to have an impact on sectors in the second category. Food and Drink, for instance, was highly protected. The survey results suggest that most firms in this sector feel that the positive effects from the SMP have outweighed the negative effects. Similar beliefs were held by firms in Textiles and Clothing and Machinery and Equipment.

As expected, there is a positive relationship between firm size and success. For example, over 40% of the largest firms (over 1,000 employees) claim that the SMP has been a success for their sector in the EU.

'The single market programme has been a success for your sector in your country'

Overall, most firms feel that the SMP has had little impact. However, a number of firms do identify the influence of the SMP. For the EU states, on average, 41.8% of firms believe the SMP has had no impact. However, sensitivity to the SMP was expected to be different in the northern countries than in the southern countries and this is borne out to some extent in the results. Roughly 60% of Italian firms believe that the programme has been successful domestically for their particular sector. In Germany, Spain and France, any reported negative effects outweigh the positive effects.

A similar picture emerges from the sector-by-sector comparison. Any positive and negative effects appear to be evenly distributed, although the impact of the SMP upon Office Machines and Computers appears to be generally beneficial.

Firm size appears to be independent of the SMP's potential for success for the various sectors in each country.

'Additional measures are needed to eliminate obstacles to EU trade'

It is important to note that the nature of trade between Member States is not the same for all sectors. In certain sectors it is inter-industrial, while in others it is intra-industry, countries trading between themselves in similar products without any dominating flow in one direction. Therefore, the structural adjustments which follow the removal of non-tariff barriers were expected to vary in kind and possibly in magnitude in these two cases. On balance most firms feel that there is no need for any further measures to remove such obstacles. However, firms in France and Italy believe more measures are needed.

In the sector-by-sector comparison, over 30% of firms in Food Products and Beverages and Chemicals and Chemical Products believe that additional measures are necessary. This is surprising in the last industry because the Chemicals industry in the EU was previously affected by only moderate non-tariff barriers.

Finally, firm size does not appear to influence firms' responses.

Conclusions

In most cases the survey results suggest little impact from the SMP. This could be because the effects will take longer to emerge. However, it also reflects a potential weakness of the survey method: firms may not attribute changes in their business environment to the SMP, even if this has played a significant role. This is especially likely to be the case for smaller firms, for whom it is difficult to determine whether changes in their own experience reflect industry-wide effects or changes specific to the firm.

Of course, the survey tells us about firms' opinions and not necessarily about the effect of the SMP. The responses simply tell us what firms **think** the effect was. Moreover, in many cases firms may not be in a position to judge the impact of the SMP.

4.3.3. DRI survey analysis

This section summarizes the results of the *Survey of the Trade Associations' Perception of the Effects of the Single Market*, carried out by DRI (1995). The purpose of the survey was to gain a broader understanding of the EU business sector's views of the impact of the SMP. The method of investigation primarily involved interviews of key representatives of the European trade associations.

Ferrous and Non-ferrous Metals

The impact of the SMP on the Ferrous Metals industry is difficult to quantify. Firstly, one needs to isolate the effects of the SMP on the industry from the impact of previous EU policy measures. Secondly, as the DRI study points out, national policies such as state aids continue to distort competition and hinder integration of the steel market. Policy responses could, therefore, include the removal of any remaining internal barriers and the harmonization of environmental regulations.

The overall impact of the SMP on the Non-ferrous Metals industry appears positive. In particular, the industry has welcomed initiatives aimed at harmonizing business conditions across the EU. This is particularly true in areas such as the liberalization of energy markets, the approximation of tax rates (VAT) and the harmonization of environmental protection and safety regulations. However, as DRI emphasizes, excessive regulatory costs need to be avoided and policy objectives achieved in an optimal way. Therefore, any additional harmonization of regulations may require an accompanying cost-benefit analysis. The only concern for the Non-ferrous Metals industry appears to be the high tariff barriers employed by a few markets, mainly in Asia, and the public procurement rules which exist in the US.

Non-metallic Mineral Products

The overall impact of the SMP on this industry has been deemed both important and beneficial. It is important to note, though, that certain sectors of this industry such as Cement, Glass, and Ceramic Goods had a strong orientation towards European integration prior to the completion of the SMP. Yet, these sectors believe the programme can not only consolidate but also bolster the integration and development processes. In the whole of the industry, issues such as the free movement of goods, technical harmonization and competition policy are of importance. The relevance of other factors will, of course, depend on the subsector in question.

A combination of energy, environmental, and competition policy issues are likely to impact strongly on highly energy-intensive industries operating in strongly competitive international markets. These industries have expressed their concern about EU legislation increasing prices to domestic producers, whereby suppliers from other countries (free from EU rules) could then gain access to European markets.

Metal Products

Thus far, the impact of the SMP on this industry has only been indirect. The programme has provided the impetus for firms to think in European terms but any direct effects on, say, sales or costs have yet to be seen. Part of the problem is the remaining internal barriers to trade. However, the industry has also expressed its concern about legislative initiatives which are designed to harmonize technical standards. Over-regulation may be a danger in such measures. Therefore, the Metal Products industry has emphasized the need for succinct and clear regulations which are also compatible with any existing international standards.

Mechanical Engineering

Despite the early stage of its integration process, this sector has benefited from the SMP, particularly in the area of technical harmonization. However, a number of problems have prevented the SMP from being truly realized. The key problem relates to the requirements of the national transpositions of the Machinery Directive. Not only do these standards differ from those in the Directive itself, but also across Member States. If requirements do vary across Member States, then free movement of products throughout the EU will be hindered. Therefore, the checking of the transposition of European legislation and the development of harmonized standards are important future issues for the Mechanical Engineering sector. The lack of licensed testing bodies represents another problem for the sector. For some kinds of machinery, inspection by licensed bodies is one condition of the Machine Directive, but end-users are even insisting on such measures.

Finally, concern is expressed about the planned agreement on public procurement between the EU and the US. The sector fears this will work in the US's favour, granting US companies wider access to EU public procurement markets.

Electrical Engineering

The overall impact of the SMP on the Electrical Engineering sector has been beneficial. However, as DRI emphasizes, a number of issues warrant attention. Firstly, any positive effects would have been bolstered if Member States had fulfilled the obligations of relevant

Directives. Secondly, problems and delays in Member States' full recognition of European standards regulation have occurred because the industry has only partially adopted the required EU legislation. Indeed, national trade barriers and trade barriers of the developed world and NICs still represent major competitive problems. With regard to the future, the industry believes legislation should be simplified and 'hidden' national trade barriers eliminated. Complete implementation of EU legislation, however, remains the main priority.

Electronics

On balance, the Electronics industry has been adversely affected by the SMP. To begin with, any improvement in the industry's competitiveness in European and world terms has been at the expense of falling employment. Secondly, within the EU, obstacles to the completion of the SMP persist. For example, technical, fiscal or legal harmonization remains deficient in many areas. Thirdly, the opening-up of public procurement markets has yet to be achieved. Finally, as DRI points out, barriers to trade still hinder exports to countries outside of the EU. Public procurement, different technical standards, discriminatory distribution systems and tariffs are the principal offenders.

Transport Equipment

As expected, the SMP has had a positive impact on the Transport Equipment sector. Free movement of goods and capital, reductions in the cost of cross-border payments, and strengthened competition within the various subsectors represent major successes of the SMP.

However, further efforts still need to be made, including increased harmonization in the areas of public procurement, indirect taxation and cross-border payments. Furthermore, the industry feels that both competition and environmental policy should be implemented with the world situation in mind. Preventing a deterioration in the competitiveness of the EU Transport industry is the crucial issue.

Food, Drink and Tobacco

The removal/reduction of internal barriers to trade has yielded globally positive benefits for the Food and Drinks sector. Any remaining barriers relate, for example, to Member States' lack of support for EU legislation with their own legislation being distinct from harmonized Community measures. Remaining areas where future progress is deemed necessary include the promotion of consumer protection and higher safety standards in food supply.

The SMP has had a limited impact on the Tobacco Products sector partly due to the nature of the product itself. However, trade intensity has been strengthened by the reduction of controls at internal frontiers. Main priorities include the reduction and harmonization of VAT taxation and duties on cigarettes.

Textiles, Clothing and Footwear

The specific impact of the SMP on the Textiles sector is difficult to assess. For example, prior to the programme's creation, the sector was already very open and hence very competitive. In addition, over-capacity is a feature of this sector and so the removal of national import quotas led to a decrease in the competitiveness of EU companies in relation to extra-EU competitors. The potential to reap economies of scale has been limited because the sector mainly consists of SMEs. Of course, the European market remains highly diversified which in part reflects the taste for variety in the EU. Given these characteristics, it is unsurprising that the SMP has had little impact in terms of volume effects.

The sector is of the opinion that future integration measures should take into account its distinct features. Training and education programmes, for example, ought to be more accessible to the SMEs and the interests of EU companies should be better represented by an improved Community external policy.

In general, the impact of the SMP on the Clothing sector has been positive. The main benefit appears to be the reduced delays and costs following the removal of controls at internal frontiers. In turn, this has stimulated EU clothing companies to specialize in higher quality and fashion-led segments of the market. On the other hand, the elimination of national import quotas has resulted in non-EU companies increasing their market share. The declining share in employment in this sector poses a great problem.

The main priorities for future action include the abolition of national non-tariff barriers to trade and measures to accelerate the fiscal harmonization process.

The impact of the SMP on the Footwear industry is hard to evaluate. Primarily, this is because the recession has offset any positive effects which may have resulted from the SMP. However, as DRI emphasizes, it is likely that the programme has increased competition in the EU, albeit at the expense of European manufacturers' profits.

The main priority for the future appears to be the abolition of any remaining non-tariff barriers.

Paper and Printing Products

The impact of the SMP on the Pulp, Paper and Board sector has been very positive. For example, cross-border EU trade flows and important restructuring have been stimulated by the SMP. However, as DRI points out, persistent currency volatility and obstacles related to technical and environmental issues represent two remaining internal barriers. Therefore, major priorities for the future include the need for currency stability and the harmonization of technical standards at the European level.

The SMP has also been beneficial for the Paper and Board Converting sector. Intra-EU trade has been stimulated and restructuring through mergers and acquisitions has also occurred. However, this restructuring has strengthened the market power of the suppliers and customers of non-integrated SMEs, thereby squeezing the latter's profits.

DRI identify direct taxation and environmental standards as the remaining internal barriers. Any external barriers have little importance because extra-EU exports do not play a major role in the sector.

Finally, the SMP has stimulated companies in the Printing and Publishing sector to think on a pan-European scale. Deregulation of telecommunications, for example, has also allowed increased exploitation of new openings. The future development of this industry, however, is being impeded by issues such as the lack of standardization of advertising regulations across Member States.

Rubber and Plastic Products

In general, the SMP has yielded positive effects for this sector. The only major problem relates to the high costs of inputs, especially labour.

DRI identify policies on the management of waste as an important remaining internal trade barrier. External trade barriers still remain, mainly in the Far East. Eliminating the former barriers ought to be assigned a high priority.

In Rubber Products, DRI identify two positive effects from the SMP: increased market potential and both producers and products securing a 'European identity'.

Any remaining internal barriers appear to be in the area of environmental legislation. Major external barriers for EU companies include the protectionism employed by the Far Eastern countries.

The sector feels that advancement on measures regarding SMEs are a policy priority. However, greater access to finance for SMEs and the stimulation of privatization within the EU are two issues which warrant future action.

Other Manufactures

Furniture, Jewellery, Musical Instruments, Toys and Sporting Goods are categorized as belonging to the Other Manufactures sector. The SMP has had no impact on the Furniture sector. One characteristic of the sector is a high level of intra-EU trade but the programme has failed to strengthen this trend. In addition, various measures, such as technical harmonization, have yet to be completed. DRI identify public procurement and safety regulations as the remaining internal barriers.

On the whole, the SMP has been beneficial to the Jewellery sector which is characterized by an international orientation. Therefore, levels of EU exports to third counties are important to the sector's performance.

In terms of internal barriers, DRI emphasize the absence of harmonized marking of precious metals. External tariff barriers also exist in some countries, such as India, and non-tariff barriers in others, such as Japan.

The sector's SMEs, however, have been placed at a competitive disadvantage by the SMP and this indicates a need for measures to counter these negative effects.

A positive impact from the SMP has been enjoyed by the Musical Instruments industry. However, as DRI point out, the position of many European SMEs has been jeopardized by the increased competition from large companies from the Far East.

The overall impact of the SMP on the Toys sector has been positive. However, the sector is adversely affected by the different interpretation and implementation of specific EU regulations and directives in the various Member States. The industry feels that removing quantitative market restrictions (quotas) ought to be the main priority.

The Sporting Goods sector has enjoyed increased intra-EU trade and greater competition from the SMP. In addition, the programme has stimulated companies to think in both global and pan-European terms.

DRI identify the inequalities in indirect taxes as one remaining internal barrier. Tariff and non-tariff barriers (import quotas, bans) in Far Eastern and South American countries should be eliminated. For the future, the main priority will be to allow SMEs easier access to finance.

Chemicals

The Chemicals industry includes Paints, Varnishes and Printing Inks, Industrial Gases, Flavours and Fragrances, Oleochemicals and Allied Products, Fertilizers, Soaps and Detergents, Personal Care Products and Non-prescription Pharmaceuticals.

The SMP has had little impact upon the Paints, Varnishes and Printing Inks sector, primarily because its products can already be freely traded within the EU. However, the industry is concerned about the possible emergence of new internal barriers.

Globally positive effects have occurred in the Industrial Gases industry due to the SMP. The full beneficial impact of the SMP, however, has yet to materialize. Of course, it is important to realize that due to the nature of the sector, the impact of the SMP will not be reflected in changing trade patterns.

Positive effects have also been attained by the Flavours and Fragrances industry. For example, internationalization of production has been facilitated by the harmonization of national laws on the authorization of food additives. Additional harmonization efforts are strongly favoured by the industry.

The Oleochemicals and Allied Products sector has benefited from the creation of the SMP. Furthermore, as DRI point out, the sector, on its own initiative, has harmonized testing methods. The only concern appears to be the possible impact of new proposals, such as hygiene requirements on imported beef tallow.

Both positive and negative effects have been felt by the Agrochemicals sector. For example, the reform of the common agricultural policy and new environmental restrictions adversely affected the sector, while the benefits of harmonization have yet to reach their full extent. For the future, progress in harmonization of authorization rules is a key priority.

The SMP has had little impact upon the Fertilizer sector. Prior to the programme's completion, the creation of the Customs Union and the subsequent removal of internal tariffs had already made a large contribution to the integration process.

Positive effects have been enjoyed by the Soaps and Detergents, Personal Care Products industry. DRI include cost savings and increased competition among these benefits. The main priority for the industry appears to be the prevention of new barriers to trade. Finally, the SMP has also yielded positive effects for the Non-prescription Pharmaceuticals sector.

Conclusion

On the whole the EU business sector feels that the SMP has had a positive effect on the economy, though much remains to be done. In the manufacturing sector, many associations believed that progress towards the single market resulted as much from improvements in the implementation of measures approved before 1986 as from new measures adopted after that date.

4.4.   Other issues relating to the *antimonde*

The implications for the SMP of foreign direct investment, structural funds spending and accession to the EU

The purpose of this section is to provide an evaluation of factors (foreign direct investment (FDI), structural funds spending and accession to the EU) whose effects would ideally have been distinguished from those of the SMP but which for various reasons, mainly data availability, could not be modelled. This section suggests the likely impact on the results if it had been possible to take these factors into account.

4.4.1.   Foreign direct investment

Buigues *et al.* (1990) identified strong growth in FDI in Spain and Portugal between 1986 and 1990. In the period 1986–88, FDI in Spain was destined primarily for the average-growth and high-demand sectors. The main beneficiaries were Chemicals, Agri-foodstuffs, Paper, and Motor Vehicles. FDI also accounted for the largest contributions to investments in high demand growth sectors. Thus, in 1990 in these sectors, 88% of investments were financed by foreign capital as opposed to 52% in average demand growth sectors, and 11% in low growth sectors.

Several factors explain these inflows of FDI. Firstly, EU membership and the prospect of the completion of the SMP gave access to a larger market and a reduction in transport and marketing costs in the EU. Also, the strong growth recorded between 1986 and 1990 made Spain and Portugal increasingly attractive to foreign investment. Finally, their advantages in terms of wage costs, the incentives offered and, in certain sectors, the absence of local competitors, have also played a part.

More recent evidence by CEPII (European Commission, 1997e) confirms that Spain's joining of the EU and the formation of the single market had a positive effect on FDI inflows in Spain, with the consequent macroeconomic benefits to saving and productive investment. The study, carried out at firm level, supports the idea that FDI has promoted the upgrading of human skills, technological advances and, although to a lesser extent, improved productivity of the firms receiving the investment. However, there is no evidence on the extent to which the benefits accruing to local subsidiaries of multinationals had spillovers for domestic firms. Thus, for example, multinational corporations may either stimulate improvements in productivity and product quality in local suppliers or conversely raise barriers to entry for local firms. In this regard, evidence on the differences of both export and import propensities of firms with foreign capital participation *vis-à-vis* domestic firms is rather ambiguous. On the one hand, FDI appears to have been helping to promote exports and, therefore, to stimulate economic growth and employment generation. On the other hand, control exerted by foreign investors may have been conducive to sourcing of inputs from abroad, so that FDI may have been constraining the development of domestic productive activity.

Ireland was also identified by Buigues *et al.* as having the potential to benefit from increased FDI. Buigues *et al.* proposed that if the removal of tariff barriers to exports to EU countries after 1973 stimulated the foreign investment in Ireland, then it seemed feasible to expect the removal of non-tariff barriers by 1992 to have further strengthened the motivation for new foreign investment in industries which, at the time, were facing non-tariff barriers to trade. Indeed, by 1990 there were some indications that this expectation would be realized with growth being particularly strong in sectors which were predominantly composed of foreign-owned multinational corporations. This growth performance was not, of course, simply a response to the potential of the single market, but was nonetheless a promising indicator. In addition, investment by new arrivals in Ireland was expected to be at least as important for growth as expansion of the existing firms.

In conclusion, there seems little doubt that FDI was boosted by the SMP. Since this effect would not be captured by the trade effects analysed in the present study, the results of this study would tend to understate the positive impact of the SMP on employment. The anecdotal evidence suggests that the FDI effect would be strongest in Spain, Portugal and Ireland. Therefore, the effect is to boost employment by more than the model would allow for.

4.4.2. Structural funds spending

In the post-SMP period, structural funds spending has also increased. This coincides with the apparent improved performance of the poorer Member States of the EU, including Spain, Portugal, Greece and Ireland (the so-called 'Objective 1 regions'). However, it is unclear whether their greater growth is due to the SMP or to the impact of enhanced structural funds spending. The *Single Market Review* VI.1: *Regional growth and convergence* (European Commission, 1997f) suggests that it is possible that differences in structural funds spending in the 1989-93 period may not explain differences in growth performances. Yet, they find no evidence to suggest that making explicit allowance for structural funds spending would explain the change in performance of poorer Member States post-1987 (although with no pre-1987 data this cannot be confirmed). Of course, this does not indicate that structural funds spending has had no impact, but it does suggest that the demand-side effects of this policy do not dominate other explanations for differences in the growth of Member States. As Fingleton *et al.* emphasize, the supply-side effects are unlikely to be evident after such a short period.

Regional growth and convergence also models the impact of the structural funds using a dummy variable approach which is not feasible for this report. Data on spending under the structural funds is not readily available, particularly in the format and dimensions necessary for this study, requiring the E3ME model to be completely re-specified and much of the work done to date to be repeated.

4.4.3. Accession to the EU

As noted in Chapter 2, there are major difficulties involved in analysing the impact of the SMP on Spain and Portugal, entrants to the EU in 1986. This is because the mechanisms induced by the completion of the single market apply during a transitional period which followed their accession to the EU. Realistically, it seems impossible to distinguish the two effects, not least because they both represent measures that increased the integration of these economies within the EU.

Unfortunately, the approach adopted in *Regional growth and convergence*, which estimated cross-section equations across countries, cannot be employed in the present report. The estimated coefficients for dummy variables in *Regional growth and convergence* reflected a comparison between the experience of countries that were in the EU for the period from 1975 and those that joined later. However, the equations for the present study are not estimated on a cross-section, but simply on the time-series for each variable in each Member State. The SMP and dummy variables are also likely to be highly collinear because the period of Spain and Portugal's accession to the EU coincided almost exactly with the start of the SMP process and also the SMP is represented in the modelling by a synthetic variable rather than a time-series that displays substantial variation. Given this possible collinearity, the time-series analysis cannot distinguish between the two effects, and any attempt to do so would require a cross-section approach, which is a completely different methodology to the one adopted in the present study.

However, we should note the results of the SMP impact on intra-EU and extra-EU exports and imports for Spain and Portugal (see Tables 3.11a–3.12b). Both countries experience greater than average increases in trade and this is particularly the case for the effect on intra-EU imports in Spain. Although we cannot isolate the SMP and EU accession effects, it is likely that the latter goes some way in explaining the large estimated increases in trade. Indeed, the start of the SMP process in 1987 may have merely bolstered any increase in trade resulting from accession to the EU.

4.5. Sensitivity analysis

4.5.1. Sensitivity to increased intra-EU trade by country and sector

This section summarizes the results of simulation exercises to examine which countries are most sensitive to increased intra-EU trade in any given sector. A separate simulation was carried out for each sector. An exogenous boost of 10% was given to each country's intra-EU imports. If each country's exports increased just sufficiently to offset the loss of sales to its domestic market, there would be no change in output. In practice, of course, this does not occur. Those countries with a more competitive sector will gain more through increased exports than they will lose from increased imports. The model represents this process: a country's response to greater demand from the rest of the EU is determined by the export equations, which include terms for competitiveness and the sensitivity of exports to changes in activity in other EU markets. Of course, these simulations are a very stylized representation of the SMP. They assume that barriers to trade are reduced equally in all countries, whereas in practice these barriers may have been much more substantial in some countries than in others. The scale of the increase in intra-EU trade (10% in every sector) is arbitrary. But the exercise does not bring out more clearly the extent to which increased trade in a given sector favours or penalizes each country.

The results are reported solely in terms of employment. For the EU as a whole, employment may change because of different productivity levels in gaining and losing countries. There are also second-round effects. For example, gaining countries have higher employment, but may have lower wages than the losing country, so that there is a net fall in pan-EU consumer spending for those employed in the sector. Or the country increasing its exports may produce at a lower price than its competitors, so that inflation in the importing country may be slightly lower, allowing consumers to spend more. Or the gaining country may source a higher proportion of its inputs from outside of the EU, so that there is an increase in extra-EU imports. All these effects are represented in the model.

Ferrous and Non-ferrous Metals (see Tables 4.19a and b)

Across the EUR-11 countries, a 10% exogenous increase in intra-EU trade leads to a 3% fall in employment (an average loss of 24,000 jobs in this sector). This fall is particularly marked in the UK with a fall in employment of 17% (17,000 jobs). Germany and Italy also experience falls in employment and in 1990 both countries held the largest shares of total EUR-12 output in this sector (32% and 20% respectively).

Taking into account changes across the total 30 EU industries, an average fall of 35,500 jobs results from the simulation exercise. The UK, Germany and Spain experience the largest losses in employment.

Non-metallic Mineral Products (see Tables 4.20a and b)

This sector only experiences negligible effects when we simulate a 10% rise in intra-EU trade, probably because its low value to weight ratio means that it is one of the less traded sectors. Employment rises by only 0.1% (a loss of 1,400 jobs). The UK experiences small increases in employment in this sector despite the fact that it was identified in Chapter 2 as one of the 'least open to trade' sectors of the UK economy prior to the SMP. It also showed relatively low levels of investment and R&D intensity, all of which suggested a greater vulnerability to greater trade under the SMP.

Taking into account changes across all 30 EU industries, an average of 2,800 jobs is gained, with exactly half of this employment being generated in Germany (which in 1990 held the largest share of total EUR-12 output in this sector (24%)).

Chemicals (see Tables 4.21a and b)

The Chemicals industry experiences a fall in employment of 4.6% (a loss of 90,000 jobs). The Netherlands and Italy experience the largest falls in employment (a loss of 28,000 and 54,000 jobs respectively). Table A3b shows that in 1990 Italy accounted for the second-largest share of EUR-12 output in this sector (20%). However, we did note in Chapter 2 that the Italian Chemicals industry may have been vulnerable to the SMP. Firstly, it tended to be specialized in lower value-added products and its international links were primarily with less-developed countries. Furthermore, Italy was one of the only countries to show a trade deficit in the Chemicals sector at the end of the 1980s, and its firms were expected to be vulnerable to the impact of the SMP due to their high concentration on the domestic market.

Taking into account changes across all 30 EU industries, the Netherlands and Italy again show the largest falls in employment, with the latter's loss of jobs increasing to an average of 66,000.

Metal Products (see Tables 4.22a and b)

The impact of a 10% rise in intra-EU trade on this sector leads to only a 0.6% fall in employment (an average loss of 16,000 jobs across the EUR-11 countries). Spain experiences the largest fall in employment in the sector (24,000 jobs, i.e. a fall of 8.6%).

Taking account of changes across the 30 EU industries, the loss of jobs rises to 32,300. The UK experiences the largest rise in employment (10,700 jobs) even though it was identified in Chapter 2 as being vulnerable to the impact of the SMP because of its relatively low levels of investment and R&D intensity in this sector.

Agricultural and Industrial Machinery (see Tables 4.23a and b)

The effects of the simulation on this sector across the EU as a whole are negligible. However, Spain experiences large falls in employment (14,200 jobs, i.e. a fall of 8.4%). This confirms the potential vulnerability of this sector in Spain as noted in Chapter 2. Italy enjoys the largest gains (5,300 jobs). Table A5b shows that in 1990 Italy accounted for the third largest share of total EUR-12 output in Agricultural and Industrial Machinery (20%).

Taking into account changes across the 30 EU industries, the average loss of jobs falls to 1,800 with the largest and smallest gains being in Italy and Spain respectively.

Office Machines (see Tables 4.24a and b)

A small negative impact on employment for the EU as a whole is experienced by the Office Machines sector (a loss of 3,600 jobs, i.e. a 0.6% fall). France shows the largest fall in employment which in 1990 held the third-largest share of EUR-12 output in Office Machines (19.5%).

Taking into account changes across the 30 EU industries, the average loss of jobs rises to 7,900, with UK employment falling by 4,900.

Electrical Goods (see Tables 4.25a and b)

A 10% exogenous increase in intra-EU trade leads to an increase of 5,000 jobs for the EU as a whole in the Electrical Goods sector, a rise of just 0.2%. Germany and the UK experience the largest employment gains (12,100 and 19,900 respectively). In 1990, both countries held the largest shares of EUR-12 manufacturing output for this sector (see Table A7b). Germany's Electrical Goods sector was identified in Chapter 2 as having a better competitive position on the EU market, with their products tending to be strongly internationally traded. The UK Electrical Goods sector was identified as having the potential to benefit from the SMP because of its relatively high investment in R&D (see Table 2.15b). By contrast, Spain and France experience falls in employment (a loss of jobs of 17,500 and 9,400 respectively). In 1990, Spain's Electrical Goods industry was characterized by high levels of inter-industry trade within the EU. Therefore, as noted in Chapter 2, the sector was expected to be vulnerable to the SMP. In 1990, France's Electrical Goods sector also appeared to be in a weak position. Foreign competition was strong and the single market was expected to boost imports from other EU countries.

Taking into account changes across the 30 EU industries, the simulated increase in intra-EU trade in the Electrical Goods sector leads to an average increase of 22,300 jobs. This employment generation is particularly marked in the UK, Germany and Italy although Spain experiences an average loss of 22,300 jobs.

Transport Equipment (see Tables 4.26a and b)

This sector shows a fall in employment for the EU as a whole when an increase in intra-EU trade is simulated (a loss of jobs of 17,400). The Netherlands and Italy experience the largest falls in employment (falls of 14% and 5% respectively). In 1990, Italy held the fourth-largest share of EUR-12 output in this sector but, as noted in Chapter 2, there were some indications of weakness in its R&D-intensive activities. Gains in employment were experienced by Belgium's Transport Equipment sector (an increase of 5,200 jobs, i.e. a rise of 7.5%). Motor Vehicles (part of Transport Equipment) was identified in Chapter 2 as being a 'traditional' Belgian industry. In the past, trade in this sector had been impeded by regulations and it was expected that Motor Vehicles would be boosted by the abolition of such trade barriers.

Taking into account changes across the 30 EU industries, employment falls by an average of 12,500 jobs with Italy experiencing the largest losses (22,000 jobs).

Food, Drink and Tobacco (see Tables 4.27a and b)

The simulation exercise has a negative effect on employment in the Food, Drink and Tobacco sector (an EU-average loss of 12,000 jobs, i.e. a 0.4% fall). The falls in employment are particularly large for Germany, which in 1990 held the third-largest share of EUR-12 output in this sector (19%). Parts of Food and Drinks were identified in Chapter 2 as being weak points of German industry. Gains in employment were experienced by the UK (an increase of 6,300 jobs, i.e. a rise of 1.3%), although this industry was considered vulnerable to greater trade under the SMP, primarily because in 1990 it was identified as one of the UK's 'least open to trade' sectors, showing relatively low levels of investment and R&D intensity.

Taking into account changes across the 30 EU industries, the simulation exercise resulted in an average loss of 24,700 jobs, with Germany experiencing the largest losses (27,400 jobs).

Textiles, Clothing and Footwear (see Tables 4.28a and b)

The impact of a 10% exogenous increase in intra-EU trade on Textiles, Clothing and Footwear is that employment rises in the sector but falls across the 30 EU industries taken as a whole. Large falls in employment in the sector are experienced by France and Spain. As noted in Chapter 2, France's Footwear and Clothing sector appeared to be in a weak position in 1990, with foreign competition being particularly strong. By contrast, Spain's sector was expected to increase its share in the EU market provided investment efforts were maintained for enhancing qualitatively both product and process.

Losses in employment were particularly marked for Denmark and the UK. Below average performance was expected for the UK's sector because over the 1980s trade had become more inter-industry orientated in the sector. However, the sector was already highly open which perhaps suggested that any structural change had already taken place. The overall position of Denmark's Textiles, Clothing and Footwear industry was unclear. For example, as noted in Chapter 2, Clothing and Accessories appeared to be in a weak competitive position in 1990 but Household Textiles were believed to hold a strong competitive position.

Taking into account changes across the 30 EU industries, the simulation exercise resulted in an average loss of 10,500 jobs, with these losses being particularly marked in Germany and the UK. However, France experienced an increase of 11,900 jobs.

Paper and Printing Products (see Tables 4.29a and b)

The effect of a 10% exogenous increase in intra-EU trade on this sector is negligible for the EU as a whole (a loss of 3,800 jobs, i.e. a fall of only 0.2%). Falls in employment are large for Luxembourg. Increases in employment were experienced by Portugal (a gain of 1,600 jobs, i.e. an increase of 3.8%).

Taking into account changes across the 30 EU industries, there is an average gain of 1,300 jobs, with Germany experiencing the largest increases (5100 jobs).

Rubber and Plastic Products (see Tables 4.30a and b)

The impact of the simulation exercise on employment in the Rubber and Plastic Products sector is negligible for the EU as a whole (a loss of 2,200 jobs, i.e. only a 0.2% fall). Portugal experienced a 7% increase in employment which is feasible given that over the 1980s, intra-industry trade had increased in the Portuguese sector within the EU.

Taking into account changes across the 30 EU industries, employment falls by an average of 4,000 although Italy experiences gains in employment.

However, it is interesting to note the simultaneous impact on the Chemicals sector which experiences an increase of 3,000 jobs. This is due to the interdependency of the Rubber and Plastic Products and Chemicals industries. The demand for chemicals in this case is a derived demand.

Table 4.19a. Sensitivity analysis: E3ME industry 8 employment in 1993 by sector (% difference from base)

		1 BE	2 DK	3 DO	4 DW	5 EL	6 ES	7 FR	8 IR	9 IT	10 IS	11 LX	12 NL	13 PO	14 UK	EU
08	Ferrous & Non-f. Metals	-2.8	37.7	0.0	-1.8	0.0	2.5	-2.5	2.8	-0.1	0.0	-1.0	0.5	-4.3	-21.2	-3.1
09	Non-metallic Min. Pr.	0.1	0.0	0.0	0.0	0.0	-0.1	0.0	0.0	0.0	0.0	0.0	0.0	0.0	0.0	0.0
10	Chemicals	0.0	0.0	0.0	0.0	0.0	-0.1	0.0	0.0	0.0	0.0	0.0	0.0	0.0	-0.1	0.0
11	Metal Products	0.3	0.1	0.0	-0.1	0.0	-0.1	0.1	-1.1	0.0	0.0	-0.1	0.0	0.0	-0.2	-0.1
12	Agri. & Indust. Mach.	0.0	0.1	0.0	-0.1	0.0	-0.3	0.0	0.0	0.0	0.0	0.0	0.0	0.0	0.0	0.0
13	Office Machines	0.0	0.0	0.0	0.0	0.0	0.0	0.0	0.0	0.0	0.0	0.0	0.0	0.0	0.0	0.0
14	Electrical Goods	0.1	0.0	0.0	0.0	0.0	-0.1	0.0	0.0	0.0	0.0	0.0	0.0	-0.1	0.0	0.0
15	Transport Equipment	0.0	0.0	0.0	0.0	0.0	0.0	0.0	0.0	0.0	0.0	0.0	0.0	0.0	0.0	0.0
16	Food, Drink & Tobacco	0.0	0.0	0.0	0.0	0.0	0.0	0.0	0.0	0.0	0.0	0.0	0.0	0.0	0.0	0.0
17	Tex., Cloth. & Footw.	0.0	0.0	0.0	0.0	0.0	0.0	0.0	0.0	0.0	0.0	0.0	0.0	0.0	0.0	0.0
18	Paper & Printing Pr.	0.1	0.0	0.0	0.0	0.0	0.0	0.0	0.0	0.0	0.0	0.1	0.0	0.0	0.0	0.0
19	Rubber & Plastic Pr.	0.1	0.0	0.0	0.0	0.0	0.0	0.0	0.0	0.0	0.0	0.0	0.0	0.0	0.0	0.0
21	Other Manufactures	0.1	0.1	0.0	0.0	0.0	-0.3	0.0	0.0	0.0	0.0	0.0	0.0	0.0	0.0	0.0
	Total (32 sectors)	0.0	0.1	0.0	0.0	0.0	-0.1	0.0	0.0	0.0	0.0	0.0	0.0	0.0	-0.1	0.0

Source: Cambridge Econometrics projection S08-S00/da1/S08.

Table 4.19b. Sensitivity analysis: E3ME industry 8 employment in 1993 by sector (difference from base in thousands)

		1 BE	2 DK	3 DO	4 DW	5 EL	6 ES	7 FR	8 IR	9 IT	10 IS	11 LX	12 NL	13 PO	14 UK	EU
08	Ferrous & Non-f. Metals	-1.2	1.0	0.0	-4.2	0.0	1.7	-3.2	0.1	-0.2	0.0	-0.1	0.1	-1.0	-17.0	-23.8
09	Non-metallic Min. Pr.	0.0	0.0	0.0	0.0	0.0	-0.3	0.0	0.0	0.0	0.0	0.0	0.0	0.0	0.0	-0.3
10	Chemicals	0.0	0.0	0.0	0.1	0.0	-0.1	0.0	0.0	0.0	0.0	0.0	0.0	0.0	-0.2	-0.1
11	Metal Products	0.2	0.0	0.0	-0.9	0.0	-0.2	0.1	0.0	0.0	0.0	0.0	0.0	0.0	-0.7	-1.5
12	Agri. & Indust. Mach.	0.0	0.1	0.0	-0.6	0.0	-0.6	0.0	0.0	0.1	0.0	0.0	0.0	0.0	0.0	-1.1
13	Office Machines	0.0	0.0	0.0	0.0	0.0	0.0	0.0	0.0	0.0	0.0	0.0	0.0	0.0	0.0	0.0
14	Electrical Goods	0.0	0.0	0.0	-0.3	0.0	-0.3	0.0	0.0	0.0	0.0	0.0	0.0	0.0	-0.1	-0.9
15	Transport Equipment	0.0	0.0	0.0	0.0	0.0	-0.1	0.0	0.0	0.0	0.0	0.0	0.0	0.0	0.1	0.0
16	Food, Drink & Tobacco	0.0	0.0	0.0	0.0	0.0	0.0	0.0	0.0	0.0	0.0	0.0	0.0	0.0	0.0	0.0
17	Tex., Cloth. & Footw.	0.0	0.0	0.0	0.0	0.0	0.1	0.0	0.0	0.0	0.0	0.0	0.0	0.0	0.0	0.1
18	Paper & Printing Pr.	0.0	0.0	0.0	0.0	0.0	0.0	0.1	0.0	0.0	0.0	0.0	0.0	0.0	-0.1	0.0
19	Rubber & Plastic Pr.	0.0	0.0	0.0	0.1	0.0	0.0	0.0	0.0	0.0	0.0	0.0	0.0	0.0	-0.1	0.0
21	Other Manufactures	0.0	0.0	0.0	0.0	0.0	-0.8	0.0	0.0	0.0	0.0	0.0	0.0	0.0	0.0	-0.7
	Total (32 sectors)	-0.3	1.4	0.0	-7.7	0.0	-13.2	0.2	0.0	-0.7	0.0	-0.1	0.1	-0.5	-14.8	-35.5

Source: Cambridge Econometrics projection S08-S00/da1/S08.

Table 4.20a. Sensitivity analysis: E3ME industry 9 employment in 1993 by sector (% difference from base)

		1 BE	2 DK	3 DO	4 DW	5 EL	6 ES	7 FR	8 IR	9 IT	10 IS	11 LX	12 NL	13 PO	14 UK	EU
08	Ferrous & Non-f. Metals	0.0	0.0	0.0	0.0	0.0	0.0	0.0	0.0	0.0	0.0	0.0	0.0	0.0	0.0	0.0
09	Non-metallic Min. Pr.	-1.7	1.1	0.0	0.1	0.0	0.2	-0.1	2.0	0.1	0.0	-0.2	0.0	0.2	0.3	0.1
10	Chemicals	0.0	0.0	0.0	0.0	0.0	0.0	0.0	0.0	0.0	0.0	0.0	0.0	0.0	0.0	0.0
11	Metal Products	0.0	0.0	0.0	0.0	0.0	0.0	0.0	0.8	0.0	0.0	0.0	0.0	0.0	0.0	0.0
12	Agri. & Indust. Mach.	0.0	-0.1	0.0	0.0	0.0	0.0	0.0	0.0	0.0	0.0	0.0	0.0	0.0	0.0	0.0
13	Office Machines	0.0	0.0	0.0	0.0	0.0	0.0	0.0	0.0	0.0	0.0	0.0	0.0	0.0	0.0	0.0
14	Electrical Goods	0.0	0.0	0.0	0.0	0.0	0.0	0.0	0.0	0.0	0.0	0.0	0.0	0.0	0.0	0.0
15	Transport Equipment	0.0	0.0	0.0	0.0	0.0	0.0	0.0	0.0	0.0	0.0	0.0	0.0	0.0	0.0	0.0
16	Food, Drink & Tobacco	0.0	0.0	0.0	0.0	0.0	0.0	0.0	0.0	0.0	0.0	0.0	0.0	0.0	0.0	0.0
17	Tex., Cloth. & Footw.	0.0	0.0	0.0	0.0	0.0	0.0	0.0	0.0	0.0	0.0	0.0	0.0	0.0	0.0	0.0
18	Paper & Printing Pr.	0.0	0.0	0.0	0.0	0.0	0.0	0.0	0.0	0.0	0.0	0.0	0.0	0.0	0.0	0.0
19	Rubber & Plastic Pr.	0.0	0.0	0.0	0.0	0.0	0.0	0.0	-0.1	0.0	0.0	0.0	0.0	0.0	0.0	0.0
21	Other Manufactures	0.0	0.0	0.0	0.0	0.0	0.0	0.0	0.1	0.0	0.0	0.0	0.0	0.0	0.0	0.0
	Total (32 sectors)	0.0	0.0	0.0	0.0	0.0	0.0	0.0	0.0	0.0	0.0	0.0	0.0	0.0	0.0	0.0

Source: Cambridge Econometrics projection S09-S00/da1/S09.

Table 4.20b. Sensitivity analysis: E3ME industry 9 employment in 1993 by sector (difference from base in thousands)

		1 BE	2 DK	3 DO	4 DW	5 EL	6 ES	7 FR	8 IR	9 IT	10 IS	11 LX	12 NL	13 PO	14 UK	EU
08	Ferrous & Non-f. Metals	0.0	0.0	0.0	0.0	0.0	0.0	0.0	0.0	0.0	0.0	0.0	0.0	0.0	0.0	0.0
09	Non-metallic Min. Pr.	-0.6	0.1	0.0	0.3	0.0	0.4	-0.1	0.4	0.3	0.0	0.0	0.0	0.2	0.6	1.4
10	Chemicals	0.0	0.0	0.0	0.0	0.0	0.0	0.0	0.0	0.0	0.0	0.0	0.0	0.0	0.0	0.0
11	Metal Products	0.0	0.0	0.0	0.1	0.0	0.0	0.0	0.0	0.0	0.0	0.0	0.0	0.0	0.0	0.1
12	Agri. & Indust. Mach.	0.0	0.0	0.0	0.3	0.0	0.0	0.0	0.0	0.0	0.0	0.0	0.0	0.0	0.0	0.3
13	Office Machines	0.0	0.0	0.0	0.0	0.0	0.0	0.0	0.0	0.0	0.0	0.0	0.0	0.0	0.0	0.0
14	Electrical Goods	0.0	0.0	0.0	0.1	0.0	0.0	0.0	0.0	0.0	0.0	0.0	0.0	0.0	0.0	0.0
15	Transport Equipment	0.0	0.0	0.0	0.0	0.0	0.0	0.0	0.0	0.0	0.0	0.0	0.0	0.0	0.0	0.0
16	Food, Drink & Tobacco	0.0	0.0	0.0	0.0	0.0	0.0	0.0	0.0	0.0	0.0	0.0	0.0	0.0	0.0	0.0
17	Tex., Cloth. & Footw.	0.0	0.0	0.0	0.0	0.0	0.0	0.0	0.0	0.0	0.0	0.0	0.0	0.0	0.0	0.0
18	Paper & Printing Pr.	0.0	0.0	0.0	0.0	0.0	0.0	0.0	0.0	0.0	0.0	0.0	0.0	0.0	0.0	0.0
19	Rubber & Plastic Pr.	0.0	0.0	0.0	0.0	0.0	0.0	0.0	0.0	0.0	0.0	0.0	0.0	0.0	0.0	0.0
21	Other Manufactures	0.0	0.0	0.0	0.0	0.0	0.0	0.0	0.0	0.0	0.0	0.0	0.0	0.0	0.0	0.0
	Total (32 sectors)	-0.8	-0.1	0.0	1.4	0.0	0.4	0.3	0.4	0.5	0.0	0.0	0.0	0.2	0.5	2.8

Source: Cambridge Econometrics projection S09-S00/da1/S09.

Table 4.21a. Sensitivity analysis: E3ME industry 10 employment in 1993 by sector (% difference from base)

		1 BE	2 DK	3 DO	4 DW	5 EL	6 ES	7 FR	8 IR	9 IT	10 IS	11 LX	12 NL	13 PO	14 UK	EU
08	Ferrous & Non-f. Metals	0.0	-0.1	0.0	0.1	0.0	0.0	0.0	0.0	0.0	0.0	0.0	0.0	-0.1	-0.1	0.0
09	Non-metallic Min. Pr.	0.0	-0.1	0.0	0.0	0.0	-0.1	0.0	0.0	-0.1	0.0	0.0	0.0	-0.2	0.0	-0.1
10	Chemicals	0.2	1.5	0.0	-0.5	0.0	-1.8	-1.0	0.8	-14.9	0.0	0.0	-19.7	-1.1	0.3	-4.6
11	Metal Products	0.0	0.1	0.0	0.1	0.0	0.0	-0.1	0.7	0.0	0.0	0.0	0.0	0.0	-0.1	0.0
12	Agri. & Indust. Mach.	0.0	0.1	0.0	0.1	0.0	0.0	0.0	0.0	0.0	0.0	0.0	0.0	0.0	0.0	0.0
13	Office Machines	0.0	0.0	0.0	0.1	0.0	0.0	0.0	0.0	0.0	0.0	0.0	0.0	0.0	0.0	0.0
14	Electrical Goods	0.0	0.0	0.0	0.0	0.0	0.0	0.0	0.0	0.0	0.0	0.0	0.0	0.0	0.0	0.0
15	Transport Equipment	0.0	0.0	0.0	0.0	0.0	0.0	0.0	0.0	-0.1	0.0	0.0	-0.1	0.0	0.0	0.0
16	Food, Drink & Tobacco	0.0	0.0	0.0	0.0	0.0	0.0	0.0	0.0	-0.2	0.0	0.0	0.1	0.0	0.0	0.0
17	Tex., Cloth. & Footw.	0.0	0.0	0.0	0.0	0.0	0.0	0.0	0.0	0.0	0.0	0.0	0.0	0.0	0.0	0.0
18	Paper & Printing Pr.	0.1	0.0	0.0	0.0	0.0	0.0	0.0	0.1	-0.2	0.0	-0.3	0.0	0.0	0.0	0.0
19	Rubber & Plastic Pr.	0.1	0.1	0.0	-0.1	0.0	-0.1	0.0	0.1	0.0	0.0	0.0	0.1	-0.1	0.0	0.0
21	Other Manufactures	0.0	0.0	0.0	0.0	0.0	0.0	0.0	0.0	0.0	0.0	0.0	-0.2	0.0	0.0	0.0
	Total (32 sectors)	0.0	0.0	0.0	0.0	0.0	0.0	0.0	0.0	-0.3	0.0	0.0	-0.5	0.0	0.0	-0.1

Source: Cambridge Econometrics projection S10-S00/da1/S10.

Table 4.21b. Sensitivity analysis: E3ME industry 10 employment in 1993 by sector (difference from base in thousands)

		1 BE	2 DK	3 DO	4 DW	5 EL	6 ES	7 FR	8 IR	9 IT	10 IS	11 LX	12 NL	13 PO	14 UK	EU
08	Ferrous & Non-f. Metals	0.0	0.0	0.0	0.2	0.0	0.0	0.0	0.0	0.0	0.0	0.0	0.0	0.0	-0.1	0.1
09	Non-metallic Min. Pr.	0.0	0.0	0.0	0.0	0.0	-0.3	0.0	0.0	-0.5	0.0	0.0	0.0	-0.2	0.0	-0.9
10	Chemicals	0.1	0.3	0.0	-2.9	0.0	-3.0	-2.7	0.1	-54.2	0.0	0.0	-28.2	-0.3	0.9	-89.8
11	Metal Products	0.0	0.1	0.0	1.2	0.0	-0.1	-0.1	0.0	0.1	0.0	0.0	0.0	0.0	-0.3	0.8
12	Agri. & Indust. Mach.	0.0	0.1	0.0	1.2	0.0	-0.1	0.0	0.0	0.0	0.0	0.0	0.0	0.0	0.0	1.2
13	Office Machines	0.0	0.0	0.0	0.1	0.0	0.0	0.0	0.0	0.0	0.0	0.0	0.0	0.0	0.0	0.1
14	Electrical Goods	0.0	0.0	0.0	0.3	0.0	0.0	0.0	0.0	0.0	0.0	0.0	0.0	0.0	0.0	0.2
15	Transport Equipment	0.0	0.0	0.0	0.0	0.0	0.0	0.0	0.0	-0.4	0.0	0.0	-0.1	0.0	0.0	-0.5
16	Food, Drink & Tobacco	0.0	0.0	0.0	0.0	0.0	0.0	0.0	0.0	-0.5	0.0	0.0	0.2	0.0	0.0	-0.4
17	Tex., Cloth. & Footw.	0.0	0.0	0.0	0.0	0.0	0.0	0.1	0.0	0.0	0.0	0.0	0.0	0.0	0.0	0.0
18	Paper & Printing Pr.	0.1	0.0	0.0	0.0	0.0	0.0	0.0	0.0	-0.3	0.0	0.0	0.0	0.0	-0.1	-0.3
19	Rubber & Plastic Pr.	0.0	0.0	0.0	-0.2	0.0	-0.2	0.0	0.0	0.0	0.0	0.0	0.0	0.0	0.0	-0.3
21	Other Manufactures	0.0	0.0	0.0	0.1	0.0	-0.1	0.0	0.0	-0.2	0.0	0.0	-0.1	0.0	0.0	-0.3
	Total (32 sectors)	0.0	0.2	0.0	0.3	0.0	-5.2	-4.0	0.2	-66.1	0.0	0.0	-26.4	-0.6	1.6	-100.0

Source: Cambridge Econometrics projection S10-S00/da1/S10.

Table 4.22a. Sensitivity analysis: E3ME industry 11 employment in 1993 by sector (% difference from base)

		1 BE	2 DK	3 DO	4 DW	5 EL	6 ES	7 FR	8 IR	9 IT	10 IS	11 LX	12 NL	13 PO	14 UK	EU
08	Ferrous & Non-f. Metals	0.0	0.3	0.0	0.0	0.0	0.4	0.0	0.0	0.0	0.0	0.0	-0.1	0.1	-0.1	0.0
09	Non-metallic Min. Pr.	0.0	0.0	0.0	0.0	0.0	-0.1	0.0	0.0	0.0	0.0	0.0	0.0	0.0	0.0	0.0
10	Chemicals	0.0	0.0	0.0	0.0	0.0	0.0	0.0	0.0	0.0	0.0	0.0	0.0	0.0	0.0	0.0
11	Metal Products	1.0	2.7	0.0	-0.1	0.0	-8.6	-1.0	-5.2	0.1	0.0	-1.6	-0.3	0.0	2.4	-0.6
12	Agri. & Indust. Mach.	0.0	0.0	0.0	0.0	0.0	-0.2	0.0	0.0	0.0	0.0	0.0	0.0	0.0	0.0	0.0
13	Office Machines	0.0	0.0	0.0	0.0	0.0	0.0	0.0	0.0	0.0	0.0	0.0	0.0	0.0	0.0	0.0
14	Electrical Goods	0.0	0.0	0.0	0.0	0.0	0.0	0.0	0.0	0.0	0.0	0.0	0.0	0.0	0.0	0.0
15	Transport Equipment	0.0	0.0	0.0	0.0	0.0	0.0	0.0	0.0	0.0	0.0	0.0	0.0	0.0	0.0	0.0
16	Food, Drink & Tobacco	0.0	0.0	0.0	0.0	0.0	0.0	0.0	0.0	0.0	0.0	0.0	0.0	0.0	0.0	0.0
17	Tex., Cloth. & Footw.	0.0	0.0	0.0	0.0	0.0	0.0	0.0	0.0	0.0	0.0	0.0	0.0	0.0	0.0	0.0
18	Paper & Printing Pr.	0.0	0.0	0.0	0.0	0.0	0.0	0.0	0.0	0.0	0.0	0.0	0.0	0.0	0.0	0.0
19	Rubber & Plastic Pr.	0.0	0.0	0.0	0.0	0.0	0.0	0.0	0.0	0.0	0.0	0.0	0.0	0.0	0.0	0.0
21	Other Manufactures	0.0	0.0	0.0	0.0	0.0	-0.2	0.0	0.0	0.0	0.0	0.0	-0.1	0.0	0.0	0.0
	Total (32 sectors)	0.0	0.0	0.0	0.0	0.0	-0.2	0.0	0.0	0.0	0.0	0.0	0.0	0.0	0.0	0.0

Source: Cambridge Econometrics projection S11-S00/da1/S11.

Table 4.22b. Sensitivity analysis: E3ME industry 11 employment in 1993 by sector (difference from base in thousands)

		1 BE	2 DK	3 DO	4 DW	5 EL	6 ES	7 FR	8 IR	9 IT	10 IS	11 LX	12 NL	13 PO	14 UK	EU
08	Ferrous & Non-f. Metals	0.0	0.0	0.0	0.0	0.0	0.3	0.0	0.0	0.0	0.0	0.0	0.0	0.0	-0.1	0.3
09	Non-metallic Min. Pr.	0.0	0.0	0.0	0.0	0.0	-0.1	0.0	0.0	0.0	0.0	0.0	0.0	0.0	0.0	-0.1
10	Chemicals	0.0	0.0	0.0	0.0	0.0	-0.1	0.0	0.0	0.1	0.0	0.0	0.0	0.0	0.1	0.1
11	Metal Products	0.6	0.6	0.0	-0.9	0.0	-24.0	-2.9	-0.2	0.3	0.0	-0.1	-0.3	0.0	10.7	-16.2
12	Agri. & Indust. Mach.	0.0	0.0	0.0	0.1	0.0	-0.3	0.0	0.0	0.1	0.0	0.0	0.0	0.0	0.0	-0.1
13	Office Machines	0.0	0.0	0.0	0.0	0.0	0.0	0.0	0.0	0.0	0.0	0.0	0.0	0.0	0.0	0.0
14	Electrical Goods	0.0	0.0	0.0	-0.1	0.0	-0.2	0.0	0.0	0.0	0.0	0.0	0.0	0.0	0.0	-0.3
15	Transport Equipment	0.0	0.0	0.0	0.0	0.0	0.0	0.0	0.0	0.0	0.0	0.0	0.0	0.0	0.0	0.0
16	Food, Drink & Tobacco	0.0	0.0	0.0	0.0	0.0	0.0	0.0	0.0	0.0	0.0	0.0	0.0	0.0	0.0	0.0
17	Tex., Cloth. & Footw.	0.0	0.0	0.0	0.0	0.0	0.0	0.0	0.0	0.0	0.0	0.0	0.0	0.0	0.0	0.0
18	Paper & Printing Pr.	0.0	0.0	0.0	0.0	0.0	0.0	0.0	0.0	0.0	0.0	0.0	0.0	0.0	0.0	0.0
19	Rubber & Plastic Pr.	0.0	0.0	0.0	0.0	0.0	-0.1	0.0	0.0	0.0	0.0	0.0	0.0	0.0	0.0	0.0
21	Other Manufactures	0.0	0.0	0.0	0.0	0.0	-0.8	0.0	0.0	0.0	0.0	0.0	0.0	0.0	0.0	-0.8
	Total (32 sectors)	0.6	0.6	0.0	-1.1	0.0	-32.3	-2.9	-0.1	1.9	0.0	-0.1	-0.3	0.0	11.0	-22.8

Source: Cambridge Econometrics projection S11-S00/da1/S11.

Table 4.23a Sensitivity analysis: E3ME industry 12 employment in 1993 by sector (% difference from base)

		1 BE	2 DK	3 DO	4 DW	5 EL	6 ES	7 FR	8 IR	9 IT	10 IS	11 LX	12 NL	13 PO	14 UK	EU
08	Ferrous & Non-f. Metals	0.0	-1.1	0.0	0.0	0.0	0.1	0.0	0.0	0.0	0.0	0.0	0.0	-0.2	0.1	0.0
09	Non-metallic Min. Pr.	0.0	0.0	0.0	0.0	0.0	0.0	0.0	0.0	0.0	0.0	0.0	0.0	0.0	0.0	0.0
10	Chemicals	0.0	0.0	0.0	0.0	0.0	0.0	0.0	0.0	0.1	0.0	0.0	0.0	0.0	0.0	0.0
11	Metal Products	0.0	0.6	0.0	0.0	0.0	-0.2	0.1	0.0	0.0	0.0	-0.1	0.0	0.0	0.1	0.0
12	Agri. & Indust. Mach.	0.0	5.1	0.0	0.3	0.0	-8.4	0.0	0.0	1.2	0.0	-11.0	0.0	-2.5	0.0	-0.1
13	Office Machines	0.0	0.0	0.0	0.0	0.0	0.0	0.0	0.0	0.0	0.0	0.0	0.0	0.0	0.0	0.0
14	Electrical Goods	0.0	0.1	0.0	-0.1	0.0	-0.1	0.0	0.0	0.0	0.0	0.0	0.0	-0.1	0.0	0.0
15	Transport Equipment	0.0	0.0	0.0	0.0	0.0	0.0	0.0	0.0	0.0	0.0	0.0	0.0	0.0	0.0	0.0
16	Food, Drink & Tobacco	0.0	0.0	0.0	0.0	0.0	0.0	0.0	0.0	0.0	0.0	0.0	0.0	0.0	0.0	0.0
17	Tex., Cloth. & Footw.	0.0	0.0	0.0	0.0	0.0	0.0	0.0	0.0	0.0	0.0	0.0	0.0	0.0	0.0	0.0
18	Paper & Printing Pr.	0.0	0.0	0.0	0.0	0.0	0.0	0.0	0.0	0.1	0.0	-0.1	0.0	0.0	0.0	0.0
19	Rubber & Plastic Pr.	0.0	0.1	0.0	0.0	0.0	-0.1	0.0	0.0	0.0	0.0	0.0	0.0	0.0	0.0	0.0
21	Other Manufactures	0.0	0.1	0.0	0.0	0.0	0.0	0.0	0.0	0.0	0.0	0.0	0.0	0.0	0.0	0.0
	Total (32 sectors)	0.0	0.2	0.0	0.0	0.0	-0.1	0.0	0.0	0.0	0.0	-0.2	0.0	0.0	0.0	0.0

Source: Cambridge Econometrics projection S12-S00/da1/S12.

Table 4.23b. Sensitivity analysis: E3ME industry 12 employment in 1993 by sector (difference from base in thousands)

		1 BE	2 DK	3 DO	4 DW	5 EL	6 ES	7 FR	8 IR	9 IT	10 IS	11 LX	12 NL	13 PO	14 UK	EU
08	Ferrous & Non-f. Metals	0.0	0.0	0.0	0.0	0.0	0.1	0.0	0.0	0.0	0.0	0.0	0.0	-0.1	0.1	0.1
09	Non-metallic Min. Pr.	0.0	0.0	0.0	0.0	0.0	-0.1	0.0	0.0	0.1	0.0	0.0	0.0	0.0	0.0	0.0
10	Chemicals	0.0	0.0	0.0	0.0	0.0	0.0	0.0	0.0	0.5	0.0	0.0	0.0	0.0	0.0	0.6
11	Metal Products	0.0	0.1	0.0	0.3	0.0	-0.4	0.2	0.0	-0.1	0.0	0.0	0.0	0.0	0.6	0.7
12	Agri. & Indust. Mach.	0.0	3.3	0.0	3.3	0.0	-14.2	0.0	0.0	5.3	0.0	-0.3	0.0	-1.0	0.0	-3.6
13	Office Machines	0.0	0.0	0.0	0.0	0.0	0.0	0.0	0.0	0.0	0.0	0.0	0.0	0.0	0.0	0.1
14	Electrical Goods	0.0	0.0	0.0	-0.6	0.0	-0.3	-0.1	0.0	0.0	0.0	0.0	0.0	0.0	0.0	-1.0
15	Transport Equipment	0.0	0.0	0.0	0.0	0.0	0.0	0.0	0.0	0.1	0.0	0.0	0.0	0.0	0.0	0.0
16	Food, Drink & Tobacco	0.0	0.0	0.0	0.0	0.0	0.0	0.0	0.0	0.0	0.0	0.0	0.0	0.0	0.0	0.0
17	Tex., Cloth. & Footw.	0.0	0.0	0.0	0.0	0.0	0.1	0.0	0.0	0.0	0.0	0.0	0.0	0.0	0.0	0.1
18	Paper & Printing Pr.	0.0	0.0	0.0	0.3	0.0	0.0	0.0	0.0	0.1	0.0	0.0	0.0	0.0	0.0	0.4
19	Rubber & Plastic Pr.	0.0	0.0	0.0	-0.1	0.0	-0.1	0.0	0.0	0.0	0.0	0.0	0.0	0.0	0.0	-0.2
21	Other Manufactures	0.0	0.1	0.0	0.1	0.0	-0.1	0.0	0.0	0.0	0.0	0.0	0.0	0.0	0.0	0.2
	Total (32 sectors)	0.0	3.8	0.0	2.9	0.0	-16.7	1.4	0.0	11.5	0.0	-0.3	0.0	0.3	-1.2	1.8

Source: Cambridge Econometrics projection S12-S00/da1/S12.

Table 4.24a. Sensitivity analysis: E3ME industry 13 employment in 1993 by sector (% difference from base)

		1 BE	2 DK	3 DO	4 DW	5 EL	6 ES	7 FR	8 IR	9 IT	10 IS	11 LX	12 NL	13 PO	14 UK	EU
08	Ferrous & Non-f. Metals	0.0	-0.5	0.0	0.0	0.0	0.0	0.0	0.0	0.0	0.0	0.0	0.0	0.0	-0.1	0.0
09	Non-metallic Min. Pr.	0.0	0.0	0.0	0.0	0.0	0.0	0.0	0.0	0.0	0.0	0.0	0.0	0.0	0.0	0.0
10	Chemicals	0.0	0.0	0.0	0.0	0.0	0.0	0.0	0.0	0.0	0.0	0.0	0.0	0.0	0.0	0.0
11	Metal Products	0.0	0.0	0.0	0.0	0.0	0.0	0.0	1.7	0.0	0.0	0.0	0.0	0.0	0.0	0.0
12	Agri. & Indust. Mach.	0.0	0.1	0.0	0.0	0.0	0.0	0.0	0.0	0.0	0.0	0.0	0.0	0.0	0.0	0.0
13	Office Machines	0.0	3.7	0.0	-0.2	0.0	0.0	-2.8	0.0	1.6	0.0	0.0	0.0	0.0	-1.3	-0.6
14	Electrical Goods	0.0	0.1	0.0	0.0	0.0	-0.1	0.0	0.5	0.0	0.0	0.0	0.0	0.0	0.0	0.0
15	Transport Equipment	0.0	0.0	0.0	0.0	0.0	0.0	0.0	0.0	0.0	0.0	0.0	0.0	0.0	0.0	0.0
16	Food, Drink & Tobacco	0.0	0.0	0.0	0.0	0.0	0.0	0.0	0.0	0.0	0.0	0.0	0.0	0.0	0.0	0.0
17	Tex., Cloth. & Footw.	0.0	-0.1	0.0	0.0	0.0	0.0	0.0	0.0	0.0	0.0	0.0	0.0	0.0	0.0	0.0
18	Paper & Printing Pr.	0.0	0.0	0.0	0.0	0.0	0.0	0.0	0.1	0.0	0.0	0.0	0.0	0.0	0.0	0.0
19	Rubber & Plastic Pr.	0.0	0.1	0.0	0.0	0.0	0.0	0.0	0.2	0.0	0.0	0.0	0.0	0.0	0.0	0.0
21	Other Manufactures	0.0	0.0	0.0	0.0	0.0	0.0	0.0	0.1	0.0	0.0	0.0	0.0	0.0	0.0	0.0
	Total (32 sectors)	0.0	0.0	0.0	0.0	0.0	0.0	0.0	0.0	0.0	0.0	0.0	0.0	0.0	0.0	0.0

Source: Cambridge Econometrics projection S13-S00/da1/S13K.

Table 4.24b. Sensitivity analysis: E3ME industry 13 employment in 1993 by sector (difference from base in thousands)

		1 BE	2 DK	3 DO	4 DW	5 EL	6 ES	7 FR	8 IR	9 IT	10 IS	11 LX	12 NL	13 PO	14 UK	EU
08	Ferrous & Non-f. Metals	0.0	0.0	0.0	0.0	0.0	0.0	0.0	0.0	0.0	0.0	0.0	0.0	0.0	-0.1	-0.1
09	Non-metallic Min. Pr.	0.0	0.0	0.0	0.0	0.0	0.0	0.0	0.0	0.0	0.0	0.0	0.0	0.0	0.0	0.0
10	Chemicals	0.0	0.0	0.0	0.0	0.0	0.0	0.0	0.0	0.0	0.0	0.0	0.0	0.0	0.0	0.0
11	Metal Products	0.0	0.0	0.0	0.0	0.0	0.0	0.0	0.1	0.0	0.0	0.0	0.0	0.0	0.1	0.1
12	Agri. & Indust. Mach.	0.0	0.0	0.0	0.0	0.0	0.0	0.0	0.0	0.0	0.0	0.0	0.0	0.0	0.0	0.1
13	Office Machines	0.0	0.5	0.0	-0.4	0.0	0.0	-3.2	0.0	1.5	0.0	0.0	0.0	0.0	-2.0	-3.6
14	Electrical Goods	0.0	0.0	0.0	0.0	0.0	-0.4	0.0	0.1	0.0	0.0	0.0	0.0	0.0	0.1	-0.2
15	Transport Equipment	0.0	0.0	0.0	0.0	0.0	0.0	0.0	0.0	0.0	0.0	0.0	0.0	0.0	0.0	0.0
16	Food, Drink & Tobacco	0.0	0.0	0.0	0.0	0.0	0.0	0.1	0.0	0.0	0.0	0.0	0.0	0.0	0.0	0.0
17	Tex., Cloth. & Footw.	0.0	0.0	0.0	0.0	0.0	0.0	0.0	0.0	0.0	0.0	0.0	0.0	0.0	0.0	0.0
18	Paper & Printing Pr.	0.0	0.0	0.0	0.0	0.0	0.0	0.0	0.0	0.0	0.0	0.0	0.0	0.0	0.0	0.0
19	Rubber & Plastic Pr.	0.0	0.0	0.0	0.0	0.0	0.0	0.1	0.0	0.0	0.0	0.0	0.0	0.0	0.0	0.1
21	Other Manufactures	0.0	0.0	0.0	0.0	0.0	0.0	0.0	0.0	0.0	0.0	0.0	0.0	0.0	0.0	0.0
	Total (32 sectors)	0.0	0.3	0.0	-0.8	0.0	-1.4	-2.9	0.3	1.8	0.0	0.0	-0.2	0.0	-4.9	-7.9

Source: Cambridge Econometrics projection S13-S00/da1/S13K.

Table 4.25a. Sensitivity analysis: E3ME industry 14 employment in 1993 by sector (% difference from base)

		1 BE	2 DK	3 DO	4 DW	5 EL	6 ES	7 FR	8 IR	9 IT	10 IS	11 LX	12 NL	13 PO	14 UK	EU
08	Ferrous & Non-f. Metals	0.0	-0.1	0.0	0.2	0.0	0.2	0.1	0.1	0.0	0.0	0.0	0.0	-0.4	0.2	0.1
09	Non-metallic Min. Pr.	0.0	-0.1	0.0	0.0	0.0	-0.1	0.0	0.0	0.0	0.0	0.0	0.0	0.0	0.0	0.0
10	Chemicals	0.0	0.0	0.0	0.0	0.0	-0.1	0.0	0.0	0.1	0.0	0.0	0.1	0.0	0.0	0.0
11	Metal Products	-0.1	0.0	0.0	0.2	0.0	-0.1	0.1	-1.0	0.0	0.0	0.0	0.0	-0.1	0.1	0.1
12	Agri. & Indust. Mach.	0.0	0.0	0.0	0.1	0.0	-0.1	0.0	0.0	0.1	0.0	0.0	0.0	0.0	0.0	0.0
13	Office Machines	0.0	0.1	0.0	0.0	0.0	0.0	0.0	0.0	0.2	0.0	0.0	0.0	0.0	0.0	0.0
14	Electrical Goods	-2.6	-2.9	0.0	2.3	0.0	-3.9	-2.0	-2.3	0.7	0.0	0.0	2.9	-5.3	2.7	0.2
15	Transport Equipment	0.0	0.0	0.0	0.0	0.0	0.0	0.0	0.0	0.0	0.0	0.0	0.0	0.0	0.0	0.0
16	Food, Drink & Tobacco	0.0	0.0	0.0	0.0	0.0	0.0	0.0	0.0	0.0	0.0	0.0	0.0	0.0	0.0	0.0
17	Tex., Cloth. & Footw.	0.0	0.0	0.0	0.0	0.0	0.0	0.0	0.0	0.0	0.0	0.0	0.0	0.0	0.0	0.0
18	Paper & Printing Pr.	0.0	0.0	0.0	0.0	0.0	0.0	0.0	0.0	0.0	0.0	0.0	0.0	0.0	0.0	0.0
19	Rubber & Plastic Pr.	0.0	0.0	0.0	-0.1	0.0	-0.2	0.1	0.0	0.0	0.0	0.0	0.0	-0.1	0.0	0.0
21	Other Manufactures	0.0	0.0	0.0	0.0	0.0	-0.1	0.0	-0.1	0.0	0.0	0.0	0.1	0.0	0.0	0.0
	Total (32 sectors)	0.0	0.0	0.0	0.1	0.0	-0.1	0.0	0.0	0.0	0.0	0.0	0.1	0.0	0.0	0.0

Source: Cambridge Econometrics projection S14-S00/da1/S14.

Table 4.25b. Sensitivity analysis: E3ME industry 14 employment in 1993 by sector (difference from base in thousands)

		1 BE	2 DK	3 DO	4 DW	5 EL	6 ES	7 FR	8 IR	9 IT	10 IS	11 LX	12 NL	13 PO	14 UK	EU
08	Ferrous & Non-f. Metals	0.0	0.0	0.0	0.4	0.0	0.1	0.1	0.0	0.0	0.0	0.0	0.0	-0.1	0.1	0.7
09	Non-metallic Min. Pr.	0.0	0.0	0.0	0.0	0.0	-0.2	0.0	0.0	0.1	0.0	0.0	0.0	0.0	0.0	0.0
10	Chemicals	0.0	0.0	0.0	0.0	0.0	-0.1	0.1	0.0	0.4	0.0	0.0	0.1	0.0	0.1	0.6
11	Metal Products	0.0	0.0	0.0	1.3	0.0	-0.4	0.4	0.0	-0.1	0.0	0.0	0.0	0.0	0.4	1.6
12	Agri. & Indust. Mach.	0.0	0.0	0.0	0.9	0.0	-0.2	0.0	0.0	0.3	0.0	0.0	0.0	0.0	0.0	1.0
13	Office Machines	0.0	0.0	0.0	0.1	0.0	0.0	0.0	0.0	0.2	0.0	0.0	0.0	0.0	0.0	0.3
14	Electrical Goods	-1.3	-0.7	0.0	19.9	0.0	-17.5	-9.4	-0.5	1.7	0.0	0.0	2.7	-1.9	12.1	5.0
15	Transport Equipment	0.0	0.0	0.0	0.0	0.0	0.0	0.1	0.0	0.1	0.0	0.0	0.0	0.0	0.0	0.1
16	Food, Drink & Tobacco	0.0	0.0	0.0	0.0	0.0	0.0	0.1	0.0	0.0	0.0	0.0	0.0	0.0	0.0	0.1
17	Tex., Cloth. & Footw.	0.0	0.0	0.0	0.0	0.0	0.1	0.1	0.0	0.0	0.0	0.0	0.0	0.0	0.0	0.2
18	Paper & Printing Pr.	0.0	0.0	0.0	0.2	0.0	0.0	0.1	0.0	0.0	0.0	0.0	0.0	0.0	0.0	0.4
19	Rubber & Plastic Pr.	0.0	0.0	0.0	-0.3	0.0	-0.3	0.1	0.0	0.0	0.0	0.0	0.0	0.0	0.0	-0.5
21	Other Manufactures	0.0	0.0	0.0	0.2	0.0	-0.3	0.1	0.0	0.0	0.0	0.0	0.0	0.0	0.0	0.1
	Total (32 sectors)	-0.9	-0.4	0.0	25.1	0.0	-22.2	-2.2	-0.5	10.3	0.0	0.0	3.1	-1.3	11.3	22.3

Source: Cambridge Econometrics projection S14-S00/da1/S14.

Table 4.26a.   Sensitivity analysis: E3ME industry 15 employment in 1993 by sector (% difference from base)

		1 BE	2 DK	3 DO	4 DW	5 EL	6 ES	7 FR	8 IR	9 IT	10 IS	11 LX	12 NL	13 PO	14 UK	EU
08	Ferrous & Non-f. Metals	0.0	-0.7	0.0	0.1	0.0	-0.1	0.0	0.0	0.0	0.0	0.0	0.0	-0.3	0.2	0.0
09	Non-metallic Min. Pr.	0.0	0.0	0.0	0.0	0.0	0.0	0.0	0.0	0.0	0.0	0.0	0.0	-0.6	0.0	0.0
10	Chemicals	0.0	0.0	0.0	0.0	0.0	0.0	0.0	0.0	-0.2	0.0	0.0	0.1	0.0	0.0	0.0
11	Metal Products	0.3	-0.1	0.0	0.2	0.0	0.3	0.1	-0.3	0.0	0.0	0.1	0.0	-0.6	0.0	0.1
12	Agri. & Indust. Mach.	0.0	-0.2	0.0	0.2	0.0	0.1	0.0	0.0	-0.1	0.0	0.1	0.0	0.0	0.0	0.0
13	Office Machines	0.0	0.0	0.0	0.0	0.0	0.0	0.0	0.0	-0.1	0.0	0.0	-0.1	-2.6	0.0	0.0
14	Electrical Goods	0.0	0.0	0.0	0.0	0.0	0.0	0.0	0.0	0.0	0.0	0.0	0.0	-1.0	0.0	0.0
15	Transport Equipment	7.5	-0.1	0.0	0.0	0.0	0.4	-0.6	3.1	-4.7	0.0	0.0	-14.4	0.0	0.7	-0.6
16	Food, Drink & Tobacco	0.0	0.0	0.0	0.0	0.0	0.0	0.0	0.0	0.0	0.0	0.0	0.1	0.0	0.0	0.0
17	Tex., Cloth. & Footw.	0.0	0.0	0.0	0.0	0.0	0.0	-0.1	0.0	0.0	0.0	0.0	0.0	0.0	0.0	0.0
18	Paper & Printing Pr.	0.1	0.0	0.0	0.0	0.0	0.0	0.1	0.0	-0.1	0.0	0.0	0.0	0.0	0.0	0.0
19	Rubber & Plastic Pr.	0.0	0.0	0.0	-0.2	0.0	0.1	-0.1	0.0	0.0	0.0	0.0	0.0	-0.1	0.0	-0.1
21	Other Manufactures	0.0	-0.1	0.0	0.1	0.0	0.0	0.0	0.0	0.0	0.0	0.0	0.0	0.0	0.0	0.0
	Total (32 sectors)	0.2	0.0	0.0	0.0	0.0	0.0	0.0	0.0	-0.1	0.0	0.0	-0.2	-0.1	0.0	0.0

Source: Cambridge Econometrics projection S15-S00/da1/S15.

Table 4.26b.   Sensitivity analysis: E3ME industry 15 employment in 1993 by sector (difference from base in thousands)

		1 BE	2 DK	3 DO	4 DW	5 EL	6 ES	7 FR	8 IR	9 IT	10 IS	11 LX	12 NL	13 PO	14 UK	EU
08	Ferrous & Non-f. Metals	0.0	0.0	0.0	0.3	0.0	0.0	0.0	0.0	0.0	0.0	0.0	0.0	-0.1	0.1	0.3
09	Non-metallic Min. Pr.	0.0	0.0	0.0	0.1	0.0	0.0	0.0	0.0	-0.1	0.0	0.0	0.0	-0.4	0.0	-0.4
10	Chemicals	0.0	0.0	0.0	0.0	0.0	0.0	0.1	0.0	-0.7	0.0	0.0	0.1	0.0	0.0	-0.6
11	Metal Products	0.2	0.0	0.0	1.7	0.0	0.8	0.4	0.0	0.1	0.0	0.0	0.0	0.0	-0.2	2.9
12	Agri. & Indust. Mach.	0.0	-0.2	0.0	1.6	0.0	0.1	0.0	0.0	-0.4	0.0	0.0	0.0	-0.2	0.0	0.9
13	Office Machines	0.0	0.0	0.0	0.1	0.0	0.0	0.1	0.0	-0.1	0.0	0.0	0.0	0.0	0.0	0.1
14	Electrical Goods	0.0	0.0	0.0	-0.2	0.0	0.1	0.1	0.0	0.0	0.0	0.0	0.0	-0.9	-0.1	-1.1
15	Transport Equipment	5.2	0.0	0.0	0.0	0.0	1.4	-3.1	-0.3	14.9	0.0	0.0	-8.9	-0.6	3.1	-17.4
16	Food, Drink & Tobacco	0.0	0.0	0.0	0.0	0.0	0.0	0.1	0.0	-0.1	0.0	0.0	0.1	0.0	0.0	0.1
17	Tex., Cloth. & Footw.	0.0	0.0	0.0	-0.1	0.0	0.1	-0.3	0.0	0.0	0.0	0.0	0.0	0.0	0.0	-0.4
18	Paper & Printing Pr.	0.0	0.0	0.0	0.2	0.0	0.0	0.2	0.0	-0.1	0.0	0.0	0.0	0.0	0.0	0.4
19	Rubber & Plastic Pr.	0.0	0.0	0.0	-0.5	0.0	-0.1	-0.2	0.0	0.0	0.0	0.0	0.0	0.0	0.0	-0.8
21	Other Manufactures	0.0	0.0	0.0	0.2	0.0	0.2	0.0	0.0	0.0	0.0	0.0	0.0	0.0	0.0	0.3
	Total (32 sectors)	5.8	-0.6	0.0	6.9	0.0	4.1	2.0	-0.3	22.0	0.0	0.0	-8.8	-4.7	4.4	-12.5

Source: Cambridge Econometrics projection S15-S00/da1/S15.

Table 4.27a. Sensitivity analysis: E3ME industry 16 employment in 1993 by sector (% difference from base)

		1 BE	2 DK	3 DO	4 DW	5 EL	6 ES	7 FR	8 IR	9 IT	10 IS	11 LX	12 NL	13 PO	14 UK	EU
08	Ferrous & Non-f. Metals	0.0	-1.1	0.0	-0.1	0.0	0.0	0.0	-0.1	0.0	0.0	0.0	-0.1	0.0	-0.3	-0.1
09	Non-metallic Min. Pr.	0.0	0.1	0.0	0.0	0.0	0.0	0.0	0.1	0.0	0.0	0.0	0.0	0.0	0.0	0.0
10	Chemicals	0.0	-0.2	0.0	0.0	0.0	0.0	0.0	0.0	0.0	0.0	0.0	-0.1	0.0	0.1	0.0
11	Metal Products	-0.1	2.8	0.0	-0.2	0.0	0.0	-0.1	3.4	0.0	0.0	0.0	0.1	0.0	0.2	0.0
12	Agri. & Indust. Mach.	0.0	1.0	0.0	-0.2	0.0	0.0	0.0	0.0	0.0	0.0	0.0	0.0	0.0	0.0	0.0
13	Office Machines	0.0	-0.2	0.0	0.0	0.0	0.0	0.0	0.0	0.0	0.0	0.0	0.0	0.0	0.0	0.0
14	Electrical Goods	0.0	0.2	0.0	0.0	0.0	0.0	0.0	0.0	0.0	0.0	0.0	0.1	0.0	0.0	0.0
15	Transport Equipment	0.0	0.0	0.0	0.0	0.0	0.0	0.0	0.0	0.0	0.0	0.0	0.0	0.0	0.0	0.0
16	Food, Drink & Tobacco	-0.3	-0.7	0.0	-2.8	0.0	0.0	0.0	0.2	-0.7	0.0	0.0	-0.4	0.0	0.0	0.0
17	Tex., Cloth. & Footw.	0.0	-0.1	0.0	0.0	0.0	0.2	0.4	0.0	0.0	0.0	-25.8	0.9	0.0	1.3	-0.4
18	Paper & Printing Pr.	-0.2	-0.1	0.0	0.0	0.0	0.0	0.0	0.0	0.0	0.0	0.0	0.0	0.0	0.0	0.0
19	Rubber & Plastic Pr.	-0.1	0.4	0.0	0.1	0.0	0.0	0.0	0.3	0.0	0.0	-0.4	0.1	0.0	0.0	0.0
21	Other Manufactures	0.1	0.1	0.0	0.0	0.0	0.0	0.0	-0.1	0.0	0.0	0.0	-0.2	0.0	0.1	0.0
	Total (32 sectors)	0.0	0.1	0.0	-0.1	0.0	0.0	0.0	0.0	0.0	0.0	-0.5	0.0	0.0	0.0	0.0

Source: Cambridge Econometrics projection S16-S00/da1/S16.

Table 4.27b. Sensitivity analysis: E3ME industry 16 employment in 1993 by sector (difference from base in thousands)

		1 BE	2 DK	3 DO	4 DW	5 EL	6 ES	7 FR	8 IR	9 IT	10 IS	11 LX	12 NL	13 PO	14 UK	EU
08	Ferrous & Non-f. Metals	0.0	0.0	0.0	-0.2	0.0	0.0	0.0	0.0	0.0	0.0	0.0	0.0	0.0	-0.2	-0.5
09	Non-metallic Min. Pr.	0.0	0.0	0.0	-0.1	0.0	0.0	0.0	0.0	0.0	0.0	0.0	0.0	0.0	0.0	-0.1
10	Chemicals	0.0	0.0	0.0	0.0	0.0	0.0	-0.1	0.0	0.0	0.0	0.0	-0.1	0.0	0.3	0.0
11	Metal Products	-0.1	0.6	0.0	-1.7	0.0	-0.1	-0.3	0.1	0.0	0.0	0.0	0.2	0.0	1.0	-0.3
12	Agri. & Indust. Mach.	0.0	0.6	0.0	-1.6	0.0	-0.1	0.0	0.0	0.0	0.0	0.0	0.0	0.0	0.0	-1.1
13	Office Machines	0.0	0.0	0.0	-0.1	0.0	0.0	0.0	0.0	0.0	0.0	0.0	0.0	0.0	0.0	-0.1
14	Electrical Goods	0.0	0.0	0.0	-0.2	0.0	0.0	0.0	0.0	0.0	0.0	0.0	0.1	0.0	0.1	0.0
15	Transport Equipment	0.0	0.0	0.0	0.0	0.0	0.0	0.0	0.0	0.0	0.0	0.0	0.0	0.0	-0.1	-0.2
16	Food, Drink & Tobacco	-0.3	-0.6	0.0	-20.0	0.0	1.5	2.1	0.1	-2.0	0.0	-0.8	-0.2	0.0	-6.3	12.0
17	Tex., Cloth. & Footw.	0.0	-0.1	0.0	0.0	0.0	0.0	0.1	0.0	0.0	0.0	0.0	1.7	0.0	0.0	0.0
18	Paper & Printing Pr.	-0.1	0.0	0.0	-0.2	0.0	0.0	-0.1	0.1	0.0	0.0	0.0	0.1	0.0	0.1	-0.1
19	Rubber & Plastic Pr.	0.0	0.1	0.0	0.2	0.0	0.0	0.1	0.0	0.0	0.0	0.0	0.0	0.0	0.1	0.4
21	Other Manufactures	0.0	0.1	0.0	-0.1	0.0	-0.1	0.1	0.0	0.0	0.0	0.0	-0.2	0.0	0.0	-0.2
	Total (32 sectors)	-0.9	3.1	0.0	-27.4	0.0	-0.6	-4.2	0.5	2.9	0.0	-0.9	2.7	0.0	-0.1	24.7

Source: Cambridge Econometrics projection S16-S00/da1/S16.

Table 4.28a. Sensitivity analysis: E3ME industry 17 employment in 1993 by sector (% difference from base)

		1 BE	2 DK	3 DO	4 DW	5 EL	6 ES	7 FR	8 IR	9 IT	10 IS	11 LX	12 NL	13 PO	14 UK	EU
08	Ferrous & Non-f. Metals	0.0	-0.3	0.0	0.0	0.0	0.0	0.0	0.0	0.0	0.0	0.0	0.0	0.1	-0.1	0.0
09	Non-metallic Min. Pr.	0.0	-0.1	0.0	0.0	0.0	0.0	0.0	-0.1	0.0	0.0	0.0	0.0	0.2	0.0	0.0
10	Chemicals	0.0	0.0	0.0	0.0	0.0	-0.2	0.0	0.0	0.1	0.0	0.0	0.0	0.0	0.0	0.0
11	Metal Products	0.0	-0.4	0.0	0.0	0.0	-0.1	0.0	-0.1	0.0	0.0	0.0	0.0	0.1	0.0	0.0
12	Agri. & Indust. Mach.	0.0	0.1	0.0	-0.1	0.0	-0.3	0.0	0.0	0.0	0.0	0.0	0.0	0.0	0.0	0.0
13	Office Machines	0.0	-0.1	0.0	0.0	0.0	0.0	0.0	-0.1	0.0	0.0	0.0	0.0	0.6	0.0	0.0
14	Electrical Goods	0.0	-0.2	0.0	0.0	0.0	0.0	0.0	0.0	0.0	0.0	0.0	0.0	0.0	0.0	0.0
15	Transport Equipment	0.0	0.0	0.0	0.0	0.0	0.0	0.0	-0.1	0.0	0.0	0.0	0.0	0.1	0.0	0.0
16	Food, Drink & Tobacco	0.0	0.0	0.0	0.0	0.0	0.1	0.0	0.0	0.0	0.0	0.0	0.0	0.7	0.0	0.0
17	Tex., Cloth. & Footw.	1.0	-8.5	0.0	-0.1	0.0	2.5	2.4	-9.7	0.0	0.0	0.0	-1.6	0.1	-1.4	0.1
18	Paper & Printing Pr.	0.0	0.0	0.0	0.0	0.0	0.0	0.0	-0.1	0.0	0.0	0.0	0.0	0.1	0.0	0.0
19	Rubber & Plastic Pr.	0.0	0.0	0.0	0.0	0.0	-0.2	0.0	-0.3	0.0	0.0	0.0	0.0	0.2	0.0	0.0
21	Other Manufactures	0.0	-0.3	0.0	0.0	0.0	-0.3	0.0	0.0	0.0	0.0	0.0	0.0	-0.1	0.0	-0.1
	Total (32 sectors)	0.0	-0.3	0.0	0.0	0.0	0.0	0.1	-0.2	0.0	0.0	0.0	0.0	0.0	0.0	0.0

Source: Cambridge Econometrics projection S17-S00/da1/S17.

Table 4.28b. Sensitivity analysis: E3ME industry 17 employment in 1993 by sector (difference from base in thousands)

		1 BE	2 DK	3 DO	4 DW	5 EL	6 ES	7 FR	8 IR	9 IT	10 IS	11 LX	12 NL	13 PO	14 UK	EU
08	Ferrous & Non-f. Metals	0.0	0.0	0.0	-0.1	0.0	0.0	0.0	0.0	0.0	0.0	0.0	0.0	0.0	-0.1	-0.1
09	Non-metallic Min. Pr.	0.0	0.0	0.0	0.0	0.0	-0.1	0.0	0.0	0.0	0.0	0.0	0.0	0.1	0.0	0.0
10	Chemicals	0.0	0.0	0.0	0.0	0.0	-0.4	0.0	0.0	0.5	0.0	0.0	0.0	0.0	0.1	0.2
11	Metal Products	0.0	-0.1	0.0	-0.4	0.0	-0.3	0.1	0.0	0.0	0.0	0.0	0.0	0.0	0.1	-0.5
12	Agri. & Indust. Mach.	0.0	0.0	0.0	-0.9	0.0	-0.4	0.0	0.0	0.2	0.0	0.0	0.0	0.1	0.0	-1.0
13	Office Machines	0.0	0.0	0.0	0.0	0.0	0.0	0.0	0.0	0.0	0.0	0.0	0.0	0.0	0.0	0.0
14	Electrical Goods	0.0	0.0	0.0	-0.2	0.0	-0.1	0.0	0.0	0.0	0.0	0.0	0.0	0.2	0.0	-0.3
15	Transport Equipment	0.0	0.0	0.0	0.0	0.0	-0.1	0.0	0.0	0.1	0.0	0.0	0.0	0.0	0.0	0.0
16	Food, Drink & Tobacco	0.0	0.0	0.0	0.0	0.0	0.4	0.0	0.0	0.1	0.0	0.0	0.0	0.1	0.0	0.6
17	Tex., Cloth. & Footw.	0.8	-6.1	0.0	-0.1	0.0	7.5	10.4	-2.0	0.3	0.0	0.0	-1.0	2.2	-7.3	4.6
18	Paper & Printing Pr.	0.0	0.0	0.0	-0.1	0.0	0.0	0.0	0.0	0.0	0.0	0.0	0.0	0.0	-0.1	0.0
19	Rubber & Plastic Pr.	0.0	0.0	0.0	0.0	0.0	-0.3	0.0	0.0	0.0	0.0	0.0	0.0	0.1	0.0	-0.2
21	Other Manufactures	0.0	-0.2	0.0	0.0	0.0	-1.0	0.0	0.0	0.0	0.0	0.0	0.0	0.0	0.0	-1.3
	Total (32 sectors)	1.1	-7.5	0.0	-2.8	0.0	-3.7	11.9	-2.3	3.0	0.0	0.0	-0.9	0.5	-9.8	-10.5

Source: Cambridge Econometrics projection S17-S00/da1/S17.

Table 4.29a. Sensitivity analysis: E3ME industry 18 employment in 1993 by sector (% difference from base)

		1 BE	2 DK	3 DO	4 DW	5 EL	6 ES	7 FR	8 IR	9 IT	10 IS	11 LX	12 NL	13 PO	14 UK	EU
08	Ferrous & Non-f. Metals	0.0	-0.3	0.0	0.0	0.0	0.0	0.0	0.0	0.0	0.0	0.0	0.1	0.1	0.0	0.0
09	Non-metallic Min. Pr.	0.0	0.0	0.0	0.0	0.0	0.0	0.0	0.0	0.0	0.0	0.0	0.0	0.1	0.0	0.0
10	Chemicals	0.0	0.0	0.0	0.0	0.0	0.0	0.0	0.0	-0.1	0.0	0.0	0.0	0.0	0.0	0.0
11	Metal Products	0.0	0.2	0.0	0.1	0.0	0.0	0.0	0.0	0.0	0.0	0.0	0.1	0.0	0.0	0.0
12	Agri. & Indust. Mach.	0.0	0.4	0.0	0.2	0.0	0.0	0.0	0.0	0.0	0.0	0.0	0.0	0.1	0.0	0.1
13	Office Machines	0.0	0.0	0.0	0.0	0.0	0.0	0.0	0.0	0.0	0.0	0.0	0.0	0.0	0.0	0.0
14	Electrical Goods	0.0	0.0	0.0	0.1	0.0	0.0	0.0	0.0	0.0	0.0	0.0	0.0	0.3	0.0	0.0
15	Transport Equipment	0.0	0.0	0.0	0.0	0.0	0.0	0.0	0.0	0.0	0.0	0.0	0.0	0.0	0.0	0.0
16	Food, Drink & Tobacco	0.0	0.0	0.0	0.0	0.0	0.0	0.0	0.0	0.0	0.0	0.0	0.0	0.0	0.0	0.0
17	Tex., Cloth. & Footw.	0.0	0.0	0.0	0.0	0.0	0.0	0.0	0.0	0.0	0.0	0.0	0.0	0.0	0.0	0.0
18	Paper & Printing Pr.	0.6	-0.2	0.0	-0.1	0.0	0.0	-0.3	-3.9	-0.8	0.0-	11.1	-2.4	3.8	0.3	-0.2
19	Rubber & Plastic Pr.	0.0	0.0	0.0	0.0	0.0	0.0	0.0	0.0	0.0	0.0	0.0	0.0	0.1	0.0	0.0
21	Other Manufactures	0.0	0.0	0.0	0.0	0.0	0.0	0.0	0.0	0.0	0.0	0.0	0.1	0.0	0.0	0.0
	Total (32 sectors)	0.0	0.0	0.0	0.0	0.0	0.0	0.0	-0.1	0.0	0.0	-0.1	0.0	0.1	0.0	0.0

Source: Cambridge Econometrics projection S18-S00/da1/s18.

Table 4.29b. Sensitivity analysis: E3ME industry 18 employment in 1993 by sector (difference from base in thousands)

		1 BE	2 DK	3 DO	4 DW	5 EL	6 ES	7 FR	8 IR	9 IT	10 IS	11 LX	12 NL	13 PO	14 UK	EU
08	Ferrous & Non-f. Metals	0.0	0.0	0.0	0.1	0.0	0.0	0.0	0.0	0.0	0.0	0.0	0.0	0.0	0.0	0.2
09	Non-metallic Min. Pr.	0.0	0.0	0.0	0.0	0.0	0.0	0.0	0.0	0.0	0.0	0.0	0.0	0.1	0.0	0.1
10	Chemicals	0.0	0.0	0.0	-0.1	0.0	0.0	0.0	0.0	-0.3	0.0	0.0	0.0	0.0	0.0	-0.3
11	Metal Products	0.0	0.0	0.0	0.8	0.0	0.0	0.0	0.0	0.0	0.0	0.0	0.1	0.0	0.0	0.9
12	Agri. & Indust. Mach.	0.0	0.3	0.0	1.5	0.0	0.0	0.0	0.0	-0.1	0.0	0.0	0.0	0.0	0.0	1.8
13	Office Machines	0.0	0.0	0.0	0.0	0.0	0.0	0.0	0.0	0.0	0.0	0.0	0.0	0.0	0.0	0.0
14	Electrical Goods	0.0	0.0	0.0	0.5	0.0	0.0	0.0	0.0	0.0	0.0	0.0	0.0	0.1	0.0	0.5
15	Transport Equipment	0.0	0.0	0.0	0.0	0.0	0.0	0.0	0.0	0.0	0.0	0.0	0.0	0.0	0.0	0.0
16	Food, Drink & Tobacco	0.0	0.0	0.0	0.0	0.0	0.0	0.0	0.0	0.0	0.0	0.0	0.2	0.0	0.0	0.1
17	Tex., Cloth. & Footw.	0.0	0.0	0.0	0.0	0.0	-0.1	0.0	0.0	0.0	0.0	0.0	0.0	0.0	0.0	-0.1
18	Paper & Printing Pr.	0.3	-0.1	0.0	-1.0	0.0	0.0	-0.9	-0.7	-1.3	0.0	-0.2	-2.7	1.6	1.2	-3.8
19	Rubber & Plastic Pr.	0.0	0.0	0.0	-0.1	0.0	0.0	0.0	0.0	0.0	0.0	0.0	0.0	0.0	0.0	-0.1
21	Other Manufactures	0.0	0.0	0.0	0.1	0.0	0.0	0.0	0.0	0.0	0.0	0.0	0.1	0.0	0.0	0.1
	Total (32 sectors)	0.3	0.1	0.0	5.1	0.0	-1.0	-0.7	-0.7	-2.7	0.0	-0.2	-2.1	1.9	1.3	1.3

Source: Cambridge Econometrics projection S18-S00/da1/s18.

Table 4.30a. Sensitivity analysis: E3ME industry 19 employment in 1993 by sector (% difference from base)

		1 BE	2 DK	3 DO	4 DW	5 EL	6 ES	7 FR	8 IR	9 IT	10 IS	11 LX	12 NL	13 PO	14 UK	EU
08	Ferrous & Non-f. Metals	0.0	-0.1	0.0	-0.1	0.0	0.0	0.0	0.0	0.0	0.0	0.0	0.0	-0.1	0.0	0.0
09	Non-metallic Min. Pr.	0.0	0.0	0.0	0.0	0.0	0.0	0.0	0.0	0.0	0.0	0.0	0.0	-0.1	0.0	0.0
10	Chemicals	0.0	-0.1	0.0	0.1	0.0	0.0	0.0	0.0	0.7	0.0	0.0	-0.1	0.0	0.0	0.2
11	Metal Products	0.0	0.1	0.0	-0.2	0.0	0.0	0.0	0.2	0.0	0.0	0.0	0.0	0.0	0.0	-0.1
12	Agri. & Indust. Mach.	0.0	0.0	0.0	-0.1	0.0	0.0	0.0	0.0	0.0	0.0	0.0	0.0	-0.1	0.0	0.0
13	Office Machines	0.0	0.0	0.0	0.0	0.0	0.0	0.0	0.0	0.0	0.0	0.0	0.0	-0.2	0.0	0.0
14	Electrical Goods	0.0	0.0	0.0	0.0	0.0	0.0	0.0	0.0	0.0	0.0	0.0	0.0	0.0	0.0	0.0
15	Transport Equipment	0.0	0.0	0.0	0.0	0.0	0.0	0.0	0.0	0.0	0.0	0.0	0.0	0.0	0.0	0.0
16	Food, Drink & Tobacco	0.0	0.0	0.0	0.0	0.0	0.0	0.0	0.0	0.0	0.0	0.0	0.0	0.0	0.0	0.0
17	Tex., Cloth. & Footw.	0.0	0.0	0.0	0.0	0.0	0.0	0.0	0.0	0.0	0.0	0.0	0.0	0.0	0.0	0.0
18	Paper & Printing Pr.	0.0	0.0	0.0	0.0	0.0	0.0	0.0	0.0	0.0	0.0	0.5	0.0	0.0	0.0	0.0
19	Rubber & Plastic Pr.	1.2	7.0	0.0	1.1	0.0	-2.2	-0.8	-5.7	0.0	0.0	4.8	-3.1	-1.8	0.8	-0.2
21	Other Manufactures	0.0	0.0	0.0	0.0	0.0	0.0	0.0	0.0	0.0	0.0	0.0	0.0	0.0	0.0	0.0
	Total (32 sectors)	0.0	0.0	0.0	0.0	0.0	0.0	0.0	0.0	0.0	0.0	0.2	0.0	0.0	0.0	0.0

Source: Cambridge Econometrics projection S19-S00/da1/S19.

Table 4.30b. Sensitivity analysis: E3ME industry 19 employment in 1993 by sector (difference from base in thousands)

		1 BE	2 DK	3 DO	4 DW	5 EL	6 ES	7 FR	8 IR	9 IT	10 IS	11 LX	12 NL	13 PO	14 UK	EU
08	Ferrous & Non-f. Metals	0.0	0.0	0.0	-0.1	0.0	0.0	0.0	0.0	0.0	0.0	0.0	0.0	0.0	0.0	-0.2
09	Non-metallic Min. Pr.	0.0	0.0	0.0	0.0	0.0	0.0	0.0	0.0	0.0	0.0	0.0	0.0	0.0	0.0	0.0
10	Chemicals	0.0	0.0	0.0	0.4	0.0	0.1	0.0	0.0	2.6	0.0	0.0	-0.1	0.0	0.0	3.0
11	Metal Products	0.0	0.0	0.0	-1.6	0.0	0.1	0.0	0.0	0.0	0.0	0.0	0.0	0.0	0.1	-1.5
12	Agri. & Indust. Mach.	0.0	0.0	0.0	-1.3	0.0	0.0	0.0	0.0	0.1	0.0	0.0	0.0	0.0	0.0	-1.2
13	Office Machines	0.0	0.0	0.0	-0.1	0.0	0.0	0.0	0.0	0.0	0.0	0.0	0.0	-0.1	0.0	-0.1
14	Electrical Goods	0.0	0.0	0.0	-0.4	0.0	0.0	0.0	0.0	0.0	0.0	0.0	0.0	0.0	0.0	-0.4
15	Transport Equipment	0.0	0.0	0.0	0.0	0.0	0.0	0.0	0.0	0.0	0.0	0.0	0.0	0.0	0.0	0.0
16	Food, Drink & Tobacco	0.0	0.0	0.0	0.0	0.0	0.0	0.0	0.0	0.0	0.0	0.0	0.0	0.0	0.0	0.0
17	Tex., Cloth. & Footw.	0.0	0.0	0.0	0.0	0.0	0.0	0.0	0.0	0.0	0.0	0.0	0.0	0.0	0.0	0.0
18	Paper & Printing Pr.	0.0	0.0	0.0	-0.1	0.0	0.0	0.0	0.0	0.1	0.0	0.0	0.0	0.0	0.0	0.0
19	Rubber & Plastic Pr.	0.3	1.0	0.0	2.8	0.0	-3.4	-1.7	-0.6	0.0	0.0	0.3	-1.9	-0.5	1.4	-2.2
21	Other Manufactures	0.0	0.0	0.0	0.0	0.0	0.0	0.0	0.0	0.0	0.0	0.0	0.0	0.0	0.0	0.0
	Total (32 sectors)	0.2	1.1	0.0	-3.7	0.0	-2.6	-1.5	-0.5	4.7	0.0	0.3	-2.0	-1.0	0.9	-4.0

Source: Cambridge Econometrics projection S19-S00/da1/S19.

4.6.   Policy issues

The distinctive contribution of this study to the assessment of the single market programme is that it has adopted a comprehensive, statistical approach to measuring the sectoral effects, by using an empirical model of the EU, estimated on time-series data over the period when the SMP effects were beginning to be experienced. This approach allows an overall impact of the SMP to be calculated, both for the effects on manufacturing employment and for all employment, income and trade. Indeed, the approach requires a comprehensive assessment because it treats the economies of Member States as interacting one with another and, within these economies, each industrial sector as interacting with all the others.

The overall conclusion is that by 1993, the last year of the study period, internal trade in the EU was some 20–30% higher than it would have been without the SMP; however, in contrast, the SMP appears to have led to a small reduction in extra-EU exports and an increase in extra-EU imports as exporters in the rest of the world have benefited from the new trading conditions in the EU. Manufacturing employment has been slightly reduced, but this reduction has been more than compensated by an increase in non-manufacturing employment. The effects are dynamic and are expected to continue well after 1993. Indeed, the fact that the analysis ends in 1993 might lead to the argument that the model simulations capture predominantly short-run effects, such as those of the J-curve. If this were the case, a more positive impact on employment might be expected in the longer term as trade prices adjust more completely to the new market structure implied by the SMP.

The SMP appears to have slightly accelerated structural change by shifting employment away from manufacturing towards services. There is no evidence that the SMP has reinforced existing disparities between rich and poor Member States: if anything, there are indications that Ireland, Spain and Portugal (although not Greece) have seen larger proportionate gains in employment. However, the evidence for Spain and Portugal is not clear cut, since these economies were also benefiting from accession to the EU over the period when the SMP was taking effect.

The general increase in trade found by the study as a result of the reduction in non-tariff barriers is in agreement with much of the existing theoretical and empirical literature on the subject. The literature is, however, less of one voice as to the effects on employment. The study finds small reductions in manufacturing employment, which appears to confirm the theories of those (e.g. Wood, 1994) who argue that changing trade patterns are the cause of decline in employment through trade competition. However, the reduction is very small in relation to the increase in trade, and the fact that there is an overall increase in employment in the study results implies that the trade-pattern theory is overstated and incomplete, because it does not take into account other responses such as those of wage rates, demand patterns and the manufacture–service mix in economies. The results are more in agreement with the new theories of intra-industry trade, economies of scale and dynamic specialization in niche markets, in which the employment effects of market integration are ambiguous because employment is created in successful sectors and destroyed in others.

One major conclusion from the study is that the effects are highly uneven as between different sectors and regions. Although the study has identified and measured these effects, they are very uncertain and depend on a perforce stylized view of the way the SMP affects trade and employment.

The study concludes that, by the end of the simulation period, manufacturing employment was increased in Ireland (5.1% increase above base), Spain (1.5% increase above base) and Portugal (1.5% above base), reduced in West Germany (-2.5% below base), the Netherlands (-1.9% below base) and Greece (-0.3% below base) and broadly unaffected in other Member States. From the analysis performed in Section 2.4 the countries expected to experience a beneficial impact from the SMP were Belgium, Spain, and to a lesser extent (due to offsetting factors, i.e. R&D versus labour-intensive sectors), Denmark, France, Germany, Ireland, Italy, Luxembourg, the Netherlands and the UK. Greece and Portugal were expected not to gain from implementation of the SMP. Thus, there is some correspondence between theory and results, although those for Spain and Portugal must be taken in the context of accession to the EU, for which it did not prove possible to provide a separate impact analysis.

At the sectoral level, manufacturing employment was increased in Ferrous and Non-ferrous Metals (6.7% above base), Metal Products (1.8% above base) and Textiles, Clothing and Footwear (0.8% above base). Reductions in employment were registered in Rubber and Plastic Products (-4.2% below base), Agricultural and Industrial Machinery (-3.2% below base), Electrical Goods (-1% below base) and Transport Equipment (-0.9% below base). Some general conclusions can be drawn from the sectoral statistics presented in Section 2.4. The industries producing similar (non-differentiated) products in a competitive market were identified as Non-metallic Mineral Products, Textiles, Clothing and Footwear, and Other Manufactures, and it is these industries which were expected to be typified by low R&D and high labour intensity of production. At the other end of the spectrum, those sectors with a more imperfectly competitive market, where non-price competition was more prevalent, were identified as Chemicals, Transport Equipment, and Food, Drink and Tobacco. The analysis proposed these sectors as most likely to undertake intra-industry trade and therefore benefit from the SMP, which is expected to increase intra-industry trade flows. In general, there does not seem to be much correspondence between the predictions from theory and the results for the study, perhaps underlining the difficulty of measuring intra-industry effects at what is a relatively broad sectoral disaggregation. It can also be difficult to generalize about sectors on a pan-European basis, as the heterogeneity that exists between Member States means that a broad classification into 'segmented market' of 'differentiated product', as in Section 2.2, can be misleading.

The conclusion for competition policy is that the SMP has accelerated market change and increased efficiency, but at the same time the market power of the leading firms is likely to have increased. The conclusion for labour market policy is that they can help to realize the potential benefits of the single market by improving the capacity of the labour market to respond to new employment opportunities and to adapt to a more efficient structure.

With regard to restructuring, the following problems need to be addressed:

(a)   How to ensure that areas of quality manufacturing jobs are sustained as far as possible, e.g. in those sectors with small enterprises producing and/or using extensively low-cost technology in production. These are the principal sources of manufacturing job gains.

(b)   How to ensure that shortages of key skills do not act as bottlenecks to capitalizing on the potential dynamic benefits of the SMP and thereby also limit the opportunities for complementary employment for the less skilled in those sectors and supplied to those sectors.

(c)   How to induce more quality into the jobs generated in the service sector by shifting more activity to high value-added activities. This is a key area of policy as services act to benefit and facilitate the manufacturing sectors while also having a role in their own right. With the majority of EU manufacturing firms involved in intra-industry trade of differentiated products, services such as marketing and advertising are central elements of the competitive edge needed in a truly international environment, which the SMP aims to create within Europe.

(d)   How to cope with the national disparities (and regional elements thereof) which appear to be opening up in the wake of the transition to the Single European Market, e.g. to ease restructuring in Germany or to counter the disadvantages of Portugal.

The empirical results suggest that the main impact of the SMP has been not to destroy or create jobs overall, but to accelerate restructuring within sectors and between Member States. There are also indications that the SMP has enabled non-EU producers to increase their exports to the EU to a greater extent, so far, than EU producers have been able to increase their extra-EU exports. The main policy implications therefore relate to actions that can ease the adjustment to restructuring, and to monitor extra-EU trade performance to examine whether the hoped-for improvements to competitiveness emerge in the longer term.

APPENDIX A

Descriptive statistics by industry

This appendix contains a series of tables (A1a–A13b) describing the industrial characteristics of the countries being analysed in the study, i.e. the EUR-12 minus Greece, which was not included on grounds of data availability. Where data permits, descriptive statistics are constructed for all 14 of the E3ME manufacturing industries and are relative to the equivalent EU figure (except for the output shares and measures of inter-industry trade), i.e. a value of unity indicates a performance in line with the EU average. The tables cover the years 1980 and 1990 in order to provide a snapshot of data before and after the implementation of the single market programme.

The abbreviations used to represent the descriptive statistics have the following meanings and constructions:

RQ

Share of output (%).

IIT_EU and IIT_RW

Balassa measures of inter-industry trade. Constructed using value data on EU trade. The measures are created using the following formula:

$$IIT\_??it = \frac{X_{it} - M_{it}}{X_{it} + M_{it}}$$

where

?? = EU for intra-EU trade and RW for extra-EU trade.

X_{it} = exports of industry i in time t

M_{it} = imports of industry i in time t

This means the statistic is bounded by -1 and +1, with a value of 0 implying equal two-way trade, i.e. intra-industry trade.

IPR

Import penetration ratio. Constructed by taking the ratio of industry imports to domestic demand.

EIR

Export penetration ratio. Constructed by taking the ratio of industry exports to gross output.

WR

Wage rate. Constructed by calculating real labour costs per employee.

ULC

Unit labour costs. Defined as the ratio of industry labour costs to industry value-added.

KI

Capital intensity ratio. Defined as the ratio of investment expenditure to value-added.

RDI

R&D expenditure ratio. Defined as the ratio of R&D expenditure to value-added.

Table A1a. **Descriptive sectoral statistics (relative to EU total): industry 8, 1980**

	RQ	IPR	EIR	WR	ULC	KI	RDI
BE	6.483	1.803	2.074	1.370	1.221	0.861	1.654
DK	0.206	2.762	2.246	1.043	1.303	0.000	0.737
DW	36.145	0.762	0.763	1.095	1.022	0.737	1.243
EL	NA	NA	NA	NA	NA	NA	NA
ES	7.793	0.470	1.376	0.665	0.764	1.970	0.431
FR	17.116	1.202	1.176	1.167	0.981	0.948	1.009
IR	0.121	3.523	3.972	0.656	0.913	0.387	0.560
IT	15.785	0.829	0.463	0.942	0.767	1.601	0.523
LX	0.976	NA	4.519	1.296	1.080	0.971	0.737
NL	2.717	2.037	1.976	1.486	0.953	0.000	1.055
PO	0.489	1.386	0.581	0.303	0.822	1.006	0.349
UK	12.169	1.051	0.807	0.840	1.349	0.739	1.174

Source: E3ME Database.

Table A1b. **Descriptive sectoral statistics (relative to EU total): industry 8, 1990**

	RQ	IPR	EIR	WR	ULC	KI	RDI
BE	7.863	1.430	1.681	1.356	0.895	1.444	2.026
DK	0.196	2.309	2.197	0.956	0.803	0.000	3.743
DW	32.835	0.801	0.809	0.970	1.190	0.760	0.907
EL	NA	NA	NA	NA	NA	NA	NA
ES	7.222	0.876	0.622	0.806	0.852	0.012	0.537
FR	16.434	1.172	1.252	1.159	0.832	1.408	1.186
IR	0.219	2.142	2.064	0.966	0.917	1.183	0.658
IT	20.000	0.696	0.453	1.002	0.958	1.391	0.835
LX	1.153	9.557	4.345	1.233	1.009	0.892	1.144
NL	2.743	1.904	1.893	1.474	1.130	0.000	0.990
PO	0.697	1.306	0.887	0.502	0.729	0.037	0.303
UK	10.638	1.191	1.355	0.901	1.098	1.193	0.859

Source: E3ME Database.

Table A2a.   Descriptive sectoral statistics (relative to EU total): industry 9, 1980

	RQ	IPR	EIR	WR	ULC	KI	RDI
BE	3.361	2.214	1.999	1.485	1.235	0.853	1.670
DK	1.551	1.377	1.111	1.212	1.215	1.369	1.802
DW	26.564	0.892	0.833	1.253	1.066	0.941	1.381
EL	NA	NA	NA	NA	NA	NA	NA
ES	8.857	0.265	0.461	0.775	0.954	0.157	0.176
FR	16.507	0.800	0.716	1.250	1.022	1.123	1.480
IR	1.533	0.956	1.239	1.099	1.220	0.760	1.423
IT	21.158	0.489	0.768	0.869	0.805	1.625	0.200
LX	0.159	3.427	2.785	1.041	0.960	2.335	1.654
NL	2.917	2.609	1.854	1.226	1.071	0.000	0.412
PO	1.167	0.720	0.737	0.271	0.816	1.693	0.202
UK	16.226	1.755	1.766	0.810	1.056	0.814	1.307

Source: E3ME Database.

Table A2b.   Descriptive sectoral statistics (relative to EU total): industry 9, 1990

	RQ	IPR	EIR	WR	ULC	KI	RDI
BE	3.249	3.069	2.699	1.334	1.061	1.337	1.556
DK	1.327	1.622	1.418	1.050	0.884	0.924	1.917
DW	24.322	0.997	1.073	1.199	1.098	1.055	1.668
EL	NA	NA	NA	NA	NA	NA	NA
ES	8.255	0.871	0.814	0.664	0.823	0.083	0.199
FR	15.521	1.089	0.993	1.279	0.966	1.162	1.627
IR	1.139	1.351	1.925	1.238	1.344	0.659	0.839
IT	25.435	0.485	0.785	0.974	0.968	1.247	0.279
LX	0.317	3.571	3.551	1.033	0.725	0.310	1.575
NL	3.348	2.308	1.767	1.209	0.880	0.000	0.369
PO	1.503	1.041	1.729	0.366	0.844	0.979	0.228
UK	15.584	0.842	0.598	0.875	1.094	1.215	0.959

Source: E3ME Database.

Table A3a.   Descriptive sectoral statistics (relative to EU total): industry 10, 1980

	RQ	IPR	EIR	WR	ULC	KI	RDI
BE	3.759	3.175	2.676	1.220	1.019	1.143	1.064
DK	1.185	1.877	1.357	1.050	0.997	0.000	0.667
DW	32.823	0.804	0.954	1.272	1.108	0.921	1.342
EL	NA	NA	NA	NA	NA	NA	NA
ES	5.671	0.483	0.533	0.773	0.715	0.808	0.345
FR	17.647	0.928	0.931	1.243	1.045	0.967	0.866
IR	0.950	1.661	1.455	0.941	0.689	1.601	1.094
IT	13.327	0.895	0.535	0.819	0.927	1.546	0.557
LX	0.038	3.811	5.557	1.209	0.815	1.060	0.844
NL	6.045	2.210	2.067	1.199	1.099	0.000	1.349
PO	0.951	1.263	0.493	0.317	0.795	3.580	0.520
UK	17.605	0.760	0.902	0.685	0.922	1.069	1.052

Source: E3ME Database.

Table A3b.   Descriptive sectoral statistics (relative to EU total): industry 10, 1990

	RQ	IPR	EIR	WR	ULC	KI	RDI
BE	3.604	2.641	2.650	1.185	1.085	2.744	0.868
DK	1.157	1.750	1.642	0.819	0.725	0.000	1.123
DW	27.630	0.946	1.128	1.140	1.213	0.977	1.183
EL	NA	NA	NA	NA	NA	NA	NA
ES	6.091	0.862	0.494	0.791	0.800	0.045	0.369
FR	17.247	1.052	1.083	1.134	0.923	1.183	1.092
IR	0.853	3.491	2.930	0.738	0.458	1.170	0.828
IT	19.945	0.576	0.305	1.106	0.975	1.134	0.570
LX	0.069	4.877	7.366	0.981	0.576	2.679	0.985
NL	5.152	2.508	2.470	0.626	0.816	0.000	0.975
PO	1.119	1.136	0.703	0.379	0.763	0.133	0.881
UK	17.132	0.753	0.766	0.811	0.930	1.101	1.415

Source: E3ME Database.

Table A4a.   Descriptive sectoral statistics (relative to EU total): industry 11, 1980

	RQ	IPR	EIR	WR	ULC	KI	RDI
BE	1.858	4.432	2.646	1.069	1.154	0.901	0.798
DK	1.472	3.628	2.371	1.105	1.067	0.000	0.744
DW	35.360	0.838	1.033	1.239	1.071	0.881	1.643
EL	NA	NA	NA	NA	NA	NA	NA
ES	6.179	0.190	0.790	0.630	0.906	0.514	0.243
FR	19.571	0.702	0.622	1.224	0.988	1.000	0.777
IR	0.302	5.693	3.103	0.829	1.791	1.349	1.373
IT	17.561	0.516	1.001	0.723	0.790	1.766	0.246
LX	0.103	6.033	3.779	1.089	0.844	1.029	0.941
NL	3.852	2.535	1.391	1.248	1.121	0.000	0.823
PO	0.802	1.228	0.908	0.314	0.752	1.551	0.242
UK	12.940	1.156	1.006	0.921	1.145	0.793	1.289

Source: E3ME Database.

Table A4b.   Descriptive sectoral statistics (relative to EU total): industry 11, 1990

	RQ	IPR	EIR	WR	ULC	KI	RDI
BE	2.218	3.103	2.483	1.386	1.140	1.768	0.814
DK	1.693	2.915	2.295	0.977	0.881	0.000	0.842
DW	32.526	0.883	1.146	1.082	1.087	1.037	1.734
EL	NA	NA	NA	NA	NA	NA	NA
ES	6.026	1.295	0.832	0.627	1.003	0.031	0.407
FR	18.224	0.855	0.664	1.219	0.983	1.422	0.673
IR	0.361	4.230	3.587	0.703	1.110	1.537	0.667
IT	22.189	0.325	0.668	0.859	0.802	1.237	0.494
LX	0.141	4.819	4.333	1.067	1.016	0.516	0.752
NL	4.391	2.155	1.870	1.114	0.966	0.000	0.921
PO	0.732	2.092	1.756	0.338	0.742	0.326	0.420
UK	11.498	1.161	0.869	1.032	1.131	0.760	0.716

Source: E3ME Database.

Table A5a.   Descriptive sectoral statistics (relative to EU total): industry 12, 1980

	RQ	IPR	EIR	WR	ULC	KI	RDI
BE	2.080	3.170	2.338	1.163	1.047	1.198	0.658
DK	1.993	1.640	1.241	0.988	0.964	0.000	0.672
DW	34.778	0.730	1.125	1.282	1.106	1.009	1.922
EL	NA	NA	NA	NA	NA	NA	NA
ES	2.836	0.656	0.624	0.593	0.896	1.139	0.123
FR	15.958	0.911	0.741	1.207	0.989	1.044	0.607
IR	0.215	2.701	1.777	0.638	1.254	2.423	1.182
IT	17.676	0.791	0.853	0.878	0.736	1.372	0.108
LX	0.098	2.744	1.987	1.069	1.066	1.372	0.577
NL	1.211	3.260	2.711	1.060	1.121	0.000	0.747
PO	0.245	2.352	0.909	0.346	0.922	1.622	0.098
UK	22.910	0.820	0.907	0.746	1.027	0.804	0.777

Source: E3ME Database.

Table A5b.   Descriptive sectoral statistics (relative to EU total): industry 12, 1990

	RQ	IPR	EIR	WR	ULC	KI	RDI
BE	2.608	2.305	1.893	1.488	1.023	1.499	1.244
DK	2.106	1.518	1.271	0.888	0.846	0.000	1.213
DW	33.904	0.719	1.127	1.112	1.146	1.130	1.489
EL	NA	NA	NA	NA	NA	NA	NA
ES	2.794	1.821	0.962	0.549	0.776	0.054	0.470
FR	15.411	1.145	0.878	1.142	0.873	1.037	1.024
IR	0.231	3.334	3.100	0.523	0.810	1.098	0.489
IT	19.799	0.602	0.925	1.000	0.828	1.323	0.555
LX	0.101	3.105	2.693	1.303	0.938	1.373	0.961
NL	1.375	4.103	3.724	0.930	0.934	0.000	0.562
PO	0.215	2.592	1.630	0.307	0.850	0.963	0.515
UK	21.456	0.691	0.615	0.813	1.020	0.794	0.603

Source: E3ME Database.

Table A6a.   Descriptive sectoral statistics (relative to EU total): industry 13, 1980

	RQ	IPR	EIR	WR	ULC	KI	RDI
BE	1.559	1.602	1.580	2.123	1.172	0.223	3.116
DK	1.104	1.261	1.584	1.105	1.174	0.000	0.841
EL	NA	NA	NA	NA	NA	NA	NA
DW	36.737	0.734	0.818	1.091	1.142	1.283	0.658
ES	1.868	1.524	1.386	0.686	0.993	1.141	0.798
FR	24.282	0.840	0.898	1.927	1.008	0.661	0.982
IR	3.755	1.179	1.635	0.704	0.463	1.421	2.542
IT	10.949	1.147	1.011	0.720	0.767	2.035	0.647
LX	NA	NA	NA	NA	NA	NA	NA
NL	0.515	2.135	4.803	2.567	1.048	0.000	0.747
PO	0.341	0.829	0.531	0.000	0.000	0.000	NA
UK	18.891	1.131	1.138	0.562	0.932	0.453	1.648

Source: E3ME Database.

Table A6b.   Descriptive sectoral statistics (relative to EU total): industry 13, 1990

	RQ	IPR	EIR	WR	ULC	KI	RDI
BE	1.088	1.916	3.271	1.425	1.305	0.333	2.050
DK	0.934	2.089	2.309	0.842	1.026	0.000	0.471
DW	25.602	1.015	1.094	1.008	1.390	0.976	0.944
EL	NA	NA	NA	NA	NA	NA	NA
ES	2.782	1.319	1.230	0.537	0.697	10.025	0.906
FR	19.532	0.960	0.910	1.429	0.848	0.633	0.654
IR	4.203	2.680	2.244	0.742	0.229	0.160	0.497
IT	12.603	0.710	0.629	1.088	0.928	0.801	0.886
LX	NA	NA	NA	NA	NA	NA	NA
NL	0.301	1.799	9.375	2.247	1.091	0.000	1.827
PO	0.513	0.779	0.272	0.000	0.000	0.000	NA
UK	32.441	0.765	0.763	0.640	1.030	0.569	2.147

Source: E3ME Database.

Table A7a. **Descriptive sectoral statistics (relative to EU total): industry 14, 1980**

	RQ	IPR	EIR	WR	ULC	KI	RDI
BE	2.132	2.966	2.880	1.118	1.110	0.968	0.971
DK	0.941	1.687	1.275	1.016	1.118	0.000	0.497
DW	35.204	0.802	0.955	1.183	1.043	0.916	0.984
EL	NA	NA	NA	NA	NA	NA	NA
ES	4.296	0.348	0.399	0.740	0.833	0.975	0.268
FR	19.415	0.753	0.802	1.187	0.975	1.002	1.044
IR	0.510	2.372	1.991	0.654	0.933	4.190	2.272
IT	14.571	0.893	0.696	0.881	0.910	1.865	0.347
LX	NA	NA	NA	NA	NA	NA	NA
NL	5.004	2.540	2.477	1.386	1.026	0.000	1.356
PO	0.576	1.591	1.110	0.356	0.707	1.479	0.348
UK	17.352	1.044	1.013	0.702	1.024	0.798	1.642

Source: E3ME Database.

Table A7b. **Descriptive sectoral statistics (relative to EU total): industry 14, 1990**

	RQ	IPR	EIR	WR	ULC	KI	RDI
BE	1.999	2.451	2.579	1.249	0.992	1.318	1.509
DK	0.889	1.994	1.911	0.831	1.086	0.000	0.753
DW	34.131	0.804	0.959	1.066	1.097	0.727	0.946
EL	NA	NA	NA	NA	NA	NA	NA
ES	3.992	1.400	0.601	0.632	0.905	5.044	0.452
FR	17.521	1.074	1.062	1.039	0.893	0.795	1.209
IR	1.095	1.891	2.269	0.809	0.479	0.465	0.912
IT	16.200	0.578	0.590	1.046	0.891	1.656	0.660
LX	NA	NA	NA	NA	NA	NA	NA
NL	4.447	1.844	1.861	1.022	0.863	0.000	1.252
PO	0.559	2.386	2.439	0.368	0.757	0.838	0.592
UK	19.167	0.926	0.926	0.923	1.082	0.539	1.235

Source: E3ME Database.

Table A8a.   Descriptive sectoral statistics (relative to EU total): industry 15, 1980

	RQ	IPR	EIR	WR	ULC	KI	RDI
BE	4.127	2.394	1.890	1.462	1.009	0.982	2.123
DK	0.693	1.911	1.215	1.031	1.161	0.000	0.194
DW	33.090	0.574	0.934	1.275	0.965	1.093	0.936
EL	NA	NA	NA	NA	NA	NA	NA
ES	6.078	0.175	0.551	0.705	0.843	0.801	0.525
FR	23.687	0.820	1.007	1.178	1.057	1.079	1.182
IR	0.369	2.040	1.142	0.868	1.099	0.235	1.787
IT	11.598	0.979	0.787	0.762	0.893	1.325	0.544
LX	0.012	3.647	9.828	0.733	0.816	1.330	1.115
NL	2.477	2.193	1.545	1.015	1.046	0.000	0.421
PO	0.684	1.274	0.696	0.329	0.633	0.934	0.451
UK	17.185	1.280	1.122	0.793	1.128	0.758	1.342

Source: E3ME Database.

Table A8b.   Descriptive sectoral statistics (relative to EU total): industry 15, 1990

	RQ	IPR	EIR	WR	ULC	KI	RDI
BE	4.164	2.399	2.257	1.053	1.012	0.977	1.954
DK	0.639	2.086	1.646	0.774	0.848	0.000	0.233
DW	36.285	0.548	0.815	1.185	1.085	1.053	0.963
EL	NA	NA	NA	NA	NA	NA	NA
ES	8.147	0.851	0.726	0.826	1.023	0.599	0.917
FR	21.689	1.124	1.250	0.961	0.891	1.213	1.348
IR	0.115	2.841	3.103	1.143	1.839	2.458	0.925
IT	12.556	0.905	0.675	1.024	0.953	1.109	0.780
LX	0.012	4.105	15.400	0.651	0.803	0.455	1.023
NL	2.314	2.130	1.727	0.864	1.006	0.000	0.447
PO	0.623	2.123	1.672	0.377	0.712	0.108	0.741
UK	13.456	1.208	0.959	0.840	0.978	0.919	0.871

Source: E3ME Database.

Table A9a.   Descriptive sectoral statistics (relative to EU total): industry 16, 1980

	RQ	IPR	EIR	WR	ULC	KI	RDI
BE	3.837	2.154	2.351	1.462	1.214	1.167	0.994
DK	3.125	1.264	2.764	1.272	1.403	1.522	1.438
DW	23.127	0.986	0.705	0.999	0.964	1.087	0.848
EL	NA	NA	NA	NA	NA	NA	NA
ES	9.302	0.145	0.539	0.759	0.944	0.353	0.189
FR	18.538	0.891	0.992	1.306	1.171	1.284	0.853
IR	1.475	1.536	3.382	1.052	1.379	1.709	2.154
IT	12.708	0.935	0.399	0.856	0.920	1.565	0.187
LX	0.078	4.265	2.719	0.912	0.969	1.572	0.774
NL	5.948	2.284	3.020	1.181	1.407	0.000	3.721
PO	1.668	0.394	0.549	0.324	0.738	0.974	0.209
UK	20.195	0.968	0.666	0.940	0.802	0.708	1.499

Source: E3ME Database.

Table A9b.   Descriptive sectoral statistics (relative to EU total): industry 16, 1990

	RQ	IPR	EIR	WR	ULC	KI	RDI
BE	4.219	1.936	2.579	1.302	1.230	1.356	1.620
DK	2.954	1.457	2.651	1.292	1.379	0.927	2.021
DW	18.909	1.126	0.813	0.774	0.967	0.997	0.654
EL	NA	NA	NA	NA	NA	NA	NA
ES	8.742	0.799	0.511	0.632	0.860	1.950	0.326
FR	19.907	0.921	1.079	1.342	1.211	1.118	1.399
IR	2.287	0.952	1.977	1.310	1.186	1.636	2.002
IT	14.373	0.767	0.388	1.144	0.941	1.088	0.398
LX	0.081	3.470	2.295	0.863	0.901	0.945	1.031
NL	6.469	1.912	2.785	1.193	1.300	0.000	2.381
PO	1.866	0.578	0.467	0.306	0.571	0.737	0.423
UK	20.192	0.816	0.534	1.145	0.850	0.589	1.120

Source: E3ME Database.

Table A10a.   Descriptive sectoral statistics (relative to EU total): industry 17, 1980

	RQ	IPR	EIR	WR	ULC	KI	RDI
BE	3.516	2.401	2.480	1.462	1.134	0.839	1.780
DK	1.008	2.006	1.703	1.191	1.022	0.781	1.444
DW	18.428	1.467	0.944	1.189	1.107	0.860	2.267
EL	NA	NA	NA	NA	NA	NA	NA
ES	8.387	0.059	0.896	0.853	0.902	0.081	0.025
FR	17.224	0.896	0.875	1.414	1.081	0.761	1.332
IR	0.693	2.351	2.205	0.815	1.291	0.687	2.256
IT	32.388	0.380	0.892	0.898	0.791	1.625	0.027
LX	0.036	6.079	8.869	1.373	0.592	0.556	1.144
NL	2.108	2.883	2.606	1.514	1.283	0.000	1.258
PO	2.759	0.340	1.065	0.388	0.975	1.486	0.029
UK	13.454	1.009	0.774	0.817	1.188	0.612	2.043

Source: E3ME Database.

Table A10b.   Descriptive sectoral statistics (relative to EU total): industry 17, 1990

	RQ	IPR	EIR	WR	ULC	KI	RDI
BE	4.237	1.916	2.253	1.521	1.196	1.429	2.292
DK	0.986	1.882	1.938	1.164	1.167	1.572	1.979
DW	16.624	1.473	1.354	1.164	1.147	0.977	2.618
EL	NA	NA	NA	NA	NA	NA	NA
ES	7.008	0.764	0.522	0.685	0.962	0.502	0.127
FR	14.940	1.086	1.061	1.457	1.123	1.034	1.777
IR	0.470	2.489	3.028	0.813	1.095	1.032	0.958
IT	38.244	0.310	0.617	1.006	0.837	1.247	0.161
LX	0.072	3.141	4.196	1.784	0.633	1.649	1.536
NL	1.679	2.480	3.116	1.152	1.057	0.000	2.883
PO	3.783	1.000	1.711	0.403	0.813	1.003	0.171
UK	11.958	1.081	0.795	0.963	1.215	0.564	0.973

Source: E3ME Database.

Table A11a.   Descriptive sectoral statistics (relative to EU total): industry 18, 1980

	RQ	IPR	EIR	WR	ULC	KI	RDI
BE	2.324	3.389	3.873	1.162	1.107	1.102	0.859
DK	2.243	1.504	0.941	1.247	1.163	0.575	0.694
DW	28.699	0.900	0.907	1.031	1.015	1.617	1.342
EL	NA	NA	NA	NA	NA	NA	NA
ES	5.213	0.280	1.064	0.694	0.852	0.433	0.209
FR	18.098	0.983	0.895	1.227	1.023	0.739	0.809
IR	0.605	2.564	1.993	0.898	1.117	0.490	1.391
IT	13.748	0.658	0.654	0.856	0.824	1.365	0.217
LX	0.052	3.789	2.006	0.866	0.974	0.877	0.670
NL	5.754	1.642	1.597	1.145	1.019	0.000	0.743
PO	0.864	1.682	5.479	0.308	0.654	1.286	0.206
UK	22.400	0.913	0.754	0.958	1.069	0.639	1.468

Source: E3ME Database.

Table A11b.   Descriptive sectoral statistics (relative to EU total): industry 18, 1990

	RQ	IPR	EIR	WR	ULC	KI	RDI
BE	2.627	2.807	3.403	1.316	1.039	1.484	1.171
DK	1.876	1.498	1.065	1.073	1.251	1.047	1.224
DW	28.503	0.876	1.068	0.916	1.034	1.385	1.029
EL	NA	NA	NA	NA	NA	NA	NA
ES	4.213	1.198	0.864	0.677	0.947	0.264	0.051
FR	18.272	1.045	0.973	1.222	0.978	1.220	1.326
IR	0.340	3.147	2.376	0.772	1.578	0.945	1.186
IT	14.394	0.615	0.541	1.037	0.859	1.270	0.068
LX	0.061	3.007	1.223	1.002	0.978	0.263	0.942
NL	5.597	1.588	1.644	1.106	0.983	0.000	1.053
PO	0.950	2.649	4.891	0.382	0.595	1.293	0.068
UK	23.168	0.807	0.634	1.041	1.067	0.558	1.462

Source: E3ME Database.

Table A12a.   Descriptive sectoral statistics (relative to EU total): industry 19, 1980

	RQ	IPR	EIR	WR	ULC	KI	RDI
BE	2.429	3.418	2.934	1.242	1.067	1.337	2.282
DK	1.093	2.143	1.689	1.097	1.033	0.000	0.677
DW	31.863	0.867	0.946	1.136	1.042	1.002	0.999
EL	NA	NA	NA	NA	NA	NA	NA
ES	6.592	0.119	0.451	0.777	0.875	0.766	0.526
FR	19.158	0.974	0.997	1.159	1.055	0.809	1.947
IR	0.810	3.215	2.578	1.026	1.076	0.988	0.543
IT	17.088	0.645	0.754	0.841	0.818	1.701	0.591
LX	0.637	2.020	2.696	1.483	0.904	1.141	2.489
NL	2.438	2.937	2.047	1.180	1.095	0.000	0.539
PO	0.794	1.233	0.392	0.293	0.704	1.437	0.644
UK	17.099	0.868	0.984	0.902	1.086	0.758	0.421

Source: E3ME Database.

Table A12b.   Descriptive sectoral statistics (relative to EU total): industry 19, 1990

	RQ	IPR	EIR	WR	ULC	KI	RDI
BE	3.598	2.381	2.354	1.160	0.716	0.895	1.448
DK	1.070	1.902	1.846	0.825	0.825	0.000	0.633
DW	29.883	0.930	1.077	1.003	1.085	1.048	0.907
EL	NA	NA	NA	NA	NA	NA	NA
ES	5.444	0.570	0.694	0.696	0.804	0.362	0.586
FR	17.600	1.070	0.986	1.118	1.011	1.103	2.070
IR	0.289	5.338	6.842	0.925	1.590	3.453	0.324
IT	19.975	0.469	0.670	1.051	0.919	1.504	0.863
LX	0.717	3.099	3.331	1.648	0.892	0.605	2.153
NL	3.338	2.247	1.756	0.863	0.887	0.000	0.614
PO	0.624	1.743	0.709	0.304	0.795	0.943	0.748
UK	17.461	0.863	0.699	1.019	1.089	0.815	0.410

Source: E3ME Database.

Table A13a.   Descriptive sectoral statistics (relative to EU total): industry 21, 1980

	RQ	IPR	EIR	WR	ULC	KI	RDI
BE	3.714	3.633	5.131	1.279	0.827	0.802	0.996
DK	1.924	1.595	2.210	1.275	1.159	0.992	4.377
DW	27.465	0.699	0.741	1.395	1.116	0.670	1.530
EL	NA	NA	NA	NA	NA	NA	NA
ES	6.556	0.203	0.694	0.672	0.972	1.911	0.078
FR	17.379	0.802	0.764	1.232	1.000	0.825	0.864
IR	0.542	2.227	0.914	0.971	1.171	1.097	2.134
IT	22.923	0.343	0.788	0.665	0.734	1.593	0.090
LX	0.020	3.343	2.586	0.818	0.911	2.029	0.831
NL	3.862	1.960	0.839	1.661	1.134	0.000	0.399
PO	1.473	0.876	0.863	0.269	0.802	1.257	0.093
UK	14.143	1.354	1.087	0.949	1.233	0.741	1.931

Source: E3ME Database.

Table A13b.   Descriptive sectoral statistics (relative to EU total): industry 21, 1990

	RQ	IPR	EIR	WR	ULC	KI	RDI
BE	3.920	2.775	3.719	1.252	0.809	2.514	1.084
DK	2.467	1.535	2.312	1.231	1.143	1.153	4.913
DW	21.991	0.910	1.116	1.184	1.144	0.902	1.607
EL	NA	NA	NA	NA	NA	NA	NA
ES	6.790	0.377	0.353	0.645	1.099	1.791	0.292
FR	17.606	0.889	0.839	1.287	0.986	1.236	0.837
IR	0.720	1.651	1.057	0.536	0.615	1.466	1.136
IT	27.397	0.426	0.590	0.869	0.722	0.883	0.391
LX	0.028	2.429	1.763	0.914	0.758	1.098	0.795
NL	4.037	1.937	2.030	1.548	1.191	0.000	0.645
PO	1.308	0.717	0.396	0.280	0.852	0.756	0.366
UK	13.736	1.085	0.895	0.963	1.217	0.576	1.098

Source: E3ME Database.

APPENDIX B

A description of the E3ME model

B1. The purpose and design of E3ME

B1.1. The policy analysis of long-term E3 interactions

E3ME is intended to meet a need expressed by researchers and policy-makers for a framework for analysing the implications of long-term E3 policies, especially those concerning R&D and environmental regulation and taxation. The model is also capable of addressing the short-term and medium-term economic effects as well as, more broadly, the long-term effects of such policies.

E3ME combines the features of an annual short- and medium-term sectoral model estimated by formal econometric methods with those of a long-term CGE model providing analysis of the movement of the long-term equilibrium levels for key E3 indicators in response to policy changes.

B1.2. The method: long-term equations and short-term dynamic estimation

The econometric model, in contrast with most macroeconomic models currently in operation, has a complete specification of the long-term solution in the form of an estimated equation which has long-term restrictions imposed on its parameters. Economic theory, for example the recent theories of endogenous growth, informs the specification of the long-term equations and hence properties of the model; dynamic equations which embody these long-term properties are estimated by econometric methods to allow the model to provide forecasts.

The method uses recent progress in time-series econometrics, with the specification of dynamic relationships in terms of error correction models (ECM) which allow behavioural equations to be written which converge to a long-term outcome.

B1.3. The model and the research strategy

E3ME is a detailed model of the 25 NACE-CLIO sectors expanded to 32 sectors with the disaggregation of the energy and environment industries, in which the energy–environment–economy interactions are central. The model is designed to be estimated and solved for 14 regions of Europe chosen for the project (the EUR-12 Member States with Germany divided into East and West and Italy divided into north and south), although in the first operational version, Italy is not divided, Eastern Germany is excluded and Greece is not estimated due to lack of data.

This one-model approach is distinguished from the multi-model approach, which is a feature of earlier model-based research for the EU. In principle, linked models (such as the DRI or the HERMES-MIDAS system of models) could be estimated and solved consistently for all the economies involved. However, in practice, this often proves difficult, if not impossible, and considerable resources have to go into linking. Even if the consistency problem in linkage can be solved by successive iterative solutions of the component models, as reported in the DRI study (1995), there remains a more basic problem with the multi-model approach if it attempts to combine macroeconomic models with detailed industry or energy models. The problem is that the system cannot adequately tackle the simulation of 'bottom-up' policies. Normally these systems are first solved at the macroeconomic level, then the results for the macroeconomic variables are disaggregated by an industry model. However, if the policy is directed at the detailed industry level (say, a tax on the carbon content of energy use), it is very difficult (without substantial intervention by the model operator) to ensure that the implicit results for macroeconomic variables from the industry model are consistent with the explicit results from the macro-model. As an example, it is difficult to use a macro-industry two-model system to simulate the effect of exempting selective energy-intensive industries from the carbon/energy tax.

B1.4.   Summary of the characteristics of E3ME

In summary, the characteristics of E3ME are such that the model is:

(a)   elaborated at a European rather than at a national level, with the national economies being treated as regions of Europe;

(b)   dealing with energy, the environment, population and the economy in one modelling framework;

(c)   designed from the outset to address issues of central importance for economic, energy and environmental policy at the European level;

(d)   capable of providing short- and medium-term economic and industrial forecasts for business and government;

(e)   capable of analysing long-term structural change in energy demand and supply and in the economy;

(f)   focused on the contribution of research and development, and associated technological innovation, on the dynamics of growth and change.

B2.   An outline description of E3ME

B2.1.   E3ME as a regional econometric sectoral model

Figure B1 shows how the economic module will be solved as an integrated EU regional model. Most of the economic variables shown in the chart are at a 32-industry level. The whole system is solved simultaneously for all industries and all regions. The chart shows interactions at three spatial levels: the outermost area, encompassing the others, is the rest of the world; the next level is the European Union outside the region/country in question; and finally, there are the relationships within the region/country.

The chart shows three loops or circuits of economic interdependence, which are described in some detail below. These are the export loop, the output–investment loop and the income loop. Figures B2–B4 provide a schematic description of the labour market and trade functions in the model, while Tables B3–B13 give an exact description of the E3ME labour market and trade equations.

The export loop (see also Figures B3 and B4)

The export loop runs from the EU transport and distribution network to the region's exports, then to total demand. The region's imports feed into other EU regions' exports and output and finally to these other regions' demand from the EU pool and back to the exports of the region in question.

An important part of the modelling concerns international trade. The basic assumption is that, for most commodities, there is a European 'pool' into which each region supplies part of its production and from which each region satisfies part of its demands. This might be compared to national electricity supplies and demands: each power plant supplies to the national grid and each user draws power from the grid and it is not possible or necessary to link a particular supply to a particular demand.

The demand for a region's exports of a commodity is related to three factors:

(a) domestic demand for the commodity in all the other EU regions, weighted by their economic distance from the region in question;

(b) activity in the main external EU export markets, as measured by GDP or industrial production;

(c) relative prices, including the effects of exchange rate changes.

Economic distance can be measured in two ways, either by using the actual trade matrix for 1985, or by a special variable, normalized with a weight of 1 being given to activity in the home region. For the special measure of distance, the weights for the other regions are inversely proportional to the economic distances of the other regions from the exporting region. Regional imports are related to demand and relative prices by commodity and region. In addition, measures of innovation (e.g. patenting activity or spending on R&D) have been introduced into the trade equations to pick up an important long-term dynamic effect on economic development.

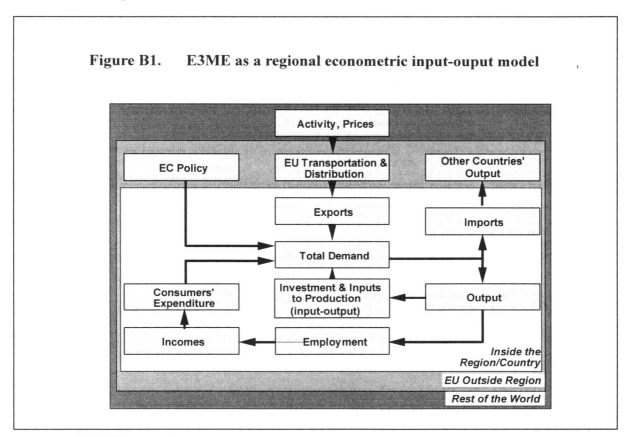

Figure B1.   E3ME as a regional econometric input-ouput model

Source: Cambridge Econometrics.

The output–investment loop

The output–investment loop includes industrial demand for goods and services and runs from total demand to output and then to investment and back to total demand. For each region, total demand for the gross output of goods and services are formed from industrial demand, consumers' expenditure, government demand, investment (fixed domestic capital formation and stockbuilding) and exports. These totals are divided between imports and output depending on relative prices, levels of activity and utilization of capacity. Industrial demand represents the inputs of goods and services from other industries required for current production, and is calculated using input–output coefficients. Eurostat national input–output tables for 1985 are taken for each of the regions. The coefficients are calculated as inputs of commodities from whatever source, including imports, per unit of gross industrial output.

Forecast changes in output are important determinants of investment in the model. Investment in new equipment and new buildings are one of the ways that companies adjust to the new challenges introduced by energy and environmental policies, so the quality of the data and the way it is modelled are of great importance to the performance of the whole model. Regional investment by investing industry is determined in the model as intertemporal choices depending on capacity output and investment prices. When investment by user industry is determined, it is converted, using coefficients derived from input–output tables, into demands on the industries producing the investment goods and services, mainly engineering and construction. These demands then constitute one of the components of total demand.

Gross fixed investment, enhanced by R&D expenditure in constant prices, is accumulated to provide a measure of the technological capital stock. There are problems with the usual definition of the capital stock, partly because there are no satisfactory data on economic scrapping. The accumulation measure is designed to get round the worst of these problems. Investment is central to the determination of long-term growth and the model embodies a theory of endogenous growth which underlies the long-term behaviour of the trade and employment equations.

The income loop (see Figure B2)

In the income loop, industrial output generates employment and incomes, which leads to further consumer expenditure, adding to total demand. Changes in output are used to determine changes in employment, along with changes in real wage costs, interest rates and energy costs. With wage rates explained by price levels and conditions in the labour market, the wage and salary payments by industry can be calculated from the industrial employment levels. These are some of the largest payments to the personal sector, but not the only ones. There are also payments of interest and dividends, transfers from government in the form of state pensions, unemployment benefits and other social security benefits. The model contains provision for the modelling of receipts and payments for seven institutional sectors, including the personal sector, government, and the company sector. Payments made by the personal sector include mortgage interest payments and personal income taxes. Personal disposable income is calculated from these accounts, and deflated by the consumer price index to give real personal disposable income.

Totals of consumer spending by region are derived from consumption functions estimated from time-series data (this treatment is similar to that of the HERMES model). These equations relate consumption to regional personal disposable income, a measure of wealth for the personal sector, inflation and interest rates. In the subsequent allocation of this spending by commodity, the approach makes the most of the disaggregated data on consumer expenditure by region available from Eurostat. Again, sets of equations have been estimated from time-series data relating the spending per capita to the national spending using the Dynamic Linear Expenditure System or the Almost Ideal Demand System. In the first version of the model, E3ME11, a standardized co-integrating equation including demographic effects has been included. The substitution between categories as a result of changes in relative prices is achieved at the national level.

B3. Accounting in E3ME

B3.1. The system of accounts

The accounting structure on which E3ME11 is based is that of the Eurostat System of Accounts 1980 (ESA). One of the characteristics of the ESA and E3ME11 is the disaggregation of economic variables.

The main classifications, following the order of accounts in ESA (Eurostat, 1980), are shown in Table B1.

B3.2. Industries and commodities

The industry and commodity classifications are in terms of industries, or their principal products and are defined on the 1980 NACE as shown in Table B2.

Table B1.   Classifications in E3ME

Key

RZ	Regions	C	Consumer Expenditure
A	World Areas	H	Institutional Sectors
Y	Industries	R	Receipts and Payments
K	Investment Sectors		
V	Investment Assets		

Member RZ	Member A
1 Belgium	1 USA
2 Denmark	2 Japan
3 Germany (East)	3 Austria
4 Germany (West)	4 Switzerland
5 Greece	5 Sweden
6 Spain	6 Norway
7 France	7 Finland
8 Ireland	8 Rest of the OECD
9 Italy (north)	9 Former USSR
10 Italy (south)	10 Eastern Europe
11 Luxembourg	11 China
12 Netherlands	12 OPEC
13 Portugal	13 NICs
14 United Kingdom	14 Rest of the world

Member Y	Member K	Member V
1 Agriculture, etc.	1 Agriculture, etc.	1 Agricultural Pr., etc.
2 Coal & Coke	2 Coal & Coke	2 Machinery
3 Oil & Gas Extraction	3 Oil & Gas Extraction	3 Transport Equipment
4 Gas Distribution	4 Gas Distribution	4 Dwellings
5 Refined Oil	5 Refined Oil	5 Non-resid. Buildings
6 Electricity, etc.	6 Electricity, etc.	6 Civil Engineer. Works
7 Water Supply	7 Water Supply	7 Other Products
8 Ferrous & Non-f. Metals	8 Ferrous & Non-f. Metals	8 Unallocated
9 Non-metallic Min. Pr.	9 Non-metallic Min. Pr.	
10 Chemicals	10 Chemicals	
11 Metal Products	11 Metal Products	
12 Agri. & Indust. Mach.	12 Agri. & Indust. Mach.	
13 Office Machines	13 Office Machines	
14 Electrical Goods	14 Electrical Goods	
15 Transport Equipment	15 Transport Equipment	
16 Food, Drink & Tobacco	16 Food, Drink & Tobacco	
17 Tex., Cloth. & Footw.	17 Tex., Cloth. & Footw.	
18 Paper & Printing Pr.	18 Paper & Printing Pr.	
19 Rubber & Plastic Pr.	19 Rubber & Plastic Pr.	
20 Recycling/Ems Abatement	20 Recycling/Ems Abatement	
21 Other Manufactures	21 Other Manufactures	
22 Construction	22 Construction	
23 Distribution, etc.	23 Distribution, etc.	
24 Lodging & Catering	24 Lodging & Catering	
25 Inland Transport	25 Inland Transport	
26 Sea & Air Transport	26 Sea & Air Transport	
27 Other Transport	27 Other Transport	
28 Communications	28 Communications	
29 Bank., Finance & Ins.	29 Bank., Finance & Ins.	
30 Other Market Services	30 Other Market Services	
31 Non-market Services	31 Non-market Services	
32 Unallocated	32 Dwellings	
	33 Unallocated	

Table B1.   Classifications in E3ME (continued)

Member C		Member H		Member R	
1	Food	1	Ind. & Comm. Comp.	1	GVA
2	Drink	2	Financial Enterprises	2	Prod. subsidies
3	Tobacco	3	Central Government	3	Prod. taxes
4	Clothing & Footw.	4	Local Government	4	VAT
5	Gross rent and water	5	Social Security Funds	5	Oth. prod. taxes
6	Electricity	6	Households + Non-p. Ins.	6	Import taxes
7	Gas	7	Rest of the World	7	Comp. of employees
8	Liquid fuels	8	Wages and salaries		
9	Other fuels	9	Gross oper. surplus		
10	Furniture, etc.	10	Prop. & enter. inc.		
11	Household text., etc.	11	Actual interest		
12	Major appliances	12	Imputed interest		
13	Hardware	13	Inc. land & assets		
14	Household operation	14	Dividends		
15	Domestic services	15	Withdrawals		
16	Medical care, etc.	16	Profits (employees)		
17	Cars, etc.	17	Accident ins. tran.		
18	Petrol, etc.	18	Current transfers		
19	Rail transport	19	Income & wealth tax		
20	Buses and coaches	20	Actual social con.		
21	Air transport	21	Employers soc. con.		
22	Other transport	22	Employees soc. con.		
23	Communication	23	Self-empl. soc. con.		
24	Equipment, etc.	24	Imputed soc. con.		
25	Entertainment, etc.	25	Social benefits		
26	Exp. in rest. and hotels	26	Curr. tran. wi. govt.		
27	Misc. goods and serv.	27	Curr. tran. to n-pfm.		
28	Unallocated	28	Curr. int. cooperat.		
29	Priv. int. tran.				
30	Misc. curr. tran.				
31	Change in pen. res.				
32	Disposable income				
33	Savings				
34	Final consumption				

Table B2.   E3ME industries defined in terms of 1980 NACE

E3ME sector	3-digit code
1 Agriculture, etc.	010
2 Coal & Coke	030+050
3 Oil & Gas Extraction	071+075
4 Gas Distribution	098
5 Refined Oil	073
6 Electricity, etc.	097+099+110
7 Water Supply	095
8 Ferrous & Non-f. Metals	130
9 Non-metallic Min. Pr.	150
10 Chemicals	170
11 Metal Products	190
12 Agri. & Indust. Mach.	210
13 Office Machines	230
14 Electrical Goods	250
15 Transport Equipment	270+290
16 Food, Drink & Tobacco	310+330+350+370+390
17 Tex., Cloth. & Footw.	410+430
18 Paper & Printing Pr.	470
19 Rubber & Plastic Pr.	490
20 Recycling/Ems Abatement	
21 Other Manufactures	450+510
22 Construction	530
23 Distribution, etc.	550+570
24 Lodging & Catering	590
25 Inland Transport	610
26 Sea & Air Transport	630
27 Other Transport	650
28 Communications	670
29 Bank., Finance & Ins.	690
30 Other Market Services	710+730+750+770+790
31 Non-market Services	810+850+890+930
32 Unallocated	

B4.   Data and data sources

B4.1.   Data types

Endogenous data

All data, exogenous or endogenous to the model, are stored on one of three E3ME databases covering time-series, cross-section or energy-environment. The list of variables available in E3ME is too large to enclose in an appendix such as this, but it is enough to say that the model can provide consistent data for most variables relevant to the study, i.e. production, value-added, employment, labour productivity, consumption, investment and external trade.

Assumptions/exogenous variables

E3ME has been developed for medium- and long-term forecasting and for the purpose of analysing economic policy effects on a regionalized EU basis. Model results for these regions (i.e. mainly the Member States of the EU) partly depend on what is assumed about the developments of economic variables outside the scope of the model, i.e. exogenous variables.

The clearest case for making assumptions in the model is for variables related to the developments outside the EU which nevertheless affect Europe's economic prospects. For example, world growth (measured in terms of industrial production) in the 14 world areas adopted for E3ME will help (via trade weighting) to create world activity variables relevant for the 14 EU regions in the model to use in export equations. Similarly, developments in world wholesale prices (and exchange rates) will affect trade by changing competitiveness levels. Inflation in the world prices of traded commodities, including food, metals and minerals, will help to explain import prices.

Most exogenous variables directly related to the EU are connected with government policy such as tax rates, interest rates and government spending, so analysis of the EU members' spending plans and views on future development are necessary. Other variables included are demographic base-line assumptions for population, and the labour force, which help drive the path of consumers' expenditure and labour force participation rates.

Assumptions related to the energy sector include oil prices, oil production, etc., while specific assumptions for the different regions' electricity supply industries must also be made. Regarding the environment, the level of a potential carbon/energy tax can be entered through this method.

B4.2. Data sources

A great deal of time and resources have gone into making sure that the data in E3ME are consistent across countries and are measured in equivalent units. The majority of data come from one of four main sources:

(a) **Eurostat.** Eurostat is always the first choice which establishes a comparable basis across Member States. Even where Eurostat data are incomplete or believed to be of poor quality, the Eurostat definitions are adopted and the data are improved via other sources. This allows the inclusion of improved Eurostat data when available.

(b) **International Sources.** When Eurostat data are not available or need to be improved, internationally available sources such as the OECD or IMF are consulted. This enables any of the team members to update the database without having to collect data from national statistics. International sources are also important for data covering the world areas outside the E3ME regions.

(c) **National Sources.** National official statistics and other data sources are used to update data series, where Eurostat is lacking national sources, i.e. to calibrate the model against the most recent past.

(d) **Quest Database.** Data from the Quest database are added as a final category in order to make the total consistent with that database.

Various sets of data were calculated by other means:

Before being put on one of the E3ME databanks (time-series, cross-sectional, etc.) data go through four stages of improvement and checking to remove any inconsistencies that may arise.

Stage 1: from raw data to databank in E3ME classifications.

The first step in Stage 1 is to extract the required raw data and store them in special format files. The second step is to reclassify the data into E3ME classifications and put them on the appropriate databank.

Stage 2: from incomplete to complete matrices.

A software program has been created to fill in gaps in any E3ME time-series matrix/vector. This is done by a variety of procedures including growth rate, sharing, extrapolation and interpolation procedures.

Stage 3: when the two above stages are not appropriate.

Updating is less transparent compared with the two previous stages. The kind of procedure is typically performed when a current price series exists, but the constant price does not. In such a case, the latter is created using a related price deflator.

Stage 4: Quest macro data added for consistency.

This consists of calculating the difference between the total of the data from the previous stages and the Quest data received from the EC. The data for industry 32 (and other unallocated classifications) is equal to the difference between the macro data and the sum of the sectoral data.

B4.3. Data coverage

At present, data for a north–south Italy split and for East Germany has been unobtainable, and thus not incorporated into the model. Data for Greece is currently not considered to be of a high enough quality to include in the model. Continued validation of the data is an on-going process.

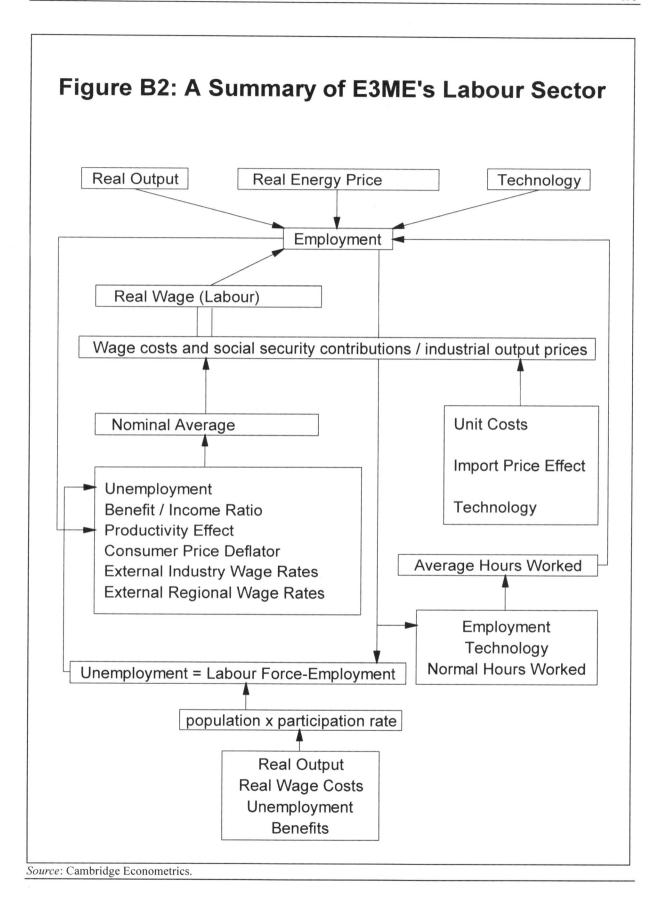

Figure B2: A Summary of E3ME's Labour Sector

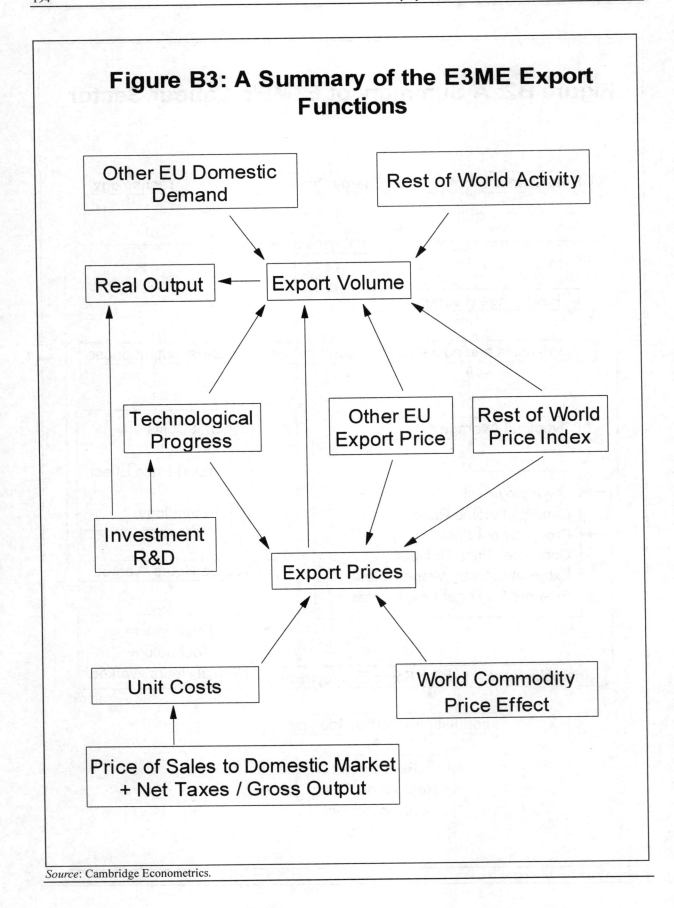

Figure B3: A Summary of the E3ME Export Functions

Source: Cambridge Econometrics.

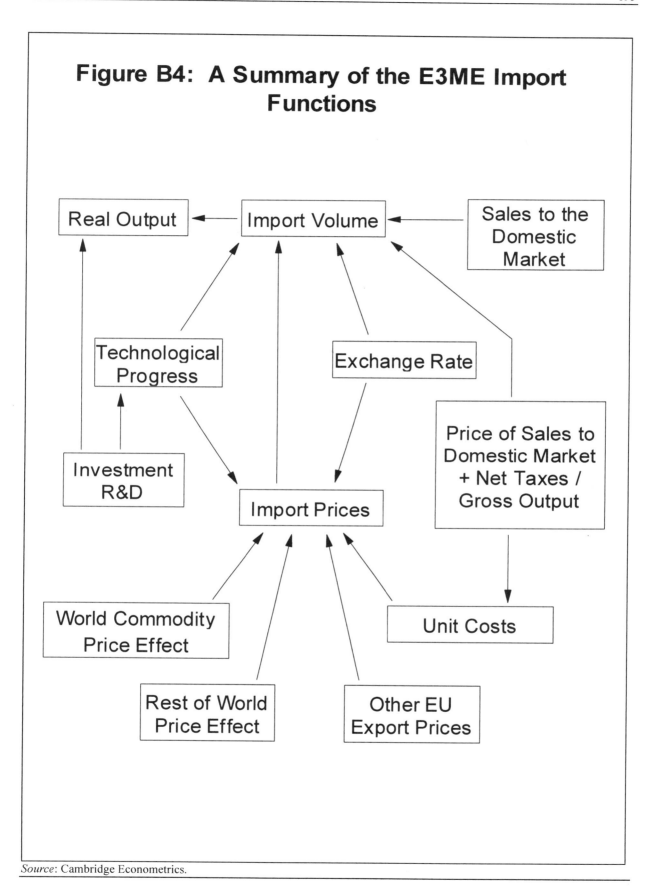

Figure B4: A Summary of the E3ME Import Functions

Real Output

Import Volume

Sales to the Domestic Market

Technological Progress

Exchange Rate

Investment R&D

Price of Sales to Domestic Market + Net Taxes / Gross Output

Import Prices

World Commodity Price Effect

Unit Costs

Rest of World Price Effect

Other EU Export Prices

Source: Cambridge Econometrics.

Table B3.   The industrial employment equations

Co-integrating long-term equation:

LN(YRE(.))   [total employment]

=	BYRE(.,9)	
+	BYRE(.,10)*LN(YR(.))	[real output]
+	BYRE(.,11)*LN(YRWC(.))	[real wage costs]
+	BYRE(.,12)*LN(YRH(.))	[hours-worked effect]
+	BYRE(.,13)*LN(PYRE(.))	[real energy price effect]
+	BYRE(.,14)*LN(YRKE(.))	[technology index]
+	ECM	[error]

Dynamic equation:

DLN(YRE(.))   [change in total employment]

=	BYRE(.,1)	
+	BYRE(.,2)*DLN(YR(.))	[real output]
+	BYRE(.,3)*DLN(YRWC(.))	[real wage costs]
+	BYRE(.,4)*DLN(YRH(.))	[hours-worked effect]
+	BYRE(.,5)*DLN(PYRE(.))	[real energy price effect]
+	BYRE(.,6)*DLN(YRKE(.))	[technology index]
+	BYRE(.,7)*DLN(YRE)(-1)	[lagged change in employment]
+	BYRE(.,8)*ECM(-1)	[lagged error correction]

Identity:

YRWC			[real wage costs]
	=	(YRLC(.)/PYR(.))/YREE(.)	
PYRE			[real energy price]
	=	PQYRE(.)/PYR(.)	
PQYRE			[nominal energy price]
PQYRE	=	(PQRD(.)*QYC(I,J)/SUM(QYC(I,J)) for I = 2,...,7	
PQRD			[price of sales to the domestic market]
PQRD	=	(VYR(.)+VYRM(.)-VYRX(.))/(YR(.)+YRM(.)-YRX(.))	

Restrictions:

BYRE(.,2),BYRE(.,10) > = 0	['right' sign]
BYRE(.,3),BYRE(.,4), BYRE(.,5), BYRE(.,11), BYRE(.,12), BYRE(.,13) < = 0	['right' sign]
0 > BYRE(.,8) > -1	['right' sign]

Definitions:

YRE	is a vector of total employment for 32 industries and 14 regions, in thousands of persons.
BYRE	is a matrix of parameters.
YR	is a vector of gross industry output for 32 industries and 14 regions, million ECU at 1985 prices.
YRH	is a vector of average hours-worked per week for 32 industries and 14 regions.
YRLC	is a vector of nominal employer wage costs (wages imputed social security contributions) for 32 industries and 14 regions, local currency prices.
YRKE	is a vector of technological progress for 32 industries and 14 regions.
PYR	is a vector of industry output prices for 32 industries and 14 regions, 1985 = 1.000, local currency.
YREE	is a vector of wage and salary earners for 14 regions, in thousands of persons.
PQYRE	is a vector of energy price for 32 industries and 14 regions, local currency.
QYC	input-output coefficient matrix (by commodity and fuel and power related).
PQRD	is a vector of price of sales to the domestic market 1985 = 1.000, local currency.
(.)	indicates that a matrix is defined across sectors.
LN	indicates natural logarithm.
DLN	indicates change in natural logarithm.
ECM	[error].

Table B4.   The industrial average earnings requirement

Co-integrating long-term equation:

LN(YRWA(.))   [gross nominal average earnings]

	=	BYRW(.,13)	
	+	BYRW(.,14)*LN(YWE(.))	[external industry wage rates]
	+	BYRW(.,15)*LN(YXE(.))	[external regional wage rates]
	+	BYRW(.,16)*(LYR(.)-LYRE(.)+LPYR(.)-LAPSC)	[productivity effect]
	+	BYRW(.,17)*LN(RUNR)	[unemployment rate effect]
	+	BYRW(.,18)*LN(RBNR)	[benefit rate effect]
	+	LAPSC-LYEC	[adjusted consumer prices]
	+	BYRW(.,20)*X8	[additional variable e.g. incomes policy dummy]
	+	ECM	[error]

Dynamic equation:

DLN(YRWA(.))   [change in gross nominal average earnings]

	=	BYRW(.,1)	
	+	BYRW(.,2)*DLN(YRWE(.))	[external industry wage rates]
	+	BYRW(.,3)*DLN(YRXE(.))	[external regional wage rates]
	+	BYRW(.,4)*D(LYR(.)-LYRE(.)+LPYR(.)-LAPSC)	[productivity effect]
	+	BYRW(.,5)*DLN(RUNR(.))	[unemployment rate effect]
	+	BYRW(.,6)*DLN(RBNR(.))	[benefit rate effect]
	+	DLAPSC(.)-DLYEC	[adjusted consumer prices]
	+	BYRW(.,8)*DX8	[additional variable e.g. incomes policy dummy]
	+	BYRW(.,9)*D(LAPSC)	[adjusted consumer prices]
	+	BYRW(.,10)*LN(YYN(.))	[nominal/actual output]
	+	BYRW(.,11)*DLN(YRWA)(-1)	[lagged change in wage rates]
	+	BYRW(.,12)*ECM(-1)	[lagged error correction]

Identity:

LAPSC	=	LN(PSCR)+LYEC+LRET	[log adjusted consumer price deflator]
LYEC	=	LN(1+(YEC/RWS))	[log employers' social security rate]
LRET		LN(RWS/(RWS-RDTX-REES))	[log reciprocal retentions ratio]
LYRWE(.)			[external industry wage rates]
	=	SUM OVER J (I NE J)(LN(YRWA(J)* YRLC(J)/SUM(YRLC(J))-LAPSC	
YRXE(.)			[external regional wage rates]
	=	LN(YRW(.))*RRDD+LN(RZEX)-LAPSC	
RBNR	=	RBEN/RWS	[the benefit rate]

Restrictions:

BYRW(.,14)+ BYRW(.,15)+ BYRW(.,16) = 1	[price homogeneity]
BYRW(.,2), BYRW(.,3), BYRW(.,4), BYRW(.,6), BYRW(.,14), BYRW(.,15),	
BYRW(.,16), BYRW(.,18) > 0	['right' sign]
BYRW(.,5), BYRW(.,17) < 0	['right' sign]
0 > BYRW (.,12) > -1	['right' sign]

Definitions:

YRWA	is a vector of nominal average earnings (contractual wage) for 32 industries and 14 regions, national currency per person-year.
BYRW	is a matrix of parameters.
YRLC	is a vector of nominal employer costs (wages and salaries plus employers' and imputed social security contributions) for 32 industries and 14 regions, local currency at current price.
RWS	is a vector of the YRWA for 14 regions.
RLC	is a vector of the YRLC for 14 regions.
LYRE	is a vector of the log of total employment for 32 industries and 14 regions, in thousands of persons.
LYR	is a vector of the log of gross industry output for 32 industries and 14 regions, national currency at 1985 prices.
PSCR	is the price deflator for total consumers' expenditure, 1985 = 1.000, local currency.
LRET	is the log of the reciprocal of the retention ratio, the proportion of earnings retained by employees after taxation and national insurance contributions.
RUNR	is the standardized unemployment rate.
RBEN	is the social benefit paid to households, million ECU at current prices.
RDTX	is the total direct tax payments made by households, million ECU at current prices.
YEC	is the total of employers' contributions to NIC, million ECU at current prices.
REES	is the total employees' contribution to NIC, million ECU at current prices.
RRDD	is the nominalized distance indicator matrix for 14 regions with zeros down the leading diagonal and rows summing to one.
(.)	indicates that a matrix is defined across sectors.
LN	indicates natural logarithm.
DLN	indicates change in natural logarithm.
ECM	[error].

Table B5.   The industrial price equations

Co-integrating long-term equation:

LN(PYRH(.))   [price of domestic sales to the domestic market]

=	BPYH(.,9)	
+	BPYH(.,10)*LN(YRUC(.))	[unit costs]
+	BPYH(.,11)*LN(PQRM(.))	[import price effect]
+	BPYH(.,12)*LN(YRKE(.))	[technology index]
+	BPYH(.,13)*LN(PQRM(3))	[oil and gas import price]
+	ECM	[error]

Dynamic equation:

DLN(PYRH(.))   [change in price of domestic sales to the domestic market]

=	BPYH(.,1)	
+	BPYH(.,2)*DLN(YRUC(.))	[unit costs]
+	BPYH(.,3)*DLN(PQRM(.))	[import price effect]
+	BPYH(.,4)*DLN(YRKE(.))	[technology index]
+	BPYH(.,5)*DLN(PQRM(.))	[oil and gas import price]
+	BPYH(.,6)*LN(YYN(.))	[normal/actual output]
+	BPYH(.,7)*DLN(PYRH)(-1)	[lagged change in prices]
+	BPYH(.,8)*ECM(-1)	[lagged error correction]

Identity:

YRUC   [unit costs]

| = | SUM(QYC(.)*PQRD(.))+(YRLC(.) + net taxes/YR(.) |

PQRH   [price of home sales by home producers]

| = | (VYR(.)+VYRM(.))/YR(.)+YRM(.)) |

PQRD   [price of home sales to the domestic market]

PQRD | = | (VYR(.)+VYRM(.)-VYRX(.))/(YR(.)+YRM(.)-YRX(.)) |

Restrictions:

BPYH(.,10)+ BPYH(.,11) = 1	[price homogeneity]
BPYH(.,2), BPYH(.,3), BPYH(.,4), BPYH(.,5), BPYH(.,10), BPYH(.,11), BPYH(.,12),	
BPYH(.,13) > = 0	['right' sign]
0 > BPYH (.,8) > -1	['right' sign]

Definitions:

BPYH	is a matrix of parameters.
PQRH	is a vector of price of home sales by home producers for 32 industries and 14 regions, 1985 = 1.000, local currency.
PQRD	is a vector of price of sales to the domestic market for 32 industries and 14 regions, 1985 = 1.000, local currency.
PQRM	is a vector of import price for 32 industries and 14 regions, 1985 = 1.000, local currency.
YR	is a vector of gross industry output for 32 industries and 14 regions, million ECU at 1985 prices.
YRKE	is a vector of technological progress for 32 industries and 14 regions.
QYC	input-output coefficient matrix (by commodity and industry).
YRLC	is a vector of nominal employer costs (wages and social security contributions) for 32 industries and 14 regions, local currency at current prices.
(.)	indicates that a matrix is defined across sectors.
LN	indicates natural logarithm.
DLN	indicates change in natural logarithm.
ECM	[error].

Table B6.   The industrial hours-worked equations

Co-integrating long-term equation:

LN(PYRH(.))			[average hours worked]
	=	BYRE(.,8)	
	+	BYRH(.,9)*LN(YRE(.))	[total employment]
	+	BYRH(.,10)*LN(YRKE(.))	[technological progress]
	+	BYRH(.,11)*LN(YNH(.))	[normal hours worked]
	+	ECM	[error]

Dynamic equation:

DLN(PYRH(.))			[change in average hours worked]
	=	BYRH(.,1)	
	+	BYRH(.,2)*DLN(YRE(.))	[total employment]
	+	BYRH(.,3)*DLN(YRKE(.))	[technological progress]
	+	BYRH(.,4)*DLN(YNH(.))	[normal hours worked]
	+	BYRH(.,5)*YYN(.)	[actual/normal output]
	+	BYRH(.,6)*DLN(YRH)(-1)	[lagged change in average hours worked]
	+	BYRH(.,7)*ECM(-1)	[lagged error correction]

Restrictions:

BYRH(.,2), BYRH(.,3), BYRH(.,9), BYRH(.,10) < = 0	['right' sign]
BYRH(.,4), BYRH(.,11) = 1	[normal hours homogeneity]
BYRH(.,5) > = 0	['right' sign]
0 > BYRH(.,7) > -1	['right' sign]

Definitions:

YRH	is a vector of average hours worked per week for 32 industries and 14 regions.
BYRH	is a matrix of parameters.
YRE	is a vector of total employment for 32 industries and 14 regions, in thousands of persons.
YRKE	is a vector of technological progress for 32 industries and 14 regions.
YNH	is a vector of normal hours worked per week for 32 industries and 14 regions.
(.)	indicates that a matrix is defined across sectors.
LN	indicates natural logarithm.
DLN	indicates change in natural logarithm.
ECM	[error].

Table B7.   The participation rate equations

Co-integrating long-term equation:

LN(LRP/(1-LRP))   [participation rate, logistic form]

	=	BLRP(9)	
	+	BLRP(10)*LN(RSQ)	[real output]
	+	BLRP(11)*LN(RWSR)	[real retained wage rates]
	+	BLRP(12)*LN(RUNR)	[unemployment rate]
	+	BLRP(13)*LN(RBNR)	[benefit rate]
	+	BLRP(14)*LN(RSER)	[economic structure]
	+	ECM	[error]

Dynamic equation:

DLN(LRP/(1-LRP))   [participaton rate, logistic form]

	=	BLRP(1)	
	+	BLRP(2)*DLN(RSQ)	[real output]
	+	BLRP(3)*DLN(RWSR)	[real retained wage rates]
	+	BLRP(4)*DLN(RUNR)	[unemployment rate]
	+	BLRP(5)*DLN(RBNR)	[benefit rate]
	+	BLRP(6)*DLN(RSER)	[economic structure]
	+	BLRP(7)*DLN(LRP/(1-LRP))(-1)	[lagged change in participation rate]
	+	BLRP(8)*ECM(-1)	[lagged error correction]

Identity:

RWSR			[real retained wage rates]
	=	REX*(RWS-RDTX-REES)/(PSCR*REEM)	
LRP			[participation rate]
	=	LAB/POP	
RUNR			[unemployment rate]
	=	UNEM/LAB	
RBNR			[benefit rate]
	=	RBEN/RWS	
RSER			[economic structure]
	=	RSERV/RMANU	

Restrictions:

BLRP(2), BLRP(3), BLRP(10), BLRP(11) > 0	['right' sign]
BLRP(4), BLRP(5), BLRP(12), BLRP(13) < 0	['right' sign]
0 > BLRP (8) > -1	['right' sign]

Definitions:

LRP	is a vector of labour force participation for 14 regions, as a proportion.
BLRP	is a matrix of parameters.
LAB	is a matrix of labour force for 14 regions, in thousands of persons.
POP	is a matrix of population of working age for 14 regions, in thousands of persons.
RSQ	is a vector of total gross industry output for14 regions, million ECU at 1985 prices.
RWS	is a vector of total nominal wages and salaries (wages and salaries excluding employers' and imputed social security contributions) for 14 regions, ECU at current prices.
UNEM	is the standardized unemployment count for 14 regions in thousands of persons.
RUNR	is the standardized unemployment rate for 14 regions.
PSCR	is a vector of total consumer price deflator for 14 regions, 1985 = 1.000, local currency.
RWS	is a vector of the total nominal average earnings for 14 regions, million ECU.
REEM	is a vector of total wage and salary earners for14 regions, in thousands of persons.
RBEN	is the social benefit paid to households, national currency at current prices.
RSERV	is total gross output of service industries for 14 regions, million ECU at 1985 prices (industries 23 to 31).
NSERV	is the total gross output of non-service industries for 14 regions, million ECU at 1985 prices (industries 1 to 22).
RSER	sectoral concentration variable for 14 regions to represent increased female participation rates.
RDTX	is the total direct tax payments made by households, million ECU at current prices.
REES	is the total of employees' contributions to NIC, million ECU at current prices.
PSCR	is the price deflator for total consumers' expenditure, 1985 = 1.000, local currency.
(.)	indicates that a matrix is defined across sectors.
LN	indicates natural logarithm.
DLN	indicates change in natural logarithm.
ECM	[error].

Table B8.   The export price equations

Co-integrating long-term equation:

LN(PQRX(.))			[export price]
	=	BPQX(.,10)	
	+	BPQX(.,11)*LN(PQRZ(.))	[other-EU export price effect]
	+	BPQX(.,12)*LN(PQRE(.))	[rest of the world price effect]
	+	BPQX(.,13)*LN(PQRF(.))	[world commodity price effect]
	+	BPQX(.,14)*LN(REX)	[exchange rate]
	+	BPQX(.,15)*LN(YRUC(.))	[unit costs]
	+	BPQX(.,16)*LN(YRKE(.))	[technological progress]
	+	ECM	[error]

Dynamic equation:

DLN(PQRX(.))			[change in export price index]
	=	BPQX(.,1)	
	+	BPQX(.,2)*DLN(PQRZ(.))	[other-EU export price effect]
	+	BPQX(.,3)*DLN(PQRE(.))	[rest of world price effect]
	+	BPQX(.,4)*DLN(PQRF(.))	[world commodity price effect]
	+	BPQX(.,5)*DLN(REX)	[exchange rate]
	+	BPQX(.,6)*DLN(YRUC(.))	[unit costs]
	+	BPQX(.,7)*DLN(YRKE(.))	[technological progress]
	+	BPQX(.,8)*DLN(PQRX(.))	[lagged change in export prices]
	+	BPQX(.,9)*ECM(-1)	[lagged error correction]

Identity:

PQRZ			[other-EU export price index]
	=	(VQRZ*RRDD)/(QRZ(.)*RRDD)	
PQRE			[rest of world price index]
	=	QAXC(.)*AWPI	
PQRF			[world commodity price index]
	=	QMC(.)*PCM	
YRUC			[unit costs]
	=	SUM(QYC(.)*PQRD(.))+(YRLC(.) + net taxes/YR(.)	
PQRD			[price of sales to the domestic market]
PQRD	=	(VYR(.)+VYRM(.)-VYRX(.))/(YR(.)+YRM(.)-YRX(.))	

Restrictions:

BPQX(.,11)+BPQX(.,12)+BPQX(.,13) = 1 – BPQX(.,15)	['right' sign]
BPQX(.,11)+BPQX(.,12)+BPQX(.,13) = BPQX(.,14)	[exchange rate symmetry]
BPQX(.,2),BPQX(.,3),BPQX(.,4),BPQX(.,5), BPQX(.,6), BPQX(.,7), BPQX(.,11),	
BPQX(.,12), BPQX(.,13), BPQX(.,14), BPQX(.,15) + BPQX(.,16) > = 0	['right' sign]
0 > BPQX(.,9) > -1	['right' sign]

Definitions:

BPQX	is a matrix of parameters.
PQRX	is a vector of export prices for 32 industries and 14 regions, 1985 = 1.000, local currency.
RRDD	is a normalized distance indicator matrix for 14 regions with zeros down the leading diagonal and rows summing to one.
QRZ	is a vector of EU industry exports for 32 industries and 14 regions, excluding the region being estimated, million ECU at 1985 prices.
VQRZ	is a vector of EU industry exports for 32 industries and 14 regions, excluding the region being estimated, million ECU at current prices.
QAXC	a converter matrix between 32 industries and 14 rest of world area classifications.
QMC	a converter matrix between 32 industries and six commodity classifications.
PCM	is a vector of commodity prices for six commodities, 1985 = 1.000, ECU.
YR	is a vector of gross industry output for 32 industries and 14 regions, million ECU at 1985 prices.
YRKE	is a vector of technological progress for 32 industries and 14 regions.
QYC	input-output coefficient matrix (by commodity and industry).
PQRD	is a vector of price of sales to the domestic market for 32 industries and 14 regions, 1985 = 1.000, local currency.
YRLC	is a vector of nominal employer costs (wages and salaries plus employers' and imputed social security contributions) for 32 industries and 14 regions, local currency at current prices.
REX	is a vector of exchange rates, local currency per ECU, 1985 = 1.000.
(.)	indicates that a matrix is defined across sectors.
LN	indicates natural logarithm.
DLN	indicates change in natural logarithm.
ECM	[error].

Table B9.   The intra-EU export volume equations

Co-integrating long-term equation:

LN(QRX(.))   [exports]

	=	BQIX(.,8)	
	+	BQIX(.,9)*LN(QXZI(.))	[other-EU domestic demand]
	+	BQIX(.,10)*LN(PQRX(.)/REX)	[export price effect]
	+	BQIX(.,11)*LN(PQRZ(.)/REX)	[other-EU export price]
	+	BQIX(.,12)*LN(YRKE(.))	[technological progress]
	+	ECM	[error]

Dynamic equation:

DLN(QIX(.))   [change in exports]

	=	BQIX(.,1)	
	+	BQIX(.,2)*DLN(QXZI(.))	[other-EU domestic demand]
	+	BPQX(.,3)*DLN(PQRX(.)/REX)	[export price effect]
	+	BQIX(.,4)*DLN(PQZX(.)/REX)	[other-EU export price]
	+	BQIX(.,5)*DLN(YRKE(.))	[technological progress]
	+	BQIX(.,6)*DLN(QIX)(-1)	[lagged change in exports]
	+	BQIX(.,7)*ECM(-1)	[lagged error correction]

Identity:

QXZI   [other-EU domestic demand]

	=	((VQR(.)+VQRM(.)-VQIX(.))*RRDD)/((QR(.)	
	+	QRM(.)-QIX(.))*RRDD)/PQRZ	
PQRX			[export price]
	=	(VQZX(.)*RRDD)/(QZX(.)*RRDD)/PQRE	
PQRZ			[other-EU export price]
	=	(VQZX(.)*RRDD)/QZX(.)*RRDD)	

Restrictions:

BQIX(.,10)+ BQIX(.,11) = 0	[price homogeneity]
BQIX(.,2), BQIX(.,4), BQIX(.,6), BQIX(.,9), BQIX(.,11), BQIX(.,12)	['right' sign]
BQIX(.,3), BQIX(.,10) < = 0	['right' sign]
0 > BQIX(.,7) > -1	['right' sign]

Definitions:

BQIX	is a matrix of parameters.
PQRX	is a vector of export prices for 32 industries and 14 regions, 1985 = 1.000, local currency.
REX	is a vector of exchange rates, local currency per ECU, 1985 = 1.000.
RRDD	is a normalized distance indicator matrix for 14 regions with zeros down the leading diagonal and rows summing to one.
QZX	is a vector of weighted EU industry exports for 32 industries and 14 regions, excluding the region being estimated, million ECU at 1985 prices.
VQZX	is a vector of weighted EU industry exports for 32 industries and 14 regions, excluding the region being estimated, million ECU at current prices.
AWPI	is a vector of wholesale prices for 14 regions outside the EU, 1985 = 1.000.
QAXC	a converter matrix between 32 industries and 14 rest of world regions classifications.
AD1	is a vector of industrial demand for the 14 regions outside the EU, 1985 = 1.000.
YRKE	is a vector of technological progress for 32 industries and 14 regions.
(.)	indicates that a matrix is defined across sectors.
LN	indicates natural logarithm.
DLN	indicates change in natural logarithm.
ECM	[error].

Table B10.   The extra-EU export volume equations

Co-integrating long-term equation:

LN(QEX(.))   [exports]

=	BQEX(.,8)	
+	BQEX(.,9)*LN(QAZI(.))	[rest of world activity index]
+	BQEX(.,10)*LN(PQEX(.)/REX)	[export price effect]
+	BQEX(.,11)*LN(PQRE(.)/REX)	[rest of world price index]
+	BQEX(.,12)*LN(YRKE(.))	[technological progress]
+	ECM	[error]

Dynamic equation:

DLN(QIX(.))   [change in exports]

=	BQEX(.,1)	
+	BQEX(.,2)*DLN(QAZI(.))	[rest of world activity index]
+	BQEX(.,3)*DLN(PQRX(.)/REX)	[export price effect]
+	BQEX(.,4)*DLN(PQRE(.)/REX)	[rest of world price index]
+	BQEX(.,5)*DLN(YRKE(.))	[technological progress]
+	BQEX(.,6)*DLN(QEX)(-1)	[lagged change in exports]
+	BQEX(.,7)*ECM(-1)	[lagged error correction]

Identity:

QAZI   [rest of world activity index]

=	QAXC(.)*AD1

PQRX   [export price]

=	(VQZX(.)*RRDD)/(QZX(.)*RRDD)/PQRE

PQRE   [rest of world price index]

=	QAXC(.)*AWPI / QAZI

Restrictions:

BQEX(.,10)+ BQEX(.,11) = 0	[price homogeneity]
BQEX(.,2), BQEX(.,4), BQEX(.,5), BQEX(.,9), BQEX(.,11), BQEX(.,12) > = 0	['right' sign]
BQEX(.,3),BQEX(.,10) < = 0	['right' sign]
0 > BQEX(.,7) > -1	['right' sign]

Definitions:

BQEX	is a matrix of parameters.
PQRX	is a vector of export prices for 32 industries and 14 regions, 1985 = 1.000, local currency.
PQRE	rest of world activity index.
REX	is a vector of exchange rates, local currency per ECU, 1985 = 1.000.
RRDD	is a normalized distance indicator matrix for 14 regions with zeros down the leading diagonal and rows summing to one.
QZX	is a vector of weighted EU industry exports for 32 industries and 14 regions, excluding the region being estimated, million ECU at 1985 prices.
VQZX	is a vector of weighted EU industry exports for 32 industries and 14 regions, excluding the region being estimated, million ECU at current prices.
AWPI	is a vector of wholesale prices for 14 regions outside the EU, 1985 = 1.000.
QAXC	a converter matrix for 32 industries and 14 rest of world regions classifications.
AD1	is a vector of industrial demand for the 14 regions outside the EU, 1985 = 1.000.
YRKE	is a vector of technological progress for 32 industries and 14 regions.
(.)	indicates that a matrix is defined across sectors.
LN	indicates natural logarithm.
DLN	indicates change in natural logarithm.
ECM	[error].

Table B11.   The import price equations

Co-integrating long-term equation:

LN(PQRM(.))			[import price]
	=	BPQM(.,10)	
	+	BPQM(.,11)*LN(PQRZ(.))	[other-EU export price effect]
	+	BPQM(.,12)*LN(PQRE(.))	[rest of the world price effect]
	+	BPQM(.,13)*LN(PQRF(.))	[world commodity price effect]
	+	BPQM(.,14)*LN(REX)	[exchange rate]
	+	BPQM(.,15)*LN(YRUC(.))	[unit costs]
	+	BPQM(.,16)*LN(YRKE(.))	[technological progress]
	+	ECM	[error]

Dynamic equation:

DLN(PQRX(.))			[change in import price index]
	=	BPQM(.,1)	
	+	BPQM(.,2)*DLN(PQRZ(.))	[other-EU export price effect]
	+	BPQM(.,3)*DLN(PQRE(.))	[rest of world price effect]
	+	BPQM(.,4)*DLN(PQRF(.))	[world commodity price effect]
	+	BPQM(.,5)*DLN(REX)	[exchange rate]
	+	BPQM(.,6)*DLN(YRUC(.))	[unit costs]
	+	BPQM(.,7)*DLN(YRKE(.))	[technological progress]
	+	BPQM(.,8)*DLN(PQRM)(-1)	[lagged change in import prices]
	+	BPQM(.,9)*ECM(-1)	[lagged error correction]

Identity:

PQRZ			[other-EU export price index]
	=	(VQRZ*RRDD)/(QRZ(.)*RRDD)	
PQRE			[rest of world price index]
	=	QAMC(.)*AWPI	
PQRF			[world commodity price index]
	=	QMC(.)*PCM	
YRUC			[unit costs]
	=	SUM(QYC(.)*PQRD(.))+(YRLC(.) + net taxes/YR(.)	
PQRD			[price of sales to the domestic market]
PQRD	=	(VYR(.)+VYRM(.)-VYRX(.))/(YR(.)+YRM(.)-YRX(.))	

Restrictions:

BPQM(.,11)+ BPQM(.,12)+ BPQM(.,13) = 1 − BPQM(.,15)	[price homogeneity]
BPQM(.,11)+ BPQM(.,12)+ BQRM(.,13) = BPQM(.,14)	[exchange rate symmetry]
BQRM(.,2), BPQM(.,3), BPQM(.,4), BPQM(.,5), BPQM(.,6), BPQM(.,11),	
BPQM(.,12), BPQM(.,13), BPQM(.,14), BPQM(.,15) >= 0	['right' sign]
BPQM(.,7), BPQM(.,16) <= 0	['right' sign]
0 > BPQM(.,9) > -1	['right' sign]

Definitions:

BPQM	is a matrix of parameters.
PQRM	is a vector of import prices for 32 industries and 14 regions, 1985 = 1.000, local currency.
RRDD	is a normalized distance indicator matrix for 14 regions with zeros down the leading diagonal and rows summing to one.
QRZ	is a vector of EU industry exports for 32 industries and 14 regions, excluding the region being estimated, million ECU at 1985 prices.
VQRZ	is a vector of EU industry exports for 32 industries and 14 regions, excluding the region being estimated, million ECU at current prices.
QAMC	a general matrix of import weights for the EU regions for 32 industries.
AWPI	is a vector of wholesale prices for 14 regions outside the EU, 1985 = 1.000, ECU.
QMC	a converter matrix between 32 industries and six commodity classifications.
PCM	is a vector of commodity prices for six commodities, 1985 = 1.000, ECU.
YR	is a vector of gross industry output for 32 industries and 14 regions, million ECU at 1985 prices.
YRKE	is a vector of technological progress for 32 industries and 14 regions.
QYC	input-output coefficient matrix (by commodity and industry).
PQRD	is a vector of price of sales to the domestic market for 32 industries and 14 regions, 1985 = 1.000, local currency.
YRLC	is a vector of nominal employer costs (wages and salaries plus employers' and imputed social security contributions) for 32 industries and 14 regions, local currency at current prices.
REX	is a vector of exchange rates, local currency per ECU, 1985 = 1.000.
(.)	indicates that a matrix is defined across sectors.
LN	indicates natural logarithm.
DLN	indicates change in natural logarithm.
ECM	[error].

Table B12.   The intra-EU import volume equations

Co-integrating long-term equation:

LN(QIM(.))			[import volumes]
	=	BQIM(.,10)	
	+	BQIM(.,11)*LN(QRDDI(.))	[sales to domestic market]
	+	BQIM(.,12)*LN(PQRM(.))	[import price effect]
	+	BQIM(.,13)*LN(PQRD(.))	[price of sales to domestic market]
	+	BQIM(.,14)*LN(REX)	[exchange rate]
	+	BQIM(.,15)*LN(YRKE(.))	[technological progress]
	+	ECM	[error]

Dynamic equation:

DLN(YRE(.))			[change in import volumes]
	=	BQIM(.,1)	
	+	BQIM(.,2)*DLN(QRDD(.))	[sales to domestic market]
	+	BQIM(.,3)*DLN(PQRM(.))	[import price effect]
	+	BQIM(.,4)*DLN(PQRD(.))	[price of sales to domestic market]
	+	BQIM(.,5)*DLN(REX)	[exchange rate]
	+	BQIM(.,6)*DLN(YKRE(.))	[technological index]
	+	BQIM(.,7)*LN(YYN(.))	[actual/normal output]
		BQIM(.,8)*DLN(QIM)(-1)	[lagged change in import volume]
	+	BQIM(.,9)*ECM(-1)	[lagged error correction]

Identity:

QRD			[sales to the domestic market]
QRD	=	YR(.)+YRM(.)-YRX(.)	
QRD			[price of sales to the domestic market]
PQRD	=	(VYR(.)+VYRM(.)-VYRX(.))/(YR(.)+YRM(.)-YRX(.))	

Restrictions:

BQIM(.,12)+ BQIM(.,13) = 0	[price homogeneity]
BQIM(.,14) = BQIM(.,12)+ BQIM(.,13)	[price and exchange rate symmetry]
BQIM(.,2), BQIM(.,4), BQIM(.,7), BQIM(.,11) > = 0	['right' sign]
BQIM(.,3), BQIM(.,5), BQIM(.,6), BQIM(.,12), BQIM(.,15) < = 0	['right' sign]
0 > BQIM(.,9) > -1	['right' sign]

Definitions:

BQIM	is a matrix of parameters.
PQRM	is a vector of import prices for 32 industries and 14 regions, 1985 = 1.000, local currency.
QRD	is a vector of sales to the domestic market for 32 industries and 14 regions, million ECU at 1985 prices.
PQRD	is a vector of price of sales to the domestic market for 32 industries and 14 regions, 1985 = 1.000, local currency.
REX	is a vector of exchange rates, local currency per ECU, 1985 = 1.000.
YRKE	is a vector technological progress for 32 industries and 14 regions.
(.)	indicates that a matrix is defined across sectors.
LN	indicates natural logarithm.
DLN	indicates change in natural logarithm.
ECM	[error].

Table B13.   The extra-EU import volume equations

Co-integrating long-term equation:

LN(QEM(.))   [import volumes]

	=	BQEM(.,10)	
	+	BQEM(.,11)*LN(QRDDI(.))	[sales to domestic market]
	+	BQEM(.,12)*LN(PQRM(.))	[import price effect]
	+	BQEM(.,13)*LN(PQRD(.))	[price of sales to domestic market]
	+	BQEM(.,14)*LN(REX)	[exchange rate]
	+	BQEM(.,15)*LN(YRKE(.))	[technological progress]
	+	ECM	[error]

Dynamic equation:

DLN(QEM(.))   [change in import volumes]

	=	BQEM(.,1)	
	+	BQEM(.,2)*DLN(QRDD(.))	[sales to domestic market]
	+	BQEM(.,3)*DLN(PQRM(.))	[import price effect]
	+	BQEM(.,4)*DLN(PQRD(.))	[price of sales to domestic market]
	+	BQEM(.,5)*DLN(REX)	[exchange rate]
	+	BQEM(.,6)*DLN(YKRE(.))	[technological progress]
	+	BQEM(.,7)*LN(YYN(.))	[actual/normal output]
	+	BQEM(.,8)*DLN(QIM)(-1)	[lagged change in import volume]
	+	BQEM(.,9)*ECM(-1)	[lagged error correction]

Identity:

QRD   [sales to the domestic market]

QRD   =   YR(.)+YRM(.)-YRX(.)

QRD   [price of sales to the domestic market]

PQRD   =   (VYR(.)+VYRM(.)-VYRX(.))/(YR(.)+YRM(.)-YRX(.))

Restrictions:

BQEM(.,12)+ BQEM(.,13) = 0   [price homogeneity]
BQEM(.,14) = BQEM(.,12)+ BQEM(.,13)   [price and exchange rate symmetry]
BQEM(.,2), BQEM(.,4), BQEM(.,7), BQEM(.,11) > = 0   ['right' sign]
BQEM(.,3), BQEM(.,5), BQEM(.,6), BQEM(.,12), BQEM(.,15) < = 0   ['right' sign]
0 > BQEM(.,9) > -1   ['right' sign]

Definitions:

BQEM	is a matrix of parameters.
PQR M	is a vector of import prices for 32 industries and 14 regions, 1985 = 1.000, local currency.
QRD	is a vector of sales to the domestic market for 32 industries and 14 regions, million ECU at 1985 prices.
PQRD	is a vector of price of sales to the domestic market for 32 industries and 14 regions, 1985 = 1.000, local currency.
REX	is a vector of exchange rates, local currency per ECU, 1985 = 1.000.
YRKE	is a vector technological progress for 32 industries and 14 regions.
(.)	indicates that a matrix is defined across sectors.
LN	indicates natural logarithm.
DLN	indicates change in natural logarithm.
ECM	[error].

The employment-output elasticities shown in Table B14 relate to the long-run coefficient of output (YR) in the employment equations, as outlined in Table B3. Parameters for all sectors and Member States are provided, and are all different by nature of the individual (instrumental variables) estimation methodology used in the E3ME model.

Table B14.   Employment – output elasticities

		BE	BK	DW	ES	FR	IR	IN	LX	NL	PO	UK
1	Agriculture etc.	0.10	0.10	0.10	1.07	1.50	0.10	0.10	1.50	0.10	0.10	0.10
2	Coal & Coke	0.82	0.00	0.15	0.10	1.07	0.10	0.44	1.50	0.00	0.10	0.10
3	Oil & Gas Extraction	0.00	1.92	0.21	1.50	1.15	1.51	1.50	0.00	0.10	0.00	1.94
4	Gas Distribution	0.00	0.84	0.10	0.10	1.50	0.00	1.50	0.00	1.50	0.00	-0.35
5	Refined Oil	0.10	0.47	0.36	0.10	1.00	0.00	0.14	0.00	0.10	0.00	1.50
6	Electricity etc.	0.10	0.91	0.10	0.10	1.43	0.44	0.10	0.86	1.50	0.10	1.06
7	Water Supply	0.10	1.29	1.79	0.79	0.90	0.00	1.50	0.82	0.01	0.00	0.10
8	Ferrous & Non-f. Metals	0.10	0.43	1.15	-0.97	1.50	0.10	0.70	0.77	0.69	0.47	0.27
9	Non-metallic Min.Pr.	1.44	1.09	1.23	0.80	1.50	0.76	0.62	0.23	-0.03	0.00	1.50
10	Chemicals	1.19	0.49	0.10	1.56	0.44	0.10	0.96	0.94	1.10	0.10	0.10
11	Metal Products	0.90	0.52	0.36	0.94	1.50	0.10	1.23	0.10	0.24	0.39	0.81
12	Agri. & Indust. Mach.	1.50	0.40	0.53	1.50	1.04	0.10	0.13	1.06	0.23	0.10	0.83
13	Office Machines	0.10	1.22	0.10	0.10	0.10	0.73	0.10	0.00	1.09	0.00	0.11
14	Electrical Goods	1.50	0.75	0.91	1.50	1.23	0.49	1.50	0.00	0.10	0.44	0.37
15	Transport Equipment	0.10	0.75	0.54	0.43	0.10	0.65	1.50	0.99	-0.01	0.18	0.41
16	Food, Drink & Tobacco	0.10	0.10	1.64	0.10	0.14	-0.22	1.29	0.10	0.10	0.10	0.10
17	Tex., Cloth. & Footw.	0.93	0.66	0.10	1.50	1.50	1.02	0.10	0.20	1.62	-0.02	0.10
18	Paper & Printing Pr.	0.90	0.10	0.10	0.10	0.10	0.48	1.50	0.98	0.62	0.10	0.43
19	Rubber & Plastic Pr.	0.10	0.48	1.16	0.10	1.10	0.10	0.10	0.10	0.44	-0.10	0.45
21	Other Manufactures	1.28	0.94	1.50	1.50	1.50	0.10	0.10	0.91	1.15	0.10	0.44
22	Construction	1.26	1.09	1.23	1.50	0.10	0.10	0.10	0.65	0.74	1.13	0.55
23	Distribution etc.	0.19	0.10	-0.24	1.50	0.50	0.10	0.95	0.19	0.10	0.70	0.10
24	Lodging & Catering	0.66	1.34	0.47	0.10	0.10	0.10	0.10	0.53	0.50	0.35	0.13
25	Inland Transport	0.50	0.57	0.14	0.31	0.38	1.50	0.60	0.15	0.92	0.10	0.10
26	Sea & Air Transport	0.10	0.24	0.10	0.17	1.11	0.10	0.10	0.00	-0.13	0.34	0.34
27	Other Transport	0.61	0.40	0.54	1.50	0.64	0.10	0.10	0.00	0.21	1.05	1.50
28	Communications	0.10	0.27	0.10	0.10	1.21	-0.09	0.44	0.75	0.10	0.10	0.42
29	Bank, Finance & Ins.	0.61	1.06	-0.28	1.50	-0.15	0.40	1.50	0.35	0.46	1.07	0.80
30	Other Market Serv.	0.91	0.77	0.60	-0.49	0.89	1.88	1.50	0.10	0.66	0.75	0.83
31	Non-market Services	0.10	0.10	0.83	0.12	1.30	1.50	1.09	1.06	0.29	0.59	0.64
32	Unallocated	0.10	0.31	0.23	0.10	0.22	0.00	0.10	0.20	0.20	0.10	0.10

Source: Cambridge Econometrics E3ME 1.1.

The employment-real wage cost elasticities shown in Table B15 relate to the long-run coefficient of real wage costs (YRWC) in the employment equations, as outlined in Table B3. Parameters for all sectors and Member States are provided, and are all different by nature of the individual (instrumental variables) estimation methodology used in the E3ME model.

Table B15.   Employment – real wage cost elasticities

		BE	BK	DW	ES	FR	IR	IN	LX	NL	PO	UK
1	Agriculture etc.	-0.71	-0.64	-1.62	-0.82	-1.50	-0.56	-0.59	-1.21	-0.48	-0.70	-0.11
2	Coal & Coke	-0.20	0.00	-0.40	-0.49	-0.50	-1.50	-0.23	-0.34	0.00	-0.40	-0.40
3	Oil & Gas Extraction	0.00	-1.50	-0.40	-0.40	-0.18	-0.77	-1.50	0.00	-0.08	0.00	-1.50
4	Gas Distribution	0.00	-0.09	0.00	-0.40	-1.50	0.00	-0.61	0.00	-0.55	0.00	-0.07
5	Refined Oil	-0.51	-0.39	-1.50	0.18	-1.50	0.00	-0.10	0.00	-0.40	0.00	-0.40
6	Electricity etc.	-0.40	-0.65	-1.85	-0.40	-1.50	-0.41	-0.45	-0.23	-0.36	-0.13	-0.67
7	Water Supply	-1.50	-0.77	-0.40	-1.50	-0.40	0.00	-0.16	-0.67	-1.50	0.00	-0.40
8	Ferrous & Non-f. Metals	-0.74	-0.92	-0.51	-1.50	-1.50	-0.43	-0.91	-1.50	-0.43	-0.01	-0.88
9	Non-metallic Min.Pr.	-1.43	-1.50	-1.14	0.50	-1.50	-0.72	-0.15	-0.40	-0.06	-0.30	-1.00
10	Chemicals	-1.50	-0.40	-0.04	-0.40	-0.52	-0.22	-0.90	-1.50	-1.50	-1.50	-0.66
11	Metal Products	-0.79	-0.78	-1.50	-0.40	-0.82	-1.50	-0.79	0.45	-0.40	-0.31	-0.24
12	Agri. & Indust. Mach.	-1.50	0.03	-0.57	-1.50	-1.13	-0.40	-0.45	-0.74	-1.50	-1.50	-0.42
13	Office Machines	-0.40	-1.82	-0.29	-1.50	-0.40	-0.40	-0.40	0.00	-0.34	0.00	-0.82
14	Electrical Goods	-1.50	-0.87	-0.85	-0.33	-0.68	-0.39	-1.77	0.00	-1.50	-0.47	-0.51
15	Transport Equipment	-0.09	-0.72	-0.40	0.16	-0.09	-0.41	-1.50	-0.12	-1.50	-0.40	-0.94
16	Food, Drink & Tobacco	-0.28	-0.35	-1.23	-0.09	-0.14	-0.40	-0.84	-0.18	-0.18	0.00	-0.56
17	Tex., Cloth. & Footw.	-1.50	-1.21	-1.58	0.38	-1.15	-1.21	0.00	-0.58	-1.50	0.00	-1.50
18	Paper & Printing Pr.	-0.79	-0.35	-0.40	-0.40	-0.26	-0.36	-1.50	-1.14	-1.50	-0.39	-0.75
19	Rubber & Plastic Pr.	-0.12	-0.53	-1.50	-0.56	-1.50	-1.50	-1.50	-0.40	-1.02	-0.69	-0.68
21	Other Manufactures	-0.91	-1.50	-1.50	-1.50	-0.40	-0.61	-0.52	-1.50	-0.08	-0.40	0.03
22	Construction	-1.50	-1.06	-1.41	-1.50	-0.45	-0.40	-0.26	-0.34	-1.50	-1.50	-0.01
23	Distribution etc.	-0.25	-0.76	0.05	-1.50	-0.40	0.04	-1.05	0.02	-1.50	0.00	0.16
24	Lodging & Catering	-0.40	-1.25	-0.79	-0.69	-0.40	-0.22	-0.40	-0.40	-0.52	-0.40	-0.40
25	Inland Transport	-0.31	-0.25	-0.86	-0.42	-0.22	-0.53	-0.15	-0.25	-0.89	-0.55	-0.40
26	Sea & Air Transport	-0.16	-0.14	-0.98	-0.70	-0.40	-0.40	-0.40	0.00	-0.40	-0.70	-0.26
27	Other Transport	-0.40	-0.16	-0.19	-1.50	-0.57	-0.08	0.00	0.00	-1.50	-0.40	-0.38
28	Communications	0.27	0.00	0.03	-0.37	-1.50	-0.07	-0.40	-0.40	-0.34	-0.40	-0.40
29	Bank, Finance & Ins.	-0.56	-4.73	0.20	-0.27	0.50	0.00	-0.40	-0.40	-0.30	-1.50	-0.40
30	Other Market Serv.	-0.36	-0.59	-0.36	-0.55	-0.57	-1.50	-1.50	-0.88	0.00	-0.52	0.00
31	Non-market Services	-0.40	-0.40	-0.40	-1.00	-1.50	-0.40	-1.50	-1.50	-1.50	-0.36	-0.61
32	Unallocated	-0.40	-0.17	-0.40	-1.74	-0.17	0.11	-1.50	-0.06	-0.40	-0.40	-0.22

Source: Cambridge Econometrics E3ME 1.1.

APPENDIX C

Labour occupational methodology

This appendix contains details of the methodology for obtaining results for employment by type of labour, i.e. by occupation. This is not as straightforward as it might seem, because the E3ME model only provides results in terms of total employment by sector.

A schematic diagram of the methodology is provided in Figure C1. Two sets of model results (with and without the SMP) are applied to a set of occupational structures for each of the E3ME manufacturing sectors which provide a breakdown by type of labour. A single cross-section for the year 1991 is used for this purpose, meaning that no effect of the SMP on intra-occupational structure (other than by the changes to sectoral employment) is allowed for.

The results give only a rough indication of the occupational effects. In particular, the dynamic effects of introducing a higher level of skill intensity in order to capitalize on the opportunities presented by the completion of the single market are not allowed for and there remain significant obstacles to the analysis of skill intensity in relation to value-added per person in an internationally comparative context. The results do, however, provide a platform for further analysis incorporating qualitative as well as other quantitative material which cannot be integrated into any model, including E3ME, or into an occupational sub-model attached to it.

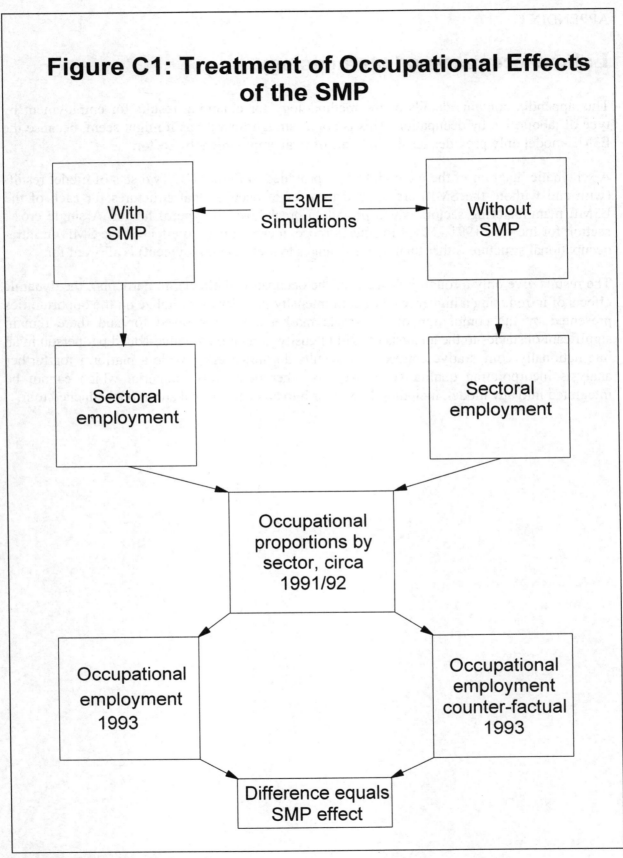

Figure C1: Treatment of Occupational Effects of the SMP

Source: Institute of Employment Research.

The occupational structures for each sector are shown in Tables C1–C13 where each row of the table gives the occupational breakdown for each country. Obviously, these embody sampling errors and classification errors, as well as substantial differences between countries. It is important to bear in mind that the occupational data are derived from the European Labour Force Surveys for each country; for manufacturing sectors the underlying samples are relatively small compared to those for the service sectors. In addition, while considerable progress has been made in the construction of a harmonized occupational classification for use in European comparisons, this is still nonetheless in its infancy. Given the lack of previous occupational data at this level in most countries, it is not possible to make an assessment of the relative importance in practice of sampling errors compared with classification errors and actual changes observed.

No detailed commentary will be provided on the tables of occupational structure. Overall, however, we can draw very broad conclusions from the changes in manufacturing occupational employment in the European Union that have taken place during the 1982–91 decade and these are given below.

Though employment in manufacturing has fallen in relative terms between 1982 and 1991, a trend which has continued to date and is expected to do so in the future, the sector continues to employ a very large number of people – about 30 million in the EUR-10 depending on the data source. Craft and Related Trades Workers (ISCO 7) was the single largest occupational group in 1991 and accounted for 25% of the work-force. Between 1982 and 1991 this occupation has become less important in relative terms although it is still dominant. Legislators, Senior Officials and Managers (ISCO 1), Professionals (ISCO 2) and Technicians and Associate Professionals (ISCO 3) have increasingly accounted for a relatively greater share of employment over the historical period and this trend has been projected to continue (Hogarth and Lindley, 1994), albeit at a more modest pace. Occupational change corresponds to the competitive pressures which the sector has faced during the 1980s and the consequent drive to improve productivity. This has resulted in a shift away from labour intensive tasks and the employment of skilled manual workers towards the greater employment of higher level skills associated with professional and technical staff. It should be borne in mind, however, that occupational change in the sector has been relatively modest during the 1980s considering the competitive pressures faced by the sector.

It will be clear from the more detailed data on occupational percentages for sectors and countries (Tables C1–C13) that there are some very striking variations in occupational structures between countries. Moreover, it is not straightforward to relate these to the value-added performance of the industry concerned. So, for example, industries with high proportions of more highly skilled workers do not necessarily coincide with industries which record the highest value-added per person ratios.

Table C1.　Occupational structure in Member States: Ferrous and Non-ferrous Metals (8), 1991

	1 Legislators, Senior Officials and Managers	2 Professionals	3 Technicians and Associate Professionals	4 Clerks	5 Service Workers etc.	6 Skilled Agricultural Workers etc.	7 Craft and Related Trades Workers	8 Plant and Machine Operators & Assemblers	9 Elementary Occupations	0 Armed Forces	Total ('000)	% of total country employment in the sector (%)
Belgium	1.1	11.7	1.4	9.6	1.7	0.0	28.5	39.1	6.9	0.0	44	4.7
Denmark	0.0	0.0	0.0	0.0	0.0	0.0	0.0	0.0	0.0	0.0	0	0.0
Germany	2.9	5.2	8.7	16.1	1.0	0.3	27.5	26.3	12.1	0.0	283	30.5
Spain	2.1	3.7	6.3	12.2	2.4	0.1	21.3	40.9	11.0	0.0	67	7.2
France	0.9	11.2	12.9	8.7	1.0	1.6	27.4	31.7	4.7	0.0	141	15.2
Italy	0.6	1.1	4.0	5.1	1.7	0.0	66.5	11.9	9.1	0.0	162	17.5
Ireland	15.5	7.0	2.5	3.0	1.3	0.0	40.0	9.4	21.4	0.0	2	0.3
Netherlands	0.0	19.3	5.8	0.0	0.0	0.0	40.5	25.5	9.0	0.0	37	4.0
Portugal	1.5	3.9	3.9	10.7	3.9	0.0	45.4	25.4	5.4	0.0	21	2.2
UK	8.5	4.9	3.9	11.4	3.6	0.0	26.1	28.8	12.6	0.0	171	18.4
EUR-10 total ('000)	27	57	64	100	16	3	321	247	92	0	929	
(%)	3.0	6.1	6.9	10.8	1.8	0.3	34.6	26.6	10.0	0.0		100.0

Source: Hogarth and Lindley (1994) with supplementary adjustments.

Table C2. Occupational structure in Member States: Non-metallic Mineral Products (9), 1991

% of total country employment in the sector

	1 Legislators, Senior Officials and Managers	2 Professionals	3 Technicians and Associate Professionals	4 Clerks	5 Service Workers etc.	6 Skilled Agricultural Workers etc.	7 Craft and Related Trades Workers	8 Plant and Machine Operators & Assemblers	9 Elementary Occupations	0 Armed Forces	Total ('000)	(%)
Belgium	3.1	7.8	2.0	12.3	0.3	0.0	40.1	25.0	9.3	0.0	35	2.3
Denmark	4.7	10.7	7.3	8.5	1.7	0.9	25.5	23.0	17.7	0.0	40	2.7
Germany	4.0	3.0	5.5	15.1	0.5	0.2	40.7	20.2	10.8	0.0	346	23.0
Spain	2.6	1.5	2.0	6.9	2.2	0.0	43.9	21.8	19.0	0.0	209	13.9
France	1.2	6.5	7.8	9.0	2.5	0.3	30.5	34.0	8.2	0.0	199	13.2
Italy	1.6	1.6	8.8	14.5	0.8	0.0	31.3	35.7	5.6	0.0	357	23.8
Ireland	5.3	3.9	4.2	8.7	0.4	0.1	35.1	32.7	9.6	0.0	18	1.2
Netherlands	0.0	0.0	0.0	0.0	0.0	0.0	54.0	34.0	12.0	0.0	45	3.0
Portugal	4.0	1.3	0.5	5.9	1.2	0.2	62.1	18.4	6.4	0.0	59	3.9
UK	10.8	3.9	3.9	11.7	5.3	0.0	24.0	28.5	11.9	0.0	194	12.9
EUR-10 total ('000)	55	48	83	172	26	2	543	416	158	0	1502	
(%)	3.6	3.2	5.5	11.5	1.7	0.1	36.2	27.7	10.5	0.0		100.0

Source: Hogarth and Lindley (1994) with supplementary adjustments.

Table C3.   **Occupational structure in Member States: Chemicals (10), 1991**

	1 Legislators, Senior Officials and Managers	2 Professionals	3 Technicians and Associate Professionals	4 Clerks	5 Service Workers etc.	6 Skilled Agricultural Workers etc.	7 Craft and Related Trades Workers	8 Plant and Machine Operators & Assemblers	9 Elementary Occupations	0 Armed Forces	Total ('000)	(%)
Belgium	4.2	21.8	5.6	21.3	1.4	0.0	12.3	25.6	7.9	0.0	66	3.5
Denmark	11.9	13.9	19.1	18.8	2.4	1.3	5.1	8.8	18.2	0.5	59	3.1
Germany	4.4	11.3	14.7	20.5	1.9	0.2	9.9	29.8	7.3	0.0	556	29.3
Spain	3.8	6.1	7.5	19.4	12.1	0.1	8.1	24.0	18.8	0.0	159	8.3
France	0.7	16.4	23.4	10.5	3.1	0.1	16.6	23.9	5.4	0.0	301	15.9
Italy	1.8	6.1	17.3	19.1	1.6	0.0	4.5	42.8	6.8	0.0	306	16.1
Ireland	10.8	8.2	13.8	15.6	0.6	0.3	14.5	26.7	9.5	0.0	13	0.7
Netherlands	8.6	20.7	6.7	12.7	0.0	0.0	27.3	17.2	6.8	0.0	95	5.0
Portugal	4.8	10.2	7.9	20.0	5.1	0.6	12.1	29.5	9.8	0.0	32	1.7
UK	11.0	10.8	8.3	14.6	8.4	0.0	16.4	22.7	7.8	0.0	314	16.5
EUR-10 Total ('000)	93	221	269	326	74	3	226	529	158	0	1900	
(%)	4.9	11.6	14.2	17.2	3.9	0.2	11.9	27.9	8.3	0.0		100.0

% of total country employment in the sector

Source: Hogarth and Lindley (1994) with supplementary adjustments.

Table C4. Occupational structure in Member States: Metal Products (11), 1991

	1 Legislators, Senior Officials and Managers	2 Professionals	3 Technicians and Associate Professionals	4 Clerks	5 Service Workers etc.	6 Skilled Agricultural Workers etc.	7 Craft and Related Trades Workers	8 Plant and Machine Operators & Assemblers	9 Elementary Occupations	0 Armed Forces	Total ('000)	% of total country employment in the sector (%)
Belgium	4.4	7.7	1.4	9.5	0.4	0.1	48.4	21.2	6.8	0.0	59	2.0
Denmark	5.3	6.2	5.9	5.6	1.0	0.0	42.1	20.2	13.7	0.0	51	1.7
Germany	3.7	2.6	6.4	14.3	0.7	0.1	41.4	21.3	9.5	0.0	1006	33.9
Spain	3.3	0.6	1.8	7.5	2.6	0.0	51.3	21.0	11.9	0.0	331	11.2
France	1.4	5.8	8.8	8.4	1.9	0.1	32.3	34.0	7.3	0.0	464	15.6
Italy	0.9	0.5	6.7	13.7	1.3	0.0	37.3	34.0	5.6	0.0	545	18.4
Ireland	6.8	2.0	3.6	7.5	0.5	0.0	70.7	2.0	6.9	0.0	17	0.6
Netherlands	7.4	7.1	2.6	11.0	0.0	0.0	38.5	24.2	9.2	0.0	120	4.0
Portugal	4.6	1.3	1.7	6.5	0.8	0.0	61.9	19.0	4.0	0.0	52	1.8
UK	10.7	3.7	3.3	11.1	4.1	0.0	35.7	22.3	9.1	0.0	321	10.8
EUR-10 total ('000)	112	87	167	344	46	1	1194	760	254	0	2966	
(%)	3.8	2.9	5.6	11.6	1.6	0.0	40.3	25.6	8.6	0.0		100.0

Source: Hogarth and Lindley (1994) with supplementary adjustments.

Table C5.   **Occupational structure in Member States: Agricultural and Industrial Machinery (12), 1991**

% of total country employment in the sector

	1 Legislators, Senior Officials and Managers	2 Professionals	3 Technicians and Associate Professionals	4 Clerks	5 Service Workers etc.	6 Skilled Agricultural Workers etc.	7 Craft and Related Trades Workers	8 Plant and Machine Operators & Assemblers	9 Elementary Occupations	0 Armed Forces	Total ('000)	(%)
Belgium	4.5	15.1	3.0	14.6	1.1	0.0	42.4	15.3	4.1	0.0	60	1.9
Denmark	6.5	10.9	9.8	8.6	1.6	0.1	34.2	18.9	9.2	0.2	88	2.8
Germany	3.6	7.5	13.4	15.1	0.7	0.1	37.1	16.0	6.5	0.0	1214	38.2
Spain	3.2	2.4	6.1	10.9	3.8	0.1	48.4	16.5	8.7	0.0	131	4.1
France	1.8	10.7	16.5	10.4	1.9	0.1	27.9	26.2	4.6	0.0	361	11.3
Italy	2.2	1.5	20.1	21.2	2.5	0.0	24.6	19.8	8.1	0.0	478	15.0
Ireland	7.8	8.2	4.9	14.7	0.3	0.0	59.1	1.8	3.2	0.0	12	0.4
Netherlands	9.4	11.9	5.2	13.1	1.4	0.0	30.6	19.3	9.0	0.0	104	3.3
Portugal	6.5	6.0	2.8	10.6	1.8	0.0	49.8	17.1	5.5	0.0	22	0.7
UK	11.7	7.4	5.9	12.4	3.9	0.0	33.2	18.5	7.0	0.0	714	22.4
EUR-10 total ('000)	169	226	385	458	64	1	1074	587	217	0	3183	
(%)	5.3	7.1	12.1	14.4	2.0	0.0	33.8	18.4	6.8	0.0		100.0

Source: Hogarth and Lindley (1994) with supplementary adjustments.

Table C6. Occupational structure in Member States: Office Machines (13), 1991

	1 Legislators, Senior Officials and Managers	2 Professionals	3 Technicians and Associate Professionals	4 Clerks	5 Service Workers etc.	6 Skilled Agricultural Workers etc.	7 Craft and Related Trades Workers	8 Plant and Machine Operators & Assemblers	9 Elementary Occupations	0 Armed Forces	Total ('000)	% of total country employment in the sector (%)
Belgium	8.6	32.3	7.1	23.6	0.5	0.0	16.3	5.9	5.7	0.0	7	1.0
Denmark	7.6	15.1	16.6	10.6	1.2	0.0	24.5	14.0	10.5	0.0	17	2.4
Germany	4.9	10.4	16.1	18.2	0.9	0.0	29.7	13.1	6.7	0.0	282	39.5
Spain	2.5	6.1	3.6	16.0	4.0	0.0	37.9	16.0	13.8	0.0	35	4.8
France	1.4	14.7	19.3	12.9	1.7	0.1	21.1	23.8	5.1	0.0	144	20.2
Italy	1.3	2.9	20.7	20.7	2.6	0.0	7.3	38.0	6.6	0.0	103	14.4
Ireland	6.2	5.4	4.6	11.1	0.5	0.2	60.3	8.2	3.6	0.0	17	2.4
Netherlands	4.0	4.6	2.2	10.8	3.0	7.2	35.8	22.5	9.9	0.0	13	1.8
Portugal	0.0	0.0	0.0	0.0	0.0	0.0	0.0	0.0	0.0	0.0	0	0.0
UK	12.6	10.6	14.9	12.3	5.0	0.0	22.4	17.3	4.9	0.0	96	13.4
EUR-10 total ('000)	33	72	114	115	15	1	177	140	47	0	714	
(%)	4.7	10.1	16.0	16.2	2.0	0.2	24.7	19.5	6.6	0.0		100.0

Source: Hogarth and Lindley (1994) with supplementary adjustments.

Table C7.   Occupational structure in Member States: Electrical Goods (14), 1991

	1 Legislators, Senior Officials and Managers	2 Professionals	3 Technicians and Associate Professionals	4 Clerks	5 Service Workers etc.	6 Skilled Agricultural Workers etc.	7 Craft and Related Trades Workers	8 Plant and Machine Operators & Assemblers	9 Elementary Occupations	0 Armed Forces	Total ('000)	(%)
Belgium	3.7	19.6	2.6	15.4	1.4	0.0	35.2	14.4	7.7	0.0	52	1.7
Denmark	7.6	17.3	11.1	10.3	1.6	0.0	26.0	16.4	9.3	0.4	41	1.3
Germany	3.8	12.3	12.9	16.5	0.9	0.0	31.0	13.5	9.1	0.0	1173	38.7
Spain	4.3	3.5	5.2	11.8	3.7	0.3	40.4	17.2	13.7	0.0	162	5.3
France	1.2	19.7	22.9	10.9	1.4	0.1	17.5	21.4	4.8	0.0	476	15.7
Italy	1.8	2.7	35.3	13.4	2.2	0.0	16.1	23.7	4.9	0.0	342	11.3
Ireland	7.1	5.1	4.4	9.6	0.8	0.0	70.1	1.3	1.6	0.0	21	0.7
Netherlands	7.7	20.4	6.7	14.8	0.0	0.0	26.8	16.9	6.8	0.0	109	3.6
Portugal	3.7	5.1	2.2	12.9	1.1	0.0	53.3	18.0	3.7	0.0	27	0.9
UK	17.0	14.1	10.4	13.7	6.3	0.0	17.6	16.2	4.7	0.0	625	20.6
EUR-10 total ('000)	185	383	469	429	72	1	764	509	215	0	3029	
(%)	6.1	12.7	15.5	14.2	2.4	0.0	25.2	16.8	7.1	0.0		100.0

% of total country employment in the sector

Source: Hogarth and Lindley (1994) with supplementary adjustments.

Table C8. Occupational structure in Member States: Transport Equipment (15), 1991

	1 Legislators, Senior Officials and Managers	2 Professionals	3 Technicians and Associate Professionals	4 Clerks	5 Service Workers etc.	6 Skilled Agricultural Workers etc.	7 Craft and Related Trades Workers	8 Plant and Machine Operators & Assemblers	9 Elementary Occupations	0 Armed Forces	Total ('000)	(%)
Belgium	2.2	6.0	1.4	9.1	1.3	0.0	54.7	18.3	6.9	0.1	76	2.7
Denmark	3.2	6.7	5.3	3.8	0.1	0.0	51.6	18.4	10.9	0.0	32	1.1
Germany	2.0	7.2	7.8	13.3	1.3	0.0	38.1	17.0	13.3	0.0	997	34.8
Spain	0.9	2.1	3.8	9.9	2.6	0.1	52.5	16.0	12.0	0.0	250	8.7
France	0.6	8.9	19.7	8.1	1.6	0.1	24.0	30.5	6.5	0.0	554	19.4
Italy	1.0	1.8	9.5	13.7	2.4	0.0	13.6	48.0	10.0	0.0	367	12.8
Ireland	5.0	5.1	3.5	5.5	0.5	0.0	70.7	5.1	4.7	0.0	6	0.2
Netherlands	0.0	12.5	4.2	10.7	0.0	0.0	38.9	24.5	9.2	0.0	75	2.6
Portugal	2.0	4.3	1.5	15.0	3.3	0.0	54.7	13.2	5.9	0.0	39	1.4
UK	7.7	9.3	5.9	10.7	3.0	0.0	33.8	21.7	7.9	0.0	466	16.3
EUR-10 total ('000)	69	194	266	325	53	1	964	699	291	0	2862	
(%)	2.4	6.8	9.3	11.4	1.9	0.0	33.7	24.4	10.2	0.0		100.0

% of total country employment in the sector

Source: Hogarth and Lindley (1994) with supplementary adjustments.

Table C9.   Occupational structure in Member States: Food, Drink and Tobacco (16), 1991

	1 Legislators, Senior Officials and Managers	2 Professionals	3 Technicians and Associate Professionals	4 Clerks	5 Service Workers etc.	6 Skilled Agricultural Workers etc.	7 Craft and Related Trades Workers	8 Plant and Machine Operators & Assemblers	9 Elementary Occupations	0 Armed Forces	Total ('000)	(%)
Belgium	4.7	5.4	3.5	12.5	2.1	0.3	12.7	48.5	10.3	0.0	104	3.2
Denmark	4.4	4.8	5.0	7.0	6.6	0.9	9.2	43.1	19.0	0.0	94	2.9
Germany	4.1	1.8	5.3	15.9	8.5	0.2	5.1	39.8	19.1	0.0	806	24.7
Spain	2.6	1.1	1.0	9.1	11.2	0.2	4.1	55.2	15.4	0.0	392	12.0
France	1.4	4.6	7.1	8.7	11.9	0.7	31.5	25.9	8.3	0.0	575	17.6
Italy	0.7	0.5	6.4	14.5	9.4	0.0	40.8	18.6	9.1	0.0	380	11.7
Ireland	8.4	3.1	6.3	13.0	3.0	0.6	45.1	10.1	10.5	0.0	43	1.3
Netherlands	8.9	6.2	4.9	9.8	3.6	0.0	33.7	21.2	11.6	0.0	180	5.5
Portugal	5.2	1.6	1.6	10.9	4.6	2.8	7.0	55.5	11.0	0.0	116	3.6
UK	9.1	3.1	3.4	11.4	11.3	0.0	17.9	32.2	11.6	0.0	571	17.5
EUR-10 total ('000)	140	89	153	390	303	11	605	1139	430	0	3260	
(%)	4.3	2.7	4.7	12.0	9.3	0.3	18.6	34.9	13.2	0.0		100.0

% of total country employment in the sector

Source: Hogarth and Lindley (1994) with supplementary adjustments.

Table C10.   Occupational structure in Member States: Textiles, Clothing and Footwear (17), 1991

	1 Legislators, Senior Officials and Managers	2 Professionals	3 Technicians and Associate Professionals	4 Clerks	5 Service Workers etc.	6 Skilled Agricultural Workers etc.	7 Craft and Related Trades Workers	8 Plant and Machine Operators & Assemblers	9 Elementary Occupations	0 Armed Forces	Total ('000)	(%)
										% of total country employment in the sector		
Belgium	3.9	2.2	1.7	7.5	0.6	0.2	56.4	23.5	4.1	0.0	92	2.6
Denmark	9.2	5.4	5.6	7.5	1.3	0.0	47.3	12.5	11.2	0.0	25	0.7
Germany	4.7	2.2	5.1	14.7	1.2	0.0	49.9	9.9	12.3	0.0	532	14.9
Spain	3.5	0.5	0.6	6.5	4.4	0.0	63.1	13.7	7.7	0.0	456	12.8
France	1.3	5.2	6.6	6.5	3.3	0.1	26.0	38.6	12.4	0.0	420	11.7
Italy	0.5	0.3	3.1	8.8	2.2	0.0	60.1	22.1	2.9	0.0	1207	33.8
Ireland	5.9	1.4	3.4	7.8	2.4	0.1	75.0	2.1	1.8	0.0	26	0.7
Netherlands	0.0	0.0	0.0	0.0	0.0	0.0	54.0	34.0	12.0	0.0	51	1.4
Portugal	4.4	0.7	0.3	5.0	0.9	0.0	69.1	15.7	3.9	0.0	299	8.4
UK	13.0	2.2	2.7	8.3	4.8	0.0	45.3	17.9	6.0	0.0	466	13.0
EUR-10 total ('000)	134	56	111	305	94	1	1916	716	240	0	3574	
(%)	3.8	1.6	3.1	8.5	2.6	0.0	53.6	20.0	6.7	0.0		100.0

Source: Hogarth and Lindley (1994) with supplementary adjustments.

Table C11.   Occupational structure in Member States: Paper and Printing Products (18), 1991

	1 Legislators, Senior Officials and Managers	2 Professionals	3 Technicians and Associate Professionals	4 Clerks	5 Service Workers etc.	6 Skilled Agricultural Workers etc.	7 Craft and Related Trades Workers	8 Plant and Machine Operators & Assemblers	9 Elementary Occupations	0 Armed Forces	Total ('000)	% of total country employment in the sector (%)
Belgium	5.9	9.9	3.3	19.3	0.4	0.0	39.4	16.3	5.5	0.0	52	2.2
Denmark	6.4	16.4	5.5	10.5	1.1	0.4	31.1	6.7	22.0	0.0	54	2.3
Germany	5.3	7.7	5.5	22.8	1.4	0.1	31.6	13.8	11.8	0.0	715	30.8
Spain	3.5	5.1	3.5	14.5	7.6	0.1	39.6	14.2	11.8	0.0	187	8.0
France	1.1	16.3	12.6	12.5	2.1	0.1	23.3	26.8	5.1	0.0	355	15.3
Italy	1.0	13.7	11.3	23.6	2.1	0.0	2.7	37.3	8.2	0.0	261	11.2
Ireland	9.5	11.8	5.2	19.6	1.0	0.2	27.0	21.8	3.9	0.0	17	0.7
Netherlands	10.1	8.7	7.9	11.5	6.7	0.0	25.8	16.3	13.1	0.0	122	5.3
Portugal	4.9	6.3	2.6	12.6	1.4	0.0	44.0	23.1	5.1	0.0	35	1.5
UK	14.7	2.9	13.5	16.2	7.9	0.0	22.6	17.7	4.6	0.0	523	22.5
EUR-10 total ('000)	150	202	207	418	89	1	597	454	203	0	2321	
(%)	6.5	8.7	8.9	18.0	3.8	0.1	25.7	19.6	8.7	0.0		100.0

Source: Hogarth and Lindley (1994) with supplementary adjustments.

Table C12. Occupational structure in Member States: Rubber and Plastic Products (19), 1991

	1 Legislators, Senior Officials and Managers	2 Professionals	3 Technicians and Associate Professionals	4 Clerks	5 Service Workers etc.	6 Skilled Agricultural Workers etc.	7 Craft and Related Trades Workers	8 Plant and Machine Operators & Assemblers	9 Elementary Occupations	0 Armed Forces	Total ('000)	Per cent
Belgium	2.9	9.2	2.1	11.0	0.2	0.0	14.8	50.1	9.7	0.0	29	2.2
Denmark	8.4	9.7	7.0	8.8	1.2	0.6	25.0	20.1	19.3	0.0	24	1.8
Germany	4.7	2.8	6.7	15.9	0.8	0.0	17.2	34.8	17.1	0.0	456	34.7
Spain	5.9	0.5	1.7	10.2	3.1	0.1	6.9	55.3	16.2	0.0	100	7.6
France	0.7	9.3	11.3	8.9	1.4	0.1	17.5	38.9	11.7	0.0	219	16.6
Italy	0.5	0.5	5.8	14.2	3.0	0.0	52.4	18.3	5.3	0.0	201	15.3
Ireland	9.2	1.7	3.5	11.4	0.8	0.0	18.5	50.2	4.8	0.0	9	0.7
Netherlands	0.0	0.0	0.7	15.2	0.0	0.0	45.0	28.3	10.8	0.0	42	3.2
Portugal	7.8	1.1	0.6	6.1	1.7	0.6	37.8	35.0	9.4	0.0	18	1.4
UK	10.8	5.3	4.6	13.1	4.7	0.0	18.1	32.4	11.0	0.0	216	16.4
EUR-10 total ('000)	58	52	81	173	27	1	306	448	168	0	1314	
(%)	4.4	3.9	6.2	13.2	2.0	0.1	23.3	34.1	12.8	0.0		100.0

% of total country employment in the sector

Source: Hogarth and Lindley (1994) with supplementary adjustments.

Table C13. Occupational structure in Member States: Other Manufactures (21), 1991

	1 Legislators, Senior Officials and Managers	2 Professionals	3 Technicians and Associate Professionals	4 Clerks	5 Service Workers etc.	6 Skilled Agricultural Workers etc.	7 Craft and Related Trades Workers	8 Plant and Machine Operators & Assemblers	9 Elementary Occupations	0 Armed Forces	Total ('000)	% of total country employment in the sector (%)
Belgium	5.9	2.8	2.1	9.8	0.5	0.2	44.3	29.7	4.7	0.1	55	2.5
Denmark	6.7	7.9	4.9	10.5	1.5	0.2	24.0	22.5	21.5	0.2	19	0.9
Germany	3.8	1.4	3.4	13.1	0.6	0.8	36.9	30.9	9.1	0.0	508	23.4
Spain	2.5	0.8	0.9	5.7	2.3	0.4	49.8	30.0	7.5	0.0	282	13.0
France	3.0	3.6	7.0	6.4	4.4	0.9	33.7	30.5	10.5	0.0	311	14.3
Italy	0.6	1.4	6.8	14.9	1.7	0.0	4.5	60.0	10.1	0.0	451	20.8
Ireland	4.1	1.8	4.2	6.5	0.7	0.0	61.8	16.3	4.6	0.0	24	1.1
Netherlands	0.6	0.0	0.4	8.6	3.6	10.8	40.2	25.3	10.6	0.0	68	3.1
Portugal	5.7	1.6	1.1	10.9	1.6	0.0	16.0	52.8	10.3	0.0	76	3.5
UK	11.0	2.3	2.9	9.5	3.8	0.0	44.1	18.8	7.6	0.0	378	17.4
EUR-10 total ('000)	90	40	87	229	49	16	703	760	197	0	2172	
(%)	4.1	1.9	4.0	10.5	2.3	0.7	32.4	35.0	9.1	0.0		100.0

Source: Hogarth and Lindley (1994) with supplementary adjustments.

APPENDIX D

Detailed model results

Appendix D is concerned with the detailed results coming from the two model simulations discussed in Chapter 4. For each simulation, employment and unit labour cost results are provided as percentage differences from base for each of the EUR-11 Member States simulated using E3ME over the period 1985–93. The Greek economy is not included in the following tables.

The detail provided allows the reader to look at a particular sector in a particular region, something not provided for in Chapter 4 which includes tables that are either averaged across regions or industries.

Table D1a.   Model simulation 1: detailed employment results – Belgium (%)

		1985	1986	1987	1988	1989	1990	1991	1992	1993
1	Agriculture etc.	0.00	0.00	0.00	0.00	0.00	0.02	0.22	0.41	0.57
2	Coal & Coke	0.00	0.00	0.00	0.00	0.00	0.00	0.00	0.00	0.00
3	Oil & Gas Extraction	0.00	0.00	0.00	0.00	0.00	0.00	0.00	0.00	0.00
4	Gas Distribution	0.00	0.00	0.00	0.00	0.00	0.00	0.00	0.00	0.00
5	Refined Oil	0.00	0.00	0.01	0.02	0.08	0.29	0.37	0.50	0.93
6	Electricity etc.	0.00	0.00	0.00	0.00	0.00	0.00	-0.02	-0.05	-0.09
7	Water Supply	0.00	0.00	0.00	0.00	0.00	0.00	0.01	0.00	0.02
8	Ferrous & Non-f. Metals	0.00	0.00	0.04	0.16	0.46	1.03	1.98	3.44	4.92
9	Non-metallic Min.Pr.	0.00	0.00	0.00	0.01	-0.19	-0.36	-0.64	-1.27	-2.37
10	Chemicals	0.00	0.00	0.00	0.00	0.00	0.01	0.02	0.05	0.10
11	Metal Products	0.00	-0.01	-0.10	-0.38	-0.80	-1.37	-2.37	-4.49	-5.35
12	Agri. & Indust. Mach.	0.00	0.00	0.00	0.00	0.00	0.00	0.00	0.00	0.00
13	Office Machines	0.00	0.00	0.00	0.00	0.00	0.00	0.00	0.00	0.00
14	Electrical Goods	0.00	0.00	0.02	0.12	0.36	0.81	-0.20	-2.50	-6.74
15	Transport Equipment	0.00	0.00	0.03	0.16	0.33	0.78	1.83	3.33	4.55
16	Food, Drink & Tobacco	0.00	0.00	0.00	0.00	0.00	0.01	0.03	0.06	0.10
17	Tex., Cloth. & Footw.	0.00	0.00	0.03	0.14	0.31	0.83	1.75	3.02	4.41
18	Paper & Printing Pr.	0.00	0.00	0.03	0.13	0.36	0.86	1.89	3.37	4.89
19	Rubber & Plastic Pr.	0.00	0.00	0.04	0.14	0.32	0.62	1.22	2.05	2.65
21	Other Manufactures	0.00	0.02	0.26	0.58	1.41	3.45	6.29	19.73	32.49
22	Construction	0.00	0.00	0.00	0.03	0.12	0.27	0.58	1.22	2.15
23	Distribution etc.	0.00	0.00	0.00	0.00	0.00	0.00	0.02	0.05	0.10
24	Lodging & Catering	0.00	0.00	0.03	0.12	0.36	0.98	2.22	4.63	6.07
25	Inland Transport	0.00	0.00	-0.01	-0.02	-0.07	-0.13	-0.24	-0.43	-0.62
26	Sea & Air Transport	0.00	0.00	-0.02	-0.06	-0.11	-0.18	-0.21	-0.04	0.87
27	Other Transport	0.00	0.00	0.01	0.03	0.09	0.15	0.31	0.48	0.64
28	Communications	0.00	0.00	0.01	0.04	0.11	0.29	0.64	1.23	1.93
29	Bank., Finance & Ins.	0.00	0.00	0.00	0.00	0.00	0.00	0.00	0.00	0.00
30	Other Market Serv.	0.00	0.00	0.00	0.03	0.14	0.37	0.78	1.56	2.93
31	Non-market Services	0.00	0.00	0.00	0.00	0.00	0.00	0.01	0.01	0.00
32	Unallocated	0.00	0.00	0.00	0.00	0.00	0.00	0.00	0.00	0.00
	Total	0.00	0.00	0.01	0.03	0.09	0.24	0.49	0.98	1.52

Source: Cambridge Econometrics Forecast C51F1Y-C51F1R/da1/F1RA.

Table D1b. Model simulation 1: detailed employment results – Denmark (%)

		1985	1986	1987	1988	1989	1990	1991	1992	1993
1	Agriculture etc.	0.00	0.00	0.01	0.02	0.06	0.16	0.36	0.98	1.43
2	Coal & Coke	0.00	0.00	0.00	0.00	0.00	0.00	0.00	0.00	0.00
3	Oil & Gas Extraction	0.00	0.00	0.00	0.00	0.00	0.00	0.00	0.00	0.00
4	Gas Distribution	0.00	0.00	0.00	0.00	0.00	0.00	0.00	0.00	0.00
5	Refined Oil	0.00	0.00	0.00	0.00	0.01	0.02	0.14	0.41	1.61
6	Electricity etc.	0.00	0.00	0.00	0.00	0.01	0.03	0.08	0.20	0.45
7	Water Supply	0.00	0.00	0.00	-0.01	-0.02	-0.04	-0.09	-0.14	-0.16
8	Ferrous & Non-f. Metals	0.00	-0.04	-0.38	-1.19	-2.78	-5.54	-	-	-
								10.28	19.70	16.17
9	Non-metallic Min.Pr.	0.00	0.00	0.01	0.04	0.10	0.22	0.29	0.37	0.36
10	Chemicals	0.00	-0.01	-0.09	-0.33	-0.85	-1.45	-2.74	-4.70	-5.49
11	Metal Products	0.00	0.01	0.08	0.38	0.99	2.27	4.62	9.94	13.80
12	Agri. & Indust. Mach.	0.00	-0.01	-0.06	-0.20	-0.45	-0.86	-1.56	-2.23	-0.62
13	Office Machines	0.00	0.00	0.02	0.06	0.18	0.38	0.82	1.90	3.30
14	Electrical Goods	0.00	0.00	0.00	-0.01	0.00	-0.09	-0.56	-1.60	-2.61
15	Transport Equipment	0.00	-0.01	-0.10	-0.34	-0.68	-1.61	-3.40	-7.92	-8.45
16	Food, Drink & Tobacco	0.00	0.00	0.00	0.01	0.04	0.07	0.12	0.15	0.17
17	Tex., Cloth. & Footw.	0.00	0.00	0.00	0.01	0.03	0.02	-0.12	-1.09	-1.31
18	Paper & Printing Pr.	0.00	0.00	0.00	0.02	0.08	0.19	0.38	0.64	0.89
19	Rubber & Plastic Pr.	0.00	-0.01	-0.11	-0.31	-0.72	-1.36	-2.44	-2.80	-2.08
21	Other Manufactures	0.00	0.00	0.04	0.14	0.36	0.69	0.95	1.86	1.96
22	Construction	0.00	0.00	0.00	-0.01	-0.03	-0.07	-0.15	-0.22	-0.24
23	Distribution etc.	0.00	0.00	0.00	-0.01	-0.02	-0.05	-0.07	-0.08	-0.03
24	Lodging & Catering	0.00	0.00	0.01	0.04	0.07	0.16	0.45	0.95	0.93
25	Inland Transport	0.00	0.00	0.00	-0.01	-0.04	-0.11	-0.24	-0.47	-0.78
26	Sea & Air Transport	0.00	0.00	0.00	-0.02	-0.04	-0.11	-0.21	-0.33	-0.37
27	Other Transport	0.00	0.00	0.00	0.01	0.05	0.14	0.30	0.66	1.24
28	Communications	0.00	0.00	0.00	0.02	0.04	0.10	0.21	0.37	0.52
29	Bank., Finance & Ins.	0.00	0.00	0.01	0.03	0.08	0.21	0.51	0.93	1.60
30	Other Market Serv.	0.00	0.00	0.00	0.01	0.02	0.04	-0.01	-0.13	-0.45
31	Non-market Services	0.00	0.00	0.00	0.01	0.03	0.08	0.15	0.27	0.40
32	Unallocated	0.00	0.00	0.00	0.00	0.00	0.00	0.00	0.00	0.00
	Total	0.00	0.00	0.00	0.00	0.01	0.04	0.06	0.14	0.29

Source: Cambridge Econometrics Forecast C51F1Y-C51F1R/da1/F1RA.

Table D1c.   Model simulation 1: detailed employment results – West Germany (%)

		1985	1986	1987	1988	1989	1990	1991	1992	1993
1	Agriculture etc.	0.00	0.00	-0.02	-0.02	0.05	0.55	0.96	1.90	1.75
2	Coal & Coke	0.00	0.00	0.00	0.00	0.00	0.00	0.00	0.00	0.00
3	Oil & Gas Extraction	0.00	0.00	0.00	0.00	0.00	0.00	0.00	0.00	0.00
4	Gas Distribution	0.00	0.00	0.00	-0.02	-0.04	-0.09	-0.12	-0.20	-0.16
5	Refined Oil	0.00	0.00	0.00	-0.01	-0.02	-0.03	0.00	-0.01	-0.10
6	Electricity etc.	0.00	0.00	0.00	0.00	0.00	0.00	0.00	0.00	0.00
7	Water Supply	0.00	0.00	0.00	0.00	0.00	0.00	0.00	0.00	0.00
8	Ferrous & Non-f. Metals	0.00	0.00	0.00	0.00	-0.01	-0.08	-0.18	-0.31	-0.38
9	Non-metallic Min.Pr.	0.00	0.00	0.00	0.00	-0.02	-0.09	-0.28	-0.55	-1.06
10	Chemicals	0.00	0.00	0.00	0.00	0.00	0.04	0.05	0.06	0.15
11	Metal Products	0.00	0.00	0.01	0.03	0.05	-0.02	-0.37	-0.98	-1.91
12	Agri. & Indust. Mach.	0.00	0.00	-0.01	-0.03	-0.08	-0.21	-0.01	-0.58	-0.95
13	Office Machines	0.00	0.00	-0.01	-0.04	-0.13	-0.23	-0.17	-0.50	-0.83
14	Electrical Goods	0.00	0.00	0.01	0.05	0.16	1.47	1.07	0.36	-0.67
15	Transport Equipment	0.00	0.00	0.00	0.00	0.00	0.00	0.00	0.00	0.00
16	Food, Drink & Tobacco	0.00	0.00	0.00	-0.01	-0.03	-0.16	0.29	-0.08	-0.16
17	Tex., Cloth. & Footw.	0.00	0.00	0.00	-0.01	-0.06	-0.22	-0.71	-1.61	-3.42
18	Paper & Printing Pr.	0.00	0.00	0.00	0.02	0.04	0.14	-0.10	0.02	0.06
19	Rubber & Plastic Pr.	0.00	0.00	0.01	0.04	0.12	0.65	1.38	2.14	2.30
21	Other Manufactures	0.00	0.00	0.00	-0.01	-0.04	-0.14	-0.28	-0.60	-1.38
22	Construction	0.00	0.00	0.00	0.00	0.01	0.03	0.08	0.06	-0.25
23	Distribution etc.	0.00	0.00	0.00	0.00	-0.01	-0.01	-0.03	-0.13	-0.57
24	Lodging & Catering	0.00	0.00	-0.01	-0.03	-0.06	-0.17	-0.49	-1.11	-1.24
25	Inland Transport	0.00	0.00	0.00	0.03	0.03	-0.49	-0.99	-2.07	-5.37
26	Sea & Air Transport	0.00	0.00	0.01	0.02	0.04	0.16	0.04	0.12	0.14
27	Other Transport	0.00	0.00	0.00	0.00	-0.01	0.01	-0.02	0.03	1.39
28	Communications	0.00	0.00	0.00	-0.01	-0.03	0.19	-0.39	-0.20	0.36
29	Bank., Finance & Ins.	0.00	0.00	0.00	0.01	0.01	-0.01	0.05	0.02	-0.50
30	Other Market Serv.	0.00	0.00	0.02	0.05	0.15	0.38	0.62	1.34	3.17
31	Non-market Services	0.00	0.00	0.00	0.00	0.00	0.00	0.00	0.00	0.00
32	Unallocated	0.00	0.00	0.00	0.00	0.00	0.00	0.00	0.00	0.00
	Total	0.00	0.00	0.00	0.01	0.02	0.11	0.11	0.11	0.10

Source: Cambridge Econometrics Forecast C51F1Y-C51F1R/da1/F1RA.

Table D1d. Model simulation 1: detailed employment results – Spain (%)

		1985	1986	1987	1988	1989	1990	1991	1992	1993
1	Agriculture etc.	0.00	0.00	0.00	0.00	0.00	0.00	0.00	0.00	0.00
2	Coal & Coke	0.00	0.00	0.00	0.00	0.00	0.00	0.00	0.00	0.00
3	Oil & Gas Extraction	0.00	0.00	0.00	0.00	0.00	0.00	0.00	0.00	0.00
4	Gas Distribution	0.00	0.00	0.01	0.04	0.07	0.15	0.27	0.38	0.32
5	Refined Oil	0.00	0.00	-0.01	-0.04	-0.10	-0.21	-0.48	-1.08	-0.92
6	Electricity etc.	0.00	0.00	0.00	0.01	0.11	0.18	0.39	0.75	0.83
7	Water Supply	0.00	0.00	0.00	0.00	0.00	0.00	0.00	0.00	0.00
8	Ferrous & Non-f. Metals	0.00	0.00	-0.02	-0.08	-0.24	-0.68	-1.26	-2.44	-4.62
9	Non-metallic Min.Pr.	0.00	0.00	0.00	0.00	0.00	-0.03	-0.07	-0.36	-0.53
10	Chemicals	0.00	0.00	0.00	0.02	0.08	0.21	0.67	1.28	1.67
11	Metal Products	0.00	0.01	0.09	0.33	0.77	1.81	3.10	4.28	4.35
12	Agri. & Indust. Mach.	0.00	0.00	0.00	-0.01	-0.03	-0.04	0.16	0.72	0.67
13	Office Machines	0.00	0.00	0.00	-0.01	-0.04	-0.12	-0.25	-0.57	-1.08
14	Electrical Goods	0.00	0.00	0.01	0.05	0.17	0.49	0.98	1.60	2.38
15	Transport Equipment	0.00	0.00	0.00	0.01	0.05	0.15	0.44	1.06	2.07
16	Food, Drink & Tobacco	0.00	0.00	0.00	0.00	0.01	0.00	0.08	0.26	0.35
17	Tex., Cloth. & Footw.	0.00	0.00	0.01	0.08	0.30	0.92	2.48	4.33	6.42
18	Paper & Printing Pr.	0.00	0.00	0.00	0.00	0.00	0.00	0.00	0.00	0.00
19	Rubber & Plastic Pr.	0.00	0.00	0.00	0.00	0.01	0.06	0.49	1.04	1.80
21	Other Manufactures	0.00	0.00	0.03	0.09	0.19	0.38	0.48	0.50	-0.13
22	Construction	0.00	0.00	0.00	0.02	0.05	0.15	0.31	0.55	0.76
23	Distribution etc.	0.00	0.00	0.00	0.01	0.04	0.13	0.30	0.61	1.05
24	Lodging & Catering	0.00	0.00	0.00	-0.01	-0.02	-0.06	-0.11	-0.16	-0.20
25	Inland Transport	0.00	0.00	0.00	0.01	0.02	0.05	0.12	0.19	0.17
26	Sea & Air Transport	0.00	0.00	0.00	0.00	0.00	0.00	0.00	0.00	0.00
27	Other Transport	0.00	0.00	0.00	0.00	0.00	0.00	0.00	0.00	0.00
28	Communications	0.00	0.00	0.00	-0.01	-0.02	-0.04	-0.06	-0.08	-0.25
29	Bank., Finance & Ins.	0.00	0.00	0.01	0.03	0.12	0.31	0.65	1.15	1.55
30	Other Market Serv.	0.00	0.00	0.00	0.00	0.00	0.00	0.00	0.00	0.00
31	Non-market Services	0.00	0.00	0.00	0.00	0.00	0.00	0.00	0.00	0.00
32	Unallocated	0.00	0.00	0.00	0.00	0.00	0.00	0.00	0.00	0.00
	Total	0.00	0.00	0.00	0.02	0.05	0.13	0.28	0.48	0.61

Source: Cambridge Econometrics Forecast C51F1Y-C51F1R/da1/F1RA.

Table D1e.   Model simulation 1: detailed employment results – France (%)

		1985	1986	1987	1988	1989	1990	1991	1992	1993
1	Agriculture etc.	0.00	0.00	0.00	-0.01	-0.02	-0.06	-0.15	-0.32	-0.74
2	Coal & Coke	0.00	0.00	0.00	0.00	0.00	0.00	0.00	0.00	0.00
3	Oil & Gas Extraction	0.00	0.00	0.00	0.00	0.00	0.00	0.00	0.00	0.00
4	Gas Distribution	0.00	0.00	0.00	0.00	0.00	0.00	0.00	-0.01	-0.13
5	Refined Oil	0.00	0.00	0.00	0.00	-0.01	-0.01	-0.01	-0.01	0.34
6	Electricity etc.	0.00	0.00	0.00	0.00	0.02	0.02	0.04	0.06	-0.20
7	Water Supply	0.00	0.00	0.01	0.03	0.08	0.18	0.42	0.68	0.90
8	Ferrous & Non-f. Metals	0.00	0.00	0.02	0.09	0.28	0.66	1.44	2.78	4.48
9	Non-metallic Min.Pr.	0.00	0.00	0.00	0.00	0.00	0.01	0.01	0.03	0.08
10	Chemicals	0.00	0.00	0.00	0.00	0.00	0.00	0.00	-0.01	-0.02
11	Metal Products	0.00	0.00	0.00	-0.04	-0.14	-0.36	-1.26	-3.07	-5.44
12	Agri. & Indust. Mach.	0.00	0.00	0.02	0.08	0.23	0.53	1.03	1.76	2.49
13	Office Machines	0.00	0.00	0.02	0.10	0.31	0.74	1.72	3.49	5.72
14	Electrical Goods	0.00	0.00	-0.01	-0.03	-0.07	-0.11	-0.29	-0.37	0.01
15	Transport Equipment	0.00	0.00	0.00	0.00	0.01	0.02	0.04	0.06	-0.02
16	Food, Drink & Tobacco	0.00	0.00	0.00	-0.01	-0.02	-0.06	-0.12	-0.22	-0.38
17	Tex., Cloth. & Footw.	0.00	0.00	0.01	0.04	0.11	0.28	0.62	1.17	1.81
18	Paper & Printing Pr.	0.00	0.00	0.00	-0.01	-0.04	-0.05	0.02	0.17	0.09
19	Rubber & Plastic Pr.	0.00	0.00	0.01	0.03	0.08	0.18	0.36	0.60	0.76
21	Other Manufactures	0.00	0.00	-0.01	-0.04	-0.13	-0.24	-0.53	-1.16	-1.79
22	Construction	0.00	0.00	0.00	-0.01	-0.02	-0.07	-0.16	-0.30	-0.86
23	Distribution etc.	0.00	0.00	0.00	-0.01	-0.04	-0.10	-0.19	-0.27	-0.52
24	Lodging & Catering	0.00	0.00	0.00	0.00	0.00	0.00	0.00	0.01	0.03
25	Inland Transport	0.00	0.00	0.00	0.01	0.01	0.00	-0.02	-0.10	-0.93
26	Sea & Air Transport	0.00	0.00	0.00	-0.01	-0.02	-0.04	-0.10	-0.22	-0.39
27	Other Transport	0.00	0.00	0.01	0.01	0.01	0.01	-0.01	-0.03	-0.32
28	Communications	0.00	0.00	0.00	0.00	-0.01	-0.02	-0.06	-0.11	-0.18
29	Bank., Finance & Ins.	0.00	0.00	0.00	0.00	-0.01	-0.03	-0.08	-0.16	-0.39
30	Other Market Serv.	0.00	0.00	-0.01	-0.02	-0.05	-0.13	-0.27	-0.54	-1.10
31	Non-market Services	0.00	0.00	0.00	0.00	0.00	0.00	-0.01	-0.03	-0.13
32	Unallocated	0.00	0.00	0.00	0.00	0.00	0.00	0.00	0.00	0.00
	Total	0.00	0.00	0.00	0.00	-0.01	-0.03	-0.08	-0.17	-0.42

Source: Cambridge Econometrics Forecast C51F1Y-C51F1R/da1/F1RA.

Table D1f.   Model simulation 1: detailed employment results – Ireland (%)

		1985	1986	1987	1988	1989	1990	1991	1992	1993
1	Agriculture etc.	0.00	0.00	0.00	0.00	0.01	0.02	0.01	0.04	0.11
2	Coal & Coke	0.00	0.00	0.00	0.00	0.00	0.00	0.00	0.00	0.00
3	Oil & Gas Extraction	0.00	0.00	0.00	0.00	0.00	0.00	0.00	0.00	0.00
4	Gas Distribution	0.00	0.00	0.00	0.00	0.00	0.00	0.00	0.00	0.00
5	Refined Oil	0.00	0.00	0.00	0.00	0.00	0.00	0.00	0.00	0.00
6	Electricity etc.	0.00	0.00	0.00	-0.01	-0.01	-0.06	0.05	-0.05	-0.29
7	Water Supply	0.00	0.00	0.00	0.00	0.00	0.00	0.00	0.00	0.00
8	Ferrous & Non-f. Metals	0.00	-0.02	-0.16	-0.12	-0.51	-0.16	1.92	-0.69	2.17
9	Non-metallic Min.Pr.	0.00	0.00	-0.01	-0.02	0.00	-0.03	-0.28	-0.11	0.32
10	Chemicals	0.00	-0.01	-0.07	-0.25	-0.41	-1.05	-1.42	-1.79	-1.65
11	Metal Products	0.00	0.01	0.18	0.96	3.10	8.20	20.31	46.69	97.02
12	Agri. & Indust. Mach.	0.00	0.00	0.00	0.00	0.00	0.00	0.00	0.00	0.00
13	Office Machines	0.00	0.00	0.00	0.00	0.00	0.00	0.00	0.00	0.00
14	Electrical Goods	0.00	0.01	0.07	0.25	0.54	1.19	1.90	2.81	2.77
15	Transport Equipment	0.00	0.00	0.01	0.01	0.02	0.14	1.20	2.11	2.18
16	Food, Drink & Tobacco	0.00	0.00	0.00	0.01	0.05	0.01	0.02	0.13	0.32
17	Tex., Cloth. & Footw.	0.00	0.02	0.25	1.21	2.65	6.66	13.43	22.16	23.48
18	Paper & Printing Pr.	0.00	0.00	0.00	0.01	0.06	0.20	0.44	1.10	2.35
19	Rubber & Plastic Pr.	0.00	0.00	0.00	0.00	0.00	0.04	0.08	0.12	-0.01
21	Other Manufactures	0.00	-0.02	-0.13	-0.42	-1.04	-2.00	-3.07	-3.93	-3.40
22	Construction	0.00	0.00	0.00	0.01	0.02	0.05	0.08	0.09	-0.06
23	Distribution etc.	0.00	0.00	0.00	0.00	0.00	0.01	0.02	0.05	0.05
24	Lodging & Catering	0.00	0.00	0.00	0.00	0.00	-0.01	-0.03	0.00	-0.01
25	Inland Transport	0.00	0.00	0.00	0.00	0.00	-0.01	-0.01	-0.04	-0.12
26	Sea & Air Transport	0.00	0.00	0.02	0.08	0.23	0.63	1.47	2.76	3.60
27	Other Transport	0.00	0.00	0.02	0.08	0.17	0.47	0.81	1.36	1.74
28	Communications	0.00	0.00	0.00	0.00	0.01	0.02	0.04	0.07	0.11
29	Bank., Finance & Ins.	0.00	0.00	0.00	0.00	0.00	0.00	0.00	-0.01	-0.02
30	Other Market Serv.	0.00	0.00	0.00	0.00	0.00	0.00	0.00	0.01	0.03
31	Non-market Services	0.00	0.00	0.00	0.01	0.02	0.04	0.05	0.08	0.09
32	Unallocated	0.00	0.00	0.00	0.00	0.00	0.00	0.00	0.00	0.00
	Total	0.00	0.00	0.01	0.04	0.10	0.26	0.57	1.01	1.33

Source: Cambridge Econometrics Forecast C51F1Y-C51F1R/da1/F1RA.

Table D1g.   Model simulation 1: detailed employment results – Italy (%)

		1985	1986	1987	1988	1989	1990	1991	1992	1993
1	Agriculture etc.	0.00	0.00	0.00	0.01	0.07	0.21	0.62	1.39	2.42
2	Coal & Coke	0.00	0.00	0.00	0.00	0.00	0.00	0.00	0.00	0.00
3	Oil & Gas Extraction	0.00	0.00	0.00	0.00	0.00	0.00	0.00	0.00	0.00
4	Gas Distribution	0.00	0.00	0.00	0.00	0.01	0.01	0.03	0.05	0.10
5	Refined Oil	0.00	0.00	0.03	0.11	0.28	0.54	0.82	0.92	-3.55
6	Electricity etc.	0.00	0.00	0.00	-0.01	-0.02	-0.05	-0.08	-0.14	-0.23
7	Water Supply	0.00	0.00	0.00	-0.01	-0.03	-0.08	-0.21	-0.35	-0.02
8	Ferrous & Non-f. Metals	0.00	0.00	0.00	0.00	0.00	-0.01	-0.03	-0.05	-0.07
9	Non-metallic Min.Pr.	0.00	0.00	0.02	0.07	0.20	0.47	0.89	1.55	2.12
10	Chemicals	0.00	0.01	0.07	0.28	0.68	1.26	2.35	3.68	1.96
11	Metal Products	0.00	0.00	-0.02	-0.04	-0.06	-0.06	-0.03	0.03	0.74
12	Agri. & Indust. Mach.	0.00	0.00	0.00	0.00	0.00	0.02	0.05	0.11	-0.10
13	Office Machines	0.00	0.00	0.01	0.03	0.10	0.19	0.46	0.72	1.16
14	Electrical Goods	0.00	0.00	0.00	0.03	0.10	0.22	0.39	0.55	-0.28
15	Transport Equipment	0.00	0.00	-0.01	-0.05	-0.15	-0.34	-0.57	-1.19	-1.81
16	Food, Drink & Tobacco	0.00	0.00	0.02	0.09	0.25	0.51	0.99	1.66	2.14
17	Tex., Cloth. & Footw.	0.00	0.00	0.00	0.00	0.00	0.00	0.00	-0.01	-0.03
18	Paper & Printing Pr.	0.00	0.00	0.02	0.07	0.23	0.55	1.16	2.06	2.56
19	Rubber & Plastic Pr.	0.00	0.00	0.00	0.00	0.01	0.03	0.05	0.06	-0.34
21	Other Manufactures	0.00	0.00	0.00	0.00	0.01	0.02	0.05	0.10	0.14
22	Construction	0.00	0.00	0.00	0.01	0.02	0.05	0.08	0.07	0.02
23	Distribution etc.	0.00	0.00	0.00	0.00	0.01	0.01	0.03	0.05	0.08
24	Lodging & Catering	0.00	0.00	0.00	0.00	-0.01	-0.02	-0.01	0.03	0.46
25	Inland Transport	0.00	0.00	0.00	0.01	0.02	0.04	0.04	0.01	-0.50
26	Sea & Air Transport	0.00	0.00	0.02	0.09	0.30	0.34	0.66	1.20	0.93
27	Other Transport	0.00	0.00	0.00	0.00	0.00	0.00	0.00	0.00	0.00
28	Communications	0.00	0.00	0.00	0.00	-0.01	-0.01	-0.01	-0.01	-0.03
29	Bank., Finance & Ins.	0.00	0.00	0.00	0.00	-0.01	-0.02	-0.06	-0.15	-0.30
30	Other Market Serv.	0.00	0.00	0.00	-0.01	-0.02	-0.05	-0.11	-0.22	-0.25
31	Non-market Services	0.00	0.00	0.00	0.00	0.00	0.01	0.01	0.00	-0.25
32	Unallocated	0.00	0.00	0.00	0.00	0.00	0.00	0.00	0.00	0.00
	Total	0.00	0.00	0.00	0.01	0.03	0.06	0.13	0.23	0.24

Source: Cambridge Econometrics Forecast C51F1Y-C51F1R/da1/F1RA.

Table D1h. Model simulation 1: detailed employment results – Luxembourg (%)

		1985	1986	1987	1988	1989	1990	1991	1992	1993
1	Agriculture etc.	0.00	0.00	-0.01	-0.02	0.01	0.05	0.29	0.66	1.12
2	Coal & Coke	0.00	0.00	0.00	0.00	0.00	0.00	0.00	0.00	0.00
3	Oil & Gas Extraction	0.00	0.00	0.00	0.00	0.00	0.00	0.00	0.00	0.00
4	Gas Distribution	0.00	0.00	0.00	0.00	0.00	0.00	0.00	0.00	0.00
5	Refined Oil	0.00	0.00	0.00	0.00	0.00	0.00	0.00	0.00	0.00
6	Electricity etc	0.00	0.00	0.00	0.00	0.00	0.00	0.00	0.00	0.00
7	Water Supply	0.00	0.00	0.00	0.00	0.00	0.00	0.00	0.00	0.00
8	Ferrous & Non-f. Metals	0.00	-0.01	-0.06	-0.24	-0.62	-1.38	-2.62	-4.09	-4.95
9	Non-metallic Min.Pr.	0.00	0.00	0.02	0.03	0.07	0.09	0.65	0.72	0.81
10	Chemicals	0.00	0.00	0.00	0.00	0.00	0.00	0.00	0.00	0.00
11	Metal Products	0.00	0.01	0.05	0.18	0.29	0.49	0.75	0.29	-0.90
12	Agri. & Indust. Mach.	0.00	0.01	0.17	0.45	1.35	2.11	4.09	3.88	-0.04
13	Office Machines	0.00	0.00	0.00	0.00	0.00	0.00	0.00	0.00	0.00
14	Electrical Goods	0.00	0.00	0.00	0.00	0.00	0.00	0.00	0.00	0.00
15	Transport Equipment	0.00	0.00	0.00	0.00	0.00	0.00	0.00	0.00	0.00
16	Food, Drink & Tobacco	0.00	0.01	0.07	0.29	0.73	1.35	2.77	4.57	5.91
17	Tex., Cloth. & Footw.	0.00	0.00	0.00	0.00	0.00	0.00	0.00	0.00	0.00
18	Paper & Printing Pr.	0.00	0.00	0.00	-0.02	-0.02	0.10	0.19	-0.03	-1.27
19	Rubber & Plastic Pr.	0.00	-0.01	-0.10	-0.55	-0.83	3.06	-3.86	-5.46	-7.78
21	Other Manufactures	0.00	0.00	0.00	0.00	0.00	0.00	0.00	0.00	0.00
22	Construction	0.00	0.00	0.00	0.00	0.00	0.00	-0.01	-0.10	-0.20
23	Distribution etc.	0.00	0.00	0.00	0.00	0.00	0.01	0.02	0.03	0.04
24	Lodging & Catering	0.00	0.00	0.00	0.01	0.01	0.03	0.06	0.03	-0.21
25	Inland Transport	0.00	0.00	0.00	0.01	0.03	0.02	-0.05	-0.16	-0.46
26	Sea & Air Transport	0.00	0.00	0.00	0.00	0.00	0.00	0.00	0.00	0.00
27	Other Transport	0.00	0.00	0.00	0.00	0.00	0.00	0.00	0.00	0.00
28	Communications	0.00	0.00	0.00	0.00	-0.01	-0.04	-0.09	-0.18	-0.40
29	Bank., Finance & Ins.	0.00	0.00	0.00	0.00	0.00	0.00	-0.01	-0.01	-0.02
30	Other Market Serv.	0.00	0.00	-0.02	-0.15	-0.57	-1.43	-2.51	-5.54	-8.91
31	Non-market Services	0.00	0.00	0.00	0.00	0.00	0.00	0.00	0.00	0.00
32	Unallocated	0.00	0.00	0.00	0.00	0.00	0.00	0.00	0.00	0.00
	Total	0.00	0.00	-0.01	-0.04	-0.11	-0.16	-0.56	-1.24	-1.98

Source: Cambridge Econometrics Forecast C51F1Y-C51F1R/da1/F1RA.

Table D1i.   Model simulation 1: detailed employment results – Netherlands (%)

		1985	1986	1987	1988	1989	1990	1991	1992	1993
1	Agriculture etc.	0.00	0.00	0.00	0.00	0.01	0.01	0.06	0.12	0.10
2	Coal & Coke	0.00	0.00	0.00	0.00	0.00	0.00	0.00	0.00	0.00
3	Oil & Gas Extraction	0.00	0.00	0.00	0.00	0.00	0.00	0.00	0.00	0.00
4	Gas Distribution	0.00	0.00	0.00	0.00	-0.01	-0.01	0.00	0.02	0.05
5	Refined Oil	0.00	0.00	-0.03	-0.13	-0.28	-0.52	-1.43	-0.22	18.52
6	Electricity etc.	0.00	0.00	0.00	0.00	0.00	0.00	0.00	0.00	0.00
7	Water Supply	0.00	0.00	0.01	0.05	0.16	0.43	0.99	1.97	2.76
8	Ferrous & Non-f. Metals	0.00	0.00	0.00	0.00	0.04	0.02	-0.05	-0.02	0.25
9	Non-metallic Min.Pr.	0.00	0.00	0.02	0.05	0.06	0.17	0.23	0.07	-0.09
10	Chemicals	0.00	0.00	-0.01	-0.02	-0.06	-0.15	-0.32	-0.67	-1.16
11	Metal Products	0.00	0.00	-0.03	-0.14	-0.34	-0.62	-1.12	-2.34	-3.42
12	Agri. & Indust. Mach.	0.00	0.00	0.00	0.00	0.00	0.00	0.00	0.00	0.00
13	Office Machines	0.00	0.00	0.00	0.00	0.00	0.00	0.00	0.00	0.00
14	Electrical Goods	0.00	-0.01	-0.16	-0.53	-1.06	-1.12	-0.28	1.31	3.51
15	Transport Equipment	0.00	0.00	0.01	0.03	0.05	0.10	0.19	-0.29	-1.31
16	Food, Drink & Tobacco	0.00	0.00	-0.01	-0.05	-0.13	-0.31	-0.53	-0.89	-1.17
17	Tex., Cloth. & Footw.	0.00	0.00	0.00	0.00	0.00	0.01	0.04	0.12	0.33
18	Paper & Printing Pr.	0.00	0.00	0.00	0.02	0.05	0.14	0.35	0.63	1.01
19	Rubber & Plastic Pr.	0.00	0.00	0.01	0.05	0.15	0.36	0.54	0.51	0.53
21	Other Manufactures	0.00	0.00	0.01	0.06	0.18	0.48	1.06	2.16	3.79
22	Construction	0.00	0.00	0.00	0.00	0.00	0.00	0.00	0.00	0.00
23	Distribution etc.	0.00	0.00	0.00	0.00	0.00	0.01	0.02	0.06	0.11
24	Lodging & Catering	0.00	0.00	0.00	0.00	-0.01	0.00	0.02	0.24	1.02
25	Inland Transport	0.00	0.00	0.01	0.03	0.05	0.05	-0.02	0.51	2.74
26	Sea & Air Transport	0.00	0.00	0.01	0.01	-0.02	-0.17	-0.72	-2.04	-4.71
27	Other Transport	0.00	0.00	0.00	0.00	0.00	0.00	0.00	0.00	0.00
28	Communications	0.00	0.00	-0.01	-0.02	-0.05	-0.11	-0.26	-0.39	-0.30
29	Bank., Finance & Ins.	0.00	0.00	0.00	0.00	0.00	0.00	0.00	0.00	0.00
30	Other Market Serv.	0.00	0.00	0.00	-0.02	-0.06	-0.14	-0.23	-0.39	-0.45
31	Non-market Services	0.00	0.00	0.00	0.00	0.00	0.00	0.00	-0.01	-0.02
32	Unallocated	0.00	0.00	0.00	0.00	0.00	0.00	0.00	0.00	0.00
	Total	0.00	0.00	0.00	-0.02	-0.04	-0.07	-0.08	-0.08	0.14

Source: Cambridge Econometrics Forecast C51F1Y-C51F1R/da1/F1RA.

Table D1j.   Model simulation 1: detailed employment results – Portugal (%)

		1985	1986	1987	1988	1989	1990	1991	1992	1993
1	Agriculture etc.	0.00	0.00	0.00	0.00	0.00	0.01	0.01	-0.01	-0.15
2	Coal & Coke	0.00	0.00	0.00	0.00	0.00	0.00	0.00	0.00	0.00
3	Oil & Gas Extraction	0.00	0.00	0.00	0.00	0.00	0.00	0.00	0.00	0.00
4	Gas Distribution	0.00	0.00	0.00	0.00	0.00	0.00	0.00	0.00	0.00
5	Refined Oil	0.00	0.00	0.00	0.00	0.00	0.00	0.00	0.00	0.00
6	Electricity etc.	0.00	0.00	0.00	0.00	0.01	0.02	0.00	0.01	0.23
7	Water Supply	0.00	0.00	0.00	0.00	0.00	0.00	0.00	0.00	0.00
8	Ferrous & Non-f. Metals	0.00	0.00	0.03	0.14	0.49	1.05	1.96	3.62	7.89
9	Non-metallic Min.Pr.	0.00	0.00	0.02	0.09	0.23	0.59	1.56	3.08	7.31
10	Chemicals	0.00	0.00	0.00	0.00	-0.01	-0.03	-0.08	-0.34	-0.77
11	Metal Products	0.00	0.00	0.00	0.01	0.07	0.20	0.48	1.01	1.53
12	Agri. & Indust. Mach.	0.00	0.00	0.00	0.00	0.01	0.06	0.27	0.94	4.13
13	Office Machines	0.00	0.00	0.00	0.00	0.00	0.00	0.00	0.00	0.00
14	Electrical Goods	0.00	0.00	0.02	0.07	0.36	0.77	0.13	1.48	14.76
15	Transport Equipment	0.00	0.00	0.00	0.00	0.01	0.04	0.10	0.12	0.09
16	Food, Drink & Tobacco	0.00	0.00	0.00	-0.01	-0.06	-0.16	-0.31	-0.61	-0.80
17	Tex., Cloth. & Footw.	0.00	0.00	0.01	0.01	0.03	0.08	0.15	0.26	0.39
18	Paper & Printing Pr.	0.00	0.00	0.01	0.03	0.08	0.20	0.45	0.95	1.70
19	Rubber & Plastic Pr.	0.00	0.00	0.02	0.07	0.21	0.47	0.91	1.52	2.83
21	Other Manufactures	0.00	0.00	0.02	0.06	0.15	0.29	0.48	0.63	0.70
22	Construction	0.00	0.00	0.00	0.00	0.01	0.04	0.24	1.05	5.95
23	Distribution etc.	0.00	0.00	0.00	0.00	0.01	0.03	0.10	0.23	0.39
24	Lodging & Catering	0.00	0.00	0.00	0.00	0.00	0.00	-0.01	0.01	0.07
25	Inland Transport	0.00	0.00	0.00	-0.02	-0.04	-0.09	-0.14	-0.15	0.61
26	Sea & Air Transport	0.00	0.00	0.00	-0.01	-0.02	-0.03	-0.05	-0.40	-1.46
27	Other Transport	0.00	0.00	0.00	-0.01	-0.05	-0.13	-0.26	-0.42	-0.70
28	Communications	0.00	0.00	0.00	0.00	0.00	0.00	-0.01	-0.01	-0.05
29	Bank., Finance & Ins.	0.00	0.00	0.00	0.00	0.00	-0.01	0.04	0.10	-0.09
30	Other Market Serv.	0.00	0.00	0.00	-0.01	-0.02	-0.04	-0.01	0.13	0.18
31	Non-market Services	0.00	0.00	0.00	0.00	0.00	-0.01	-0.01	-0.02	-0.18
32	Unallocated	0.00	0.00	0.00	0.00	0.00	0.00	0.00	0.00	0.00
	Total	0.00	0.00	0.00	0.00	0.02	0.04	0.10	0.27	1.10

Source: Cambridge Econometrics Forecast C51F1Y-C51F1R/da1/F1RA.

Table D1k.   Model simulation 1: detailed employment results – United Kingdom (%)

		1985	1986	1987	1988	1989	1990	1991	1992	1993
1	Agriculture etc.	0.00	0.00	0.00	0.00	0.01	0.01	0.03	0.01	-0.27
2	Coal & Coke	0.00	0.00	0.00	0.00	0.00	0.00	0.00	0.00	0.00
3	Oil & Gas Extraction	0.00	0.00	0.00	0.00	0.00	0.00	0.00	0.00	0.00
4	Gas Distribution	0.00	0.00	0.00	0.00	0.00	0.00	0.01	0.05	0.09
5	Refined Oil	0.00	0.00	0.00	-0.01	-0.01	-0.01	0.02	0.39	3.74
6	Electricity etc.	0.00	0.00	0.00	0.00	0.00	0.00	-0.01	-0.01	-0.02
7	Water Supply	0.00	0.00	0.00	0.00	0.01	0.01	0.01	0.00	-0.02
8	Ferrous & Non-f. Metals	0.00	-0.01	-0.07	-0.14	0.26	2.31	7.47	21.78	61.54
9	Non-metallic Min.Pr.	0.00	0.00	0.00	-0.01	-0.02	-0.03	-0.05	-0.21	-0.27
10	Chemicals	0.00	0.00	0.00	0.00	0.01	0.01	0.00	-0.18	-1.50
11	Metal Products	0.00	0.00	0.04	0.13	0.37	0.81	1.27	1.34	1.04
12	Agri. & Indust. Mach.	0.00	0.00	0.00	0.00	0.00	0.00	0.00	0.00	0.00
13	Office Machines	0.00	0.00	0.01	0.06	0.17	0.40	0.83	1.71	2.90
14	Electrical Goods	0.00	0.00	-0.01	-0.03	-0.04	-0.05	-0.06	-0.11	0.05
15	Transport Equipment	0.00	0.00	0.00	0.01	0.03	0.08	0.27	0.50	0.54
16	Food, Drink & Tobacco	0.00	0.00	0.00	-0.01	-0.03	-0.09	-0.24	-0.57	-1.26
17	Tex., Cloth. & Footw.	0.00	0.00	0.00	0.00	-0.01	-0.01	0.03	0.11	0.14
18	Paper & Printing Pr.	0.00	0.00	-0.02	-0.05	-0.13	-0.26	-0.38	-0.61	-0.52
19	Rubber & Plastic Pr.	0.00	0.00	0.00	0.01	0.03	0.07	0.05	-0.10	0.05
21	Other Manufactures	0.00	0.00	0.00	0.00	0.00	0.01	0.02	0.05	0.11
22	Construction	0.00	0.00	0.00	0.00	-0.01	-0.03	-0.06	-0.15	-1.01
23	Distribution etc.	0.00	0.00	0.00	0.00	-0.01	-0.01	-0.03	-0.05	-0.16
24	Lodging & Catering	0.00	0.00	0.00	0.00	0.00	-0.01	-0.02	-0.07	-0.17
25	Inland Transport	0.00	0.00	0.00	0.00	-0.01	-0.02	-0.04	-0.06	-0.10
26	Sea & Air Transport	0.00	0.00	0.02	0.09	0.24	0.56	1.12	1.97	2.64
27	Other Transport	0.00	0.00	0.00	0.00	-0.01	-0.04	-0.08	-0.20	-1.07
28	Communications	0.00	0.00	0.00	0.00	0.00	0.01	0.01	0.01	-0.14
29	Bank., Finance & Ins.	0.00	0.00	0.00	0.00	0.00	-0.01	-0.01	-0.03	-0.06
30	Other Market Serv.	0.00	0.00	-0.02	-0.07	-0.21	-0.45	-0.90	-1.58	-1.98
31	Non-market Services	0.00	0.00	0.00	-0.01	-0.02	-0.05	-0.11	-0.21	-0.25
32	Unallocated	0.00	0.00	0.00	0.00	0.00	0.00	0.00	0.00	0.00
	Total	0.00	0.00	0.00	-0.01	-0.03	-0.07	-0.13	-0.22	-0.30

Source: Cambridge Econometrics Forecast C51F1Y-C51F1R/da1/F1RA.

Table D1l.   Model simulation 1: detailed employment results – EUR-11 (%)

		1985	1986	1987	1988	1989	1990	1991	1992	1993
1	Agriculture etc.	0.00	0.00	0.00	0.00	0.02	0.11	0.26	0.56	0.73
2	Coal & Coke	0.00	0.00	0.00	0.00	0.00	0.00	0.00	0.00	0.00
3	Oil & Gas Extraction	0.00	0.00	0.00	0.00	0.00	0.00	0.00	0.00	0.00
4	Gas Distribution	0.00	0.00	0.00	0.00	-0.01	-0.02	-0.02	-0.02	-0.02
5	Refined Oil	0.00	0.00	0.00	0.01	0.02	0.05	0.03	0.04	0.14
6	Electricity etc.	0.00	0.00	0.00	0.00	0.01	0.01	0.02	0.05	0.01
7	Water Supply	0.00	0.00	0.00	0.01	0.01	0.03	0.06	0.09	0.10
8	Ferrous & Non-f. Metals	0.00	0.00	-0.01	-0.01	0.08	0.42	1.23	3.25	7.52
9	Non-metallic Min.Pr.	0.00	0.00	0.01	0.02	0.05	0.10	0.17	0.25	0.42
10	Chemicals	0.00	0.00	0.01	0.04	0.10	0.18	0.35	0.50	0.06
11	Metal Products	0.00	0.00	0.01	0.05	0.12	0.25	0.23	-0.03	-0.46
12	Agri. & Indust. Mach.	0.00	0.00	0.00	-0.01	-0.01	-0.03	0.09	0.03	0.07
13	Office Machines	0.00	0.00	0.01	0.02	0.06	0.13	0.41	0.78	0.64
14	Electrical Goods	0.00	0.00	-0.01	0.00	0.04	0.59	0.48	0.32	0.30
15	Transport Equipment	0.00	0.00	0.00	0.00	0.00	0.00	0.05	0.07	0.09
16	Food, Drink & Tobacco	0.00	0.00	0.00	0.00	0.01	-0.02	0.11	0.04	-0.05
17	Tex., Cloth. & Footw.	0.00	0.00	0.01	0.02	0.07	0.19	0.46	0.77	1.01
18	Paper & Printing Pr.	0.00	0.00	0.00	0.00	0.01	0.07	0.11	0.31	0.50
19	Rubber & Plastic Pr.	0.00	0.00	0.00	0.02	0.06	0.30	0.62	0.96	1.06
21	Other Manufactures	0.00	0.00	0.01	0.02	0.05	0.11	0.17	0.36	0.30
22	Construction	0.00	0.00	0.00	0.00	0.01	0.03	0.06	0.11	0.10
23	Distribution etc.	0.00	0.00	0.00	0.00	0.00	0.00	0.00	0.00	-0.10
24	Lodging & Catering	0.00	0.00	0.00	0.00	-0.01	-0.02	-0.04	-0.09	-0.02
25	Inland Transport	0.00	0.00	0.00	0.01	0.01	-0.05	-0.12	-0.24	-0.78
26	Sea & Air Transport	0.00	0.00	0.01	0.03	0.11	0.16	0.18	0.31	0.25
27	Other Transport	0.00	0.00	0.00	0.00	0.00	0.00	0.00	-0.01	0.09
28	Communications	0.00	0.00	0.00	0.00	-0.01	0.05	-0.08	-0.03	0.09
29	Bank., Finance & Ins.	0.00	0.00	0.00	0.00	0.01	0.02	0.04	0.04	-0.12
30	Other Market Serv.	0.00	0.00	0.00	-0.01	-0.04	-0.07	-0.15	-0.19	0.03
31	Non-market Services	0.00	0.00	0.00	0.00	0.00	-0.01	-0.02	-0.04	-0.10
32	Unallocated	0.00	0.00	0.00	0.00	0.00	0.00	0.00	0.00	0.00
	Total	0.00	0.00	0.00	0.00	0.01	0.04	0.06	0.09	0.10

Source: Cambridge Econometrics Forecast C51F1Y-C51F1R/da1/F1RA.

Table D2a.   Model simulation 1: detailed unit labour cost results –Belgium (%)

		1985	1986	1987	1988	1989	1990	1991	1992	1993
1	Agriculture etc.	0.00	0.00	0.47	0.95	1.70	4.38	7.81	12.36	12.76
2	Coal & Coke	0.00	0.09	1.21	4.24	12.83	29.19	60.71	94.29	106.25
3	Oil & Gas Extraction	0.00	0.00	0.00	0.00	0.00	0.00	0.00	0.00	0.00
4	Gas Distribution	0.00	0.00	0.00	0.00	0.11	0.14	0.13	0.30	0.34
5	Refined Oil	0.00	0.00	0.09	0.46	1.29	3.17	6.09	10.08	22.12
6	Electricity etc.	0.00	0.00	0.00	0.08	0.16	0.36	0.60	0.89	1.08
7	Water Supply	0.00	0.00	0.00	0.00	0.00	-0.26	-0.33	-1.27	-1.03
8	Ferrous & Non-f. Metals	0.00	0.00	0.31	1.09	3.21	6.59	11.95	20.02	24.50
9	Non-metallic Min.Pr.	0.00	0.00	0.04	0.16	0.24	0.73	1.23	2.19	2.22
10	Chemicals	0.00	0.00	-0.05	-0.10	-0.29	-0.57	-1.00	-1.55	-1.88
11	Metal Products	0.00	0.03	0.20	0.84	1.86	2.78	3.53	3.12	1.17
12	Agri. & Indust. Mach.	0.00	0.00	0.03	0.14	0.38	0.86	1.41	1.88	0.73
13	Office Machines	0.00	0.00	0.00	-0.20	-0.48	-0.86	-1.71	-2.53	-3.59
14	Electrical Goods	0.00	0.04	0.52	2.15	8.14	9.10	9.99	9.59	11.91
15	Transport Equipment	0.00	0.00	-0.05	-0.48	-1.24	-2.95	-6.68	-12.55	-18.81
16	Food, Drink & Tobacco	0.00	0.00	0.00	0.00	0.00	-0.07	-0.49	-0.93	-1.67
17	Tex., Cloth. & Footw.	0.00	0.00	-0.09	-0.39	-1.08	-2.24	-4.03	-6.70	-8.87
18	Paper & Printing Pr.	0.00	0.00	-0.03	-0.07	-0.17	-0.45	-0.90	-1.45	-1.73
19	Rubber & Plastic Pr.	0.00	0.00	0.00	-0.08	-0.22	-0.54	-1.13	-1.87	-2.49
21	Other Manufactures	0.00	0.00	-0.05	-0.06	-0.05	-0.18	0.68	14.51	25.59
22	Construction	0.00	0.00	-0.04	-0.04	-0.07	-0.07	-0.17	-0.30	-0.20
23	Distribution etc.	0.00	0.00	0.00	0.00	0.00	-0.14	-0.34	-0.63	-0.91
24	Lodging & Catering	0.00	0.00	0.00	-0.09	-0.26	-0.81	-2.09	-4.55	-7.48
25	Inland Transport	0.00	0.00	0.01	0.06	0.16	0.33	0.71	1.34	1.99
26	Sea & Air Transport	0.00	0.00	0.00	0.05	0.10	0.10	0.10	0.05	-0.58
27	Other Transport	0.00	0.00	-0.04	-0.14	-0.27	-0.46	-0.93	-1.33	-1.42
28	Communications	0.00	0.00	0.00	-0.04	-0.09	-0.20	-0.35	-0.51	-0.35
29	Bank., Finance & Ins.	0.00	0.00	0.00	0.00	-0.03	-0.17	-0.40	-0.72	-1.13
30	Other Market Serv.	0.00	-0.04	-0.08	-0.35	-0.64	-1.32	-2.54	-4.41	-4.22
31	Non-market Services	0.00	0.00	0.00	0.00	0.00	-0.01	-0.02	-0.04	-0.07
32	Unallocated	0.00	0.00	0.00	0.00	0.00	0.00	0.00	0.00	0.00
	Total	0.00	0.00	0.05	0.14	0.44	0.41	0.27	-0.03	-0.49

Source: Cambridge Econometrics Forecast C51F1Y-C51F1R/da1/F1RA.

Table D2b.   Model simulation 1: detailed unit labour cost results – Denmark (%)

		1985	1986	1987	1988	1989	1990	1991	1992	1993
1	Agriculture etc.	0.00	0.00	-0.11	-0.11	-0.11	-0.48	-1.01	-2.50	-2.70
2	Coal & Coke	0.00	0.00	0.00	0.00	0.00	0.00	0.00	0.00	0.00
3	Oil & Gas Extraction	0.00	0.00	0.00	-0.30	-0.74	-0.66	-1.36	-2.25	-2.92
4	Gas Distribution	0.00	0.00	0.00	0.00	0.00	0.26	0.22	0.24	0.26
5	Refined Oil	0.00	0.00	-0.13	-0.25	-0.79	-1.61	-4.40	-15.70	-94.29
6	Electricity etc.	0.00	0.00	0.00	-0.10	-0.22	-0.51	-0.76	-1.32	-1.68
7	Water Supply	0.00	0.00	0.00	0.00	-0.10	-0.11	-0.21	-0.42	-0.55
8	Ferrous & Non-f. Metals	0.00	0.00	-0.15	-0.46	-1.22	-2.40	-4.27	-7.77	-4.97
9	Non-metallic Min.Pr.	0.00	0.03	0.03	0.21	0.56	1.25	1.90	2.46	3.44
10	Chemicals	0.00	0.00	0.04	0.13	0.28	0.35	0.83	1.42	1.39
11	Metal Products	0.00	0.00	0.00	0.12	0.28	0.73	1.38	2.44	2.51
12	Agri. & Indust. Mach.	0.00	0.00	0.00	-0.03	-0.05	-0.06	-0.06	-0.03	0.51
13	Office Machines	0.00	0.00	-0.02	-0.12	-0.30	-0.49	-1.18	-3.09	-3.93
14	Electrical Goods	0.00	0.00	0.00	0.07	0.08	0.51	1.48	3.82	4.52
15	Transport Equipment	0.00	0.03	0.18	0.64	1.32	3.30	6.79	18.40	19.08
16	Food, Drink & Tobacco	0.00	0.00	0.00	0.06	0.19	0.31	0.60	0.78	0.72
17	Tex., Cloth. & Footw.	0.00	0.00	-0.04	-0.07	-0.22	-0.56	-1.22	-3.79	-5.38
18	Paper & Printing Pr.	0.00	0.00	0.12	0.46	1.22	2.47	4.67	6.61	8.32
19	Rubber & Plastic Pr.	0.00	0.00	0.03	0.09	0.21	0.48	0.80	1.06	1.24
21	Other Manufactures	0.00	0.00	-0.03	-0.10	-0.25	-0.56	-0.86	-3.22	-4.50
22	Construction	0.00	0.00	0.00	0.00	-0.03	-0.03	0.03	0.07	0.19
23	Distribution etc	0.00	0.00	0.00	0.00	0.00	0.00	0.12	0.10	0.07
24	Lodging & Catering	0.00	0.00	0.00	0.17	0.27	0.81	2.74	6.06	8.51
25	Inland Transport	0.00	0.00	0.00	0.02	0.02	0.07	0.22	0.42	0.74
26	Sea & Air Transport	0.00	0.00	0.00	0.08	0.24	0.43	0.85	1.37	1.32
27	Other Transport	0.00	0.00	-0.08	-0.25	-0.60	-1.30	-2.57	-4.38	-5.16
28	Communications	0.00	0.00	0.00	0.00	0.03	0.11	0.30	0.59	0.85
29	Bank., Finance & Ins.	0.00	0.00	-0.02	-0.03	-0.05	0.00	0.11	0.21	0.60
30	Other Market Serv.	0.00	0.00	0.00	0.04	0.04	0.16	0.24	0.39	0.43
31	Non-market Services	0.00	0.00	0.00	-0.01	-0.04	-0.07	-0.07	-0.12	-0.21
32	Unallocated	0.00	0.00	0.00	0.00	0.00	0.00	0.00	0.00	0.00
	Total	0.00	0.00	0.01	0.00	0.00	0.22	0.17	0.17	-9.32

Source: Cambridge Econometrics Forecast C51F1Y-C51F1R/da1/F1RA.

Table D2c.   Model simulation 1: detailed unit labour cost results – West Germany (%)

		1985	1986	1987	1988	1989	1990	1991	1992	1993
1	Agriculture etc.	0.00	0.00	0.11	0.47	1.09	2.72	5.29	8.54	9.08
2	Coal & Coke	0.00	0.00	0.00	0.00	0.00	0.57	0.49	1.10	3.07
3	Oil & Gas Extraction	0.00	0.00	0.00	0.18	0.77	1.49	2.32	5.00	9.13
4	Gas Distribution	0.00	0.00	0.00	0.00	0.00	0.09	-0.10	-0.22	0.36
5	Refined Oil	0.00	0.00	0.00	-0.09	-0.15	0.00	-0.32	3.18	117.47
6	Electricity etc	0.00	0.00	0.00	0.00	0.06	0.12	0.25	0.41	0.47
7	Water Supply	0.00	0.00	0.00	0.00	-0.06	-0.64	-2.13	-15.63	0.00
8	Ferrous & Non-f. Metals	0.00	0.00	0.00	-0.08	-0.08	-0.16	-0.23	-0.41	-1.00
9	Non-metallic Min.Pr.	0.00	0.00	0.03	0.13	0.31	0.73	0.87	1.62	2.17
10	Chemicals	0.00	0.00	0.04	0.04	0.04	0.15	0.43	0.86	1.12
11	Metal Products	0.00	0.00	-0.03	-0.05	-0.08	0.13	-0.10	-0.46	-0.15
12	Agri. & Indust. Mach.	0.00	0.00	0.00	0.02	0.09	0.67	0.84	1.35	1.15
13	Office Machines	0.00	0.00	0.03	0.11	0.38	0.86	1.57	2.79	3.47
14	Electrical Goods	0.00	0.00	0.00	-0.05	0.02	1.68	0.42	-0.26	0.28
15	Transport Equipment	0.00	0.03	0.00	0.07	0.21	1.25	1.15	2.20	4.37
16	Food, Drink & Tobacco	0.00	0.00	0.00	0.00	0.00	-0.06	0.80	0.45	0.63
17	Tex., Cloth. & Footw.	0.00	0.00	-0.04	-0.16	-0.43	-0.97	-2.07	-4.05	-5.98
18	Paper & Printing Pr.	0.00	0.00	0.03	0.10	0.21	0.20	0.74	1.66	1.88
19	Rubber & Plastic Pr.	0.00	0.00	0.00	0.06	0.18	1.20	1.85	2.93	4.11
21	Other Manufactures	0.00	0.00	0.03	0.09	0.26	0.84	0.76	0.85	0.21
22	Construction	0.00	0.00	0.00	0.00	0.03	-0.08	-0.05	0.10	-0.12
23	Distribution etc	0.00	0.00	0.02	0.00	0.04	-0.02	0.19	0.45	-0.22
24	Lodging & Catering	0.00	0.00	0.04	0.04	0.11	0.41	0.40	1.12	1.57
25	Inland Transport	0.00	0.00	-0.02	-0.02	-0.09	-0.28	-0.97	-2.40	-4.38
26	Sea & Air Transport	0.00	0.04	0.06	0.18	0.41	1.72	0.04	0.12	0.14
27	Other Transport	0.00	0.00	0.00	0.00	-0.03	0.34	-0.20	-0.79	-0.18
28	Communications	0.00	0.00	0.00	0.00	0.00	0.16	-0.35	0.07	0.58
29	Bank., Finance & Ins.	0.00	0.00	-0.03	-0.03	-0.03	-0.06	-0.30	-0.41	-0.74
30	Other Market Serv.	0.00	0.00	0.00	0.00	0.00	0.00	-0.07	0.00	1.50
31	Non-market Services	0.00	0.00	0.00	0.00	0.00	0.00	0.00	0.00	-0.01
32	Unallocated	0.00	0.00	0.00	0.00	0.00	0.00	0.00	0.00	0.00
	Total	0.00	0.00	0.00	0.00	0.00	0.36	0.04	0.12	0.13

Source: Cambridge Econometrics Forecast C51F1Y-C51F1R/da1/F1RA

Table D2d. Model simulation 1: detailed unit labour cost results – Spain (%)

		1985	1986	1987	1988	1989	1990	1991	1992	1993
1	Agriculture etc.	0.00	0.00	-0.07	-0.13	-0.18	-0.41	-0.46	-0.25	-0.47
2	Coal & Coke	0.00	0.00	-0.30	-1.24	-2.44	-6.19	-11.54	-19.91	-21.43
3	Oil & Gas Extraction	0.00	-18.22	-0.40	-2.13	-3.03	-9.30	-25.95	-57.20	-43.97
4	Gas Distribution	0.00	0.00	0.00	-0.07	-0.16	-0.35	-0.62	-1.14	-1.45
5	Refined Oil	0.00	0.00	-0.08	-0.24	-0.51	-1.13	-2.53	-5.26	-5.58
6	Electricity etc.	0.00	0.00	-0.05	-0.21	-0.40	-0.98	-1.90	-2.95	-3.23
7	Water Supply	0.00	0.00	0.00	-0.04	-0.04	-0.16	-0.42	-0.62	-0.87
8	Ferrous & Non-f. Metals	0.00	-0.07	-0.35	-1.48	-4.17	-9.55	-16.35	-28.02	-35.90
9	Non-metallic Min.Pr.	0.00	0.00	0.00	0.03	0.09	0.23	0.26	0.33	0.32
10	Chemicals	0.00	0.00	0.00	-0.06	-0.11	-0.26	0.31	-0.05	-0.25
11	Metal Products	0.00	0.00	0.00	0.05	0.05	0.07	0.17	-0.11	-0.05
12	Agri. & Indust. Mach.	0.00	0.00	-0.04	-0.04	-0.03	-0.05	0.14	0.76	0.95
13	Office Machines	0.00	0.00	-0.06	-0.11	-0.23	-0.46	-1.04	-1.67	-2.38
14	Electrical Goods	0.00	0.00	-0.03	-0.10	-0.25	-0.50	-0.56	-0.65	-0.15
15	Transport Equipment	0.00	0.00	-0.04	-0.09	-0.23	-0.73	-1.49	-2.62	-3.38
16	Food, Drink & Tobacco	0.00	0.00	0.00	0.00	0.13	0.13	0.31	1.36	2.09
17	Tex., Cloth. & Footw.	0.00	0.00	0.06	0.27	1.09	3.40	8.47	17.67	31.68
18	Paper & Printing Pr.	0.00	0.00	0.00	0.03	0.08	0.21	0.59	0.86	1.30
19	Rubber & Plastic Pr.	0.00	0.00	0.00	-0.07	-0.11	-0.20	-0.03	0.27	0.79
21	Other Manufactures	0.00	0.00	0.04	0.04	0.12	0.20	0.26	0.35	0.08
22	Construction	0.00	0.00	0.00	0.03	0.02	0.06	0.12	0.22	0.35
23	Distribution etc.	0.00	0.00	0.00	0.00	-0.08	-0.17	-0.22	-0.29	-0.07
24	Lodging & Catering	0.00	0.00	0.00	0.00	0.00	0.00	-0.07	-0.19	-0.23
25	Inland Transport	0.00	0.00	0.00	-0.03	-0.08	-0.16	-0.25	-0.31	-0.14
26	Sea & Air Transport	0.00	0.00	0.06	0.10	0.31	0.50	-2.71	-4.32	-12.57
27	Other Transport	0.00	0.00	0.00	0.03	0.14	0.36	1.11	2.00	2.99
28	Communications	0.00	0.00	-0.02	0.00	-0.03	-0.05	-0.13	-0.20	-0.03
29	Bank., Finance & Ins.	0.00	0.00	0.00	0.00	0.00	0.01	0.06	0.13	0.29
30	Other Market Serv.	0.00	0.00	0.00	0.00	0.00	0.00	-0.03	-0.07	-0.05
31	Non-market Services	0.00	0.00	0.00	0.00	0.00	0.00	0.00	0.00	0.00
32	Unallocated	0.00	0.00	0.00	0.00	0.00	0.00	0.00	0.00	0.00
	Total	0.00	0.00	0.00	0.00	0.00	-0.13	0.00	-1.29	-4.56

Source: Cambridge Econometrics Forecast C51F1Y-C51F1R/da1/F1RA.

Table D2e.   Model simulation 1: detailed unit labour cost results – France (%)

		1985	1986	1987	1988	1989	1990	1991	1992	1993
1	Agriculture etc.	0.00	0.00	0.00	-0.27	-0.80	-2.01	-4.17	-7.38	-11.29
2	Coal & Coke	0.00	0.00	0.00	-0.28	-0.65	-1.46	-3.52	-5.86	-6.85
3	Oil & Gas Extraction	0.00	0.00	0.00	0.00	0.40	0.45	0.52	3.09	5.65
4	Gas Distribution	0.00	0.00	0.00	0.00	0.00	0.00	0.00	0.00	0.00
5	Refined Oil	0.00	0.00	0.00	0.00	-0.08	-0.08	0.00	0.27	8.05
6	Electricity etc.	0.00	0.00	0.00	0.00	-0.05	-0.12	-0.22	-0.39	-0.60
7	Water Supply	0.00	0.00	0.00	0.00	0.00	0.00	0.05	0.00	0.05
8	Ferrous & Non-f. Metals	0.00	0.00	-0.06	-0.34	-0.74	-1.54	-3.15	-5.42	-5.76
9	Non-metallic Min.Pr.	0.00	0.00	-0.03	-0.03	-0.10	-0.16	-0.44	-0.57	-0.10
10	Chemicals	0.00	0.00	0.00	0.05	0.15	0.35	0.50	0.78	1.31
11	Metal Products	0.00	0.03	0.00	-0.03	-0.16	-0.52	-1.99	-4.48	-7.48
12	Agri. & Indust. Mach.	0.00	0.00	0.00	-0.10	-0.20	-0.37	-0.87	-1.37	-1.19
13	Office Machines	0.00	0.00	0.10	0.43	1.26	2.75	6.22	11.90	18.74
14	Electrical Goods	0.00	0.00	0.00	0.00	0.06	0.09	-0.22	-0.65	-0.21
15	Transport Equipment	0.00	0.00	-0.04	-0.08	-0.33	-0.69	-1.51	-2.62	-3.29
16	Food, Drink & Tobacco	0.00	0.00	0.00	-0.06	-0.13	-0.25	-0.69	-1.39	-1.93
17	Tex., Cloth. & Footw.	0.00	0.00	0.00	0.00	0.00	-0.03	-0.19	-0.47	-0.59
18	Paper & Printing Pr.	0.00	0.00	0.04	0.11	0.36	0.86	1.73	3.34	5.01
19	Rubber & Plastic Pr.	0.00	0.00	0.00	0.00	-0.03	-0.10	-0.28	-0.53	-0.32
21	Other Manufactures	0.00	0.00	0.00	0.03	0.07	0.10	0.00	-0.10	0.00
22	Construction	0.00	0.00	0.00	0.00	0.03	0.06	0.12	0.27	0.49
23	Distribution etc.	0.00	0.00	0.00	-0.02	0.00	-0.02	-0.04	-0.11	0.02
24	Lodging & Catering	0.00	0.00	0.00	0.00	0.00	0.02	0.04	0.13	0.18
25	Inland Transport	0.00	0.00	0.00	0.00	0.02	0.05	0.07	0.12	0.37
26	Sea & Air Transport	0.00	0.00	0.08	0.32	0.92	2.18	5.64	12.39	21.49
27	Other Transport	0.00	0.00	0.00	0.03	0.05	0.15	0.33	0.66	1.05
28	Communications	0.00	0.00	0.00	0.03	0.08	0.15	0.27	0.38	0.80
29	Bank., Finance & Ins.	0.00	0.00	0.00	-0.04	-0.08	-0.22	-0.34	-0.47	-0.29
30	Other Market Serv.	0.00	0.00	0.00	0.04	0.12	0.23	0.36	0.64	0.60
31	Non-market Services	0.00	0.00	0.00	0.01	0.04	0.10	0.12	0.15	0.57
32	Unallocated	0.00	0.00	0.00	0.00	0.00	0.00	0.00	0.00	0.00
	Total	0.00	0.00	0.00	0.00	0.04	0.06	0.06	0.09	0.44

Source: Cambridge Econometrics Forecast C51F1Y-C51F1R/da1/F1RA.

Table D2f. Model simulation 1: detailed unit labour cost results – Ireland (%)

		1985	1986	1987	1988	1989	1990	1991	1992	1993
1	Agriculture etc.	0.00	0.00	0.00	-0.26	-0.71	-1.26	-2.23	-3.26	-3.30
2	Coal & Coke	0.00	0.00	0.11	0.57	1.27	3.38	4.45	6.26	7.45
3	Oil & Gas Extraction	0.00	0.00	0.00	0.00	0.00	0.00	0.00	0.00	0.45
4	Gas Distribution	0.00	0.00	0.00	0.00	0.00	0.00	0.00	0.00	0.00
5	Refined Oil	0.00	0.00	0.00	0.00	0.00	0.35	0.32	0.43	0.32
6	Electricity etc.	0.00	0.00	0.00	0.00	0.04	0.11	0.00	0.09	0.36
7	Water Supply	0.00	0.00	0.00	0.00	0.00	0.00	-0.05	-0.08	-0.07
8	Ferrous & Non-f. Metals	0.00	0.02	0.22	0.52	-0.92	-1.29	-3.33	-6.55	-7.48
9	Non-metallic Min.Pr.	0.00	-0.03	-0.05	-0.09	-0.13	-0.26	-0.99	-1.04	-0.65
10	Chemicals	0.00	0.06	0.24	1.03	2.30	5.75	9.09	12.84	15.14
11	Metal Products	0.00	-0.03	-0.11	-0.20	0.15	1.54	4.80	15.60	44.67
12	Agri. & Indust. Mach.	0.00	0.00	-0.14	-0.80	-2.33	-7.85	-14.96	-24.21	-20.87
13	Office Machines	0.00	0.00	0.29	1.15	2.87	6.10	11.37	19.06	22.34
14	Electrical Goods	0.00	-0.04	-0.14	-0.46	-0.92	-1.73	-2.52	-3.93	-3.19
15	Transport Equipment	0.00	0.00	-0.02	-0.02	-0.04	-0.11	-0.66	-1.26	-1.42
16	Food, Drink & Tobacco	0.00	0.00	0.00	-0.09	-0.27	-0.62	-1.28	-1.90	-1.97
17	Tex., Cloth. & Footw.	0.00	-0.03	-0.42	-1.31	-3.19	-4.49	-7.45	-10.97	-10.58
18	Paper & Printing Pr.	0.00	0.00	-0.11	-0.40	-1.12	-2.36	-5.03	-8.56	-10.18
19	Rubber & Plastic Pr.	0.00	0.00	-0.03	-0.15	-0.31	-0.57	-1.31	-2.86	-5.27
21	Other Manufactures	0.00	0.00	0.15	0.58	1.58	3.30	5.68	8.35	9.57
22	Construction	0.00	0.00	0.00	0.00	0.00	0.00	-0.03	-0.01	0.06
23	Distribution etc.	0.00	0.00	0.00	-0.05	-0.07	-0.16	-0.34	-0.50	-0.51
24	Lodging & Catering	0.00	0.00	0.00	-0.02	-0.02	-0.06	0.04	0.50	1.44
25	Inland Transport	0.00	0.00	0.00	-0.03	-0.07	-0.13	-0.22	-0.47	-0.54
26	Sea & Air Transport	0.00	0.00	0.09	0.30	0.71	2.16	5.06	9.24	11.35
27	Other Transport	0.00	0.00	0.04	0.16	0.48	1.43	2.77	5.26	8.18
28	Communications	0.00	0.00	0.00	0.00	-0.03	-0.03	-0.06	-0.06	-0.06
29	Bank., Finance & Ins.	0.00	0.00	0.00	0.00	0.00	0.07	0.00	0.08	0.16
30	Other Market Serv.	0.00	0.00	0.00	0.03	0.03	0.05	0.10	0.27	0.50
31	Non-market Services	0.00	0.00	0.00	-0.02	-0.07	-0.17	-0.28	-0.50	-0.49
32	Unallocated	0.00	0.00	0.00	0.00	0.00	0.00	0.00	0.00	0.00
	Total	0.00	0.00	0.00	0.00	0.00	0.00	0.00	-2.75	0.00

Source: Cambridge Econometrics Forecast C51F1Y-C51F1R/da1/F1RA.

Table D2g.   Model simulation 1: detailed unit labour cost results – Italy (%)

		1985	1986	1987	1988	1989	1990	1991	1992	1993
1	Agriculture etc.	0.00	0.00	0.00	0.05	0.15	0.29	0.92	1.87	3.43
2	Coal & Coke	0.00	0.00	0.00	-33.35	-35.27	-41.10	-45.50	-52.52	-63.67
3	Oil & Gas Extraction	0.00	-0.05	-0.14	-0.60	-1.78	-4.30	-6.15	-10.07	-13.84
4	Gas Distribution	0.00	0.00	0.00	0.00	0.00	0.00	-0.11	0.00	-0.09
5	Refined Oil	0.00	0.00	0.00	-0.06	-0.06	-0.21	-0.46	-0.42	6.69
6	Electricity etc.	0.00	0.00	0.00	0.00	0.00	-0.07	-0.06	-0.11	0.04
7	Water Supply	0.00	0.00	0.00	0.00	0.00	0.00	0.06	0.11	0.23
8	Ferrous & Non-f. Metals	0.00	0.00	0.00	0.00	0.00	0.00	0.00	-0.14	-0.58
9	Non-metallic Min.Pr.	0.00	0.00	-0.04	-0.24	-0.52	-1.12	-1.80	-2.90	-2.65
10	Chemicals	0.00	0.00	0.06	0.13	0.36	0.67	1.24	1.90	-0.16
11	Metal Products	0.00	0.00	-0.08	-0.21	-0.55	-1.02	-1.81	-2.42	-2.13
12	Agri. & Indust. Mach.	0.00	0.00	0.00	0.00	0.00	-0.03	-0.15	-0.31	-1.40
13	Office Machines	0.00	0.00	-0.04	-0.04	-0.10	-0.21	-0.42	-0.76	-1.17
14	Electrical Goods	0.00	0.04	0.08	0.24	0.50	1.05	1.74	2.41	1.65
15	Transport Equipment	0.00	0.00	0.00	0.00	0.03	0.06	0.17	0.05	0.13
16	Food, Drink & Tobacco	0.00	0.00	0.00	0.08	0.08	0.07	0.19	0.29	0.64
17	Tex., Cloth. & Footw.	0.00	0.00	0.09	0.22	0.53	1.18	2.24	3.43	4.36
18	Paper & Printing Pr.	0.00	0.00	0.00	-0.04	-0.07	-0.24	-0.21	-0.60	-0.24
19	Rubber & Plastic Pr.	0.00	0.00	0.00	-0.09	-0.17	-0.37	-0.80	-1.57	-2.60
21	Other Manufactures	0.00	0.00	0.00	-0.05	-0.26	-0.45	-0.88	-1.24	-1.71
22	Construction	0.00	0.00	0.00	0.00	0.00	0.00	0.00	-0.03	0.02
23	Distribution etc.	0.00	0.00	0.00	0.00	-0.05	-0.04	-0.04	-0.07	-0.16
24	Lodging & Catering	0.00	0.00	0.00	0.00	0.06	0.06	-0.05	-0.09	-0.12
25	Inland Transport	0.00	0.00	0.00	0.00	-0.03	-0.08	-0.07	-0.16	-0.43
26	Sea & Air Transport	0.00	0.00	0.00	0.09	0.52	0.43	0.57	1.06	0.52
27	Other Transport	0.00	0.00	0.00	0.00	-0.04	-0.04	-0.07	-0.12	-0.29
28	Communications	0.00	0.00	0.00	0.00	-0.02	-0.05	-0.12	-0.18	0.10
29	Bank., Finance & Ins.	0.00	0.00	0.02	0.00	0.02	0.09	0.18	0.29	0.40
30	Other Market Serv.	0.00	0.00	0.00	0.00	0.04	0.04	0.10	0.21	0.27
31	Non-market Services	0.00	0.00	0.00	0.00	0.00	0.00	-0.01	-0.01	-0.04
32	Unallocated	0.00	0.00	0.00	0.00	0.00	0.00	0.00	0.00	0.00
	Total	0.00	0.00	0.00	0.00	0.00	-0.80	-0.32	-0.27	-0.20

Source: Cambridge Econometrics Forecast C51F1Y-C51F1R/da1/F1RA.

Table D2h. Model simulation 1: detailed unit labour cost results – Luxembourg (%)

		1985	1986	1987	1988	1989	1990	1991	1992	1993
1	Agriculture etc.	0.00	0.00	-0.27	-0.56	-1.43	-2.98	-6.56	-11.37	-16.36
2	Coal & Coke	0.00	0.00	0.00	0.00	0.00	0.00	0.00	0.00	0.00
3	Oil & Gas Extraction	0.00	0.00	0.00	0.00	0.00	0.00	0.00	0.00	0.00
4	Gas Distribution	0.00	0.00	-0.05	-0.11	-0.19	-0.46	-0.71	-0.96	-1.02
5	Refined Oil	0.00	0.00	0.00	0.00	0.00	0.00	0.00	0.00	0.00
6	Electricity etc.	0.00	0.00	-0.14	-0.32	-1.84	-3.93	-14.46	-31.74	-21.77
7	Water Supply	0.00	0.00	0.00	0.04	0.14	0.38	0.87	2.42	4.16
8	Ferrous & Non-f. Metals	0.00	-0.05	-0.52	-1.70	-4.10	-8.98	-15.62	-22.99	-25.94
9	Non-metallic Min.Pr.	0.00	0.00	-0.04	-0.09	-0.18	-0.30	-1.50	-2.13	-3.15
10	Chemicals	0.00	0.08	0.45	4.29	0.93	-1.49	-1.20	15.09	0.85
11	Metal Products	0.00	0.00	-0.26	-0.93	-1.80	-5.42	-10.73	-16.79	-20.53
12	Agri. & Indust. Mach.	0.00	0.00	-0.16	-0.36	-1.05	-1.15	-2.35	-2.46	1.78
13	Office Machines	0.00	0.00	0.00	0.00	0.00	0.00	0.00	0.00	0.00
14	Electrical Goods	0.00	0.00	0.00	0.00	0.00	0.00	0.00	0.00	0.00
15	Transport Equipment	0.00	0.09	1.64	1.52	2.40	5.69	8.24	13.82	8.39
16	Food, Drink & Tobacco	0.00	0.00	0.00	0.00	-0.07	-0.06	-0.23	-0.84	-1.21
17	Tex., Cloth. & Footw.	0.00	-0.33	-2.99	-7.34	-12.84	-20.35	-42.95	-62.16	-63.66
18	Paper & Printing Pr.	0.00	0.00	-0.03	-0.03	-0.13	-0.19	-0.36	-1.17	-5.54
19	Rubber & Plastic Pr.	0.00	0.05	0.22	0.62	2.06	9.03	7.93	10.38	12.96
21	Other Manufactures	0.00	0.00	0.00	-0.77	-2.44	-1.88	1.08	0.16	0.32
22	Construction	0.00	0.00	0.00	0.00	0.00	0.00	0.00	0.02	0.06
23	Distribution etc.	0.00	0.00	0.00	0.00	-0.14	-0.14	-0.14	-0.13	-0.13
24	Lodging & Catering	0.00	0.00	0.00	0.00	0.00	-0.17	0.00	0.16	0.46
25	Inland Transport	0.00	-0.01	-0.03	-0.11	-0.23	-0.45	-0.58	-0.45	-0.58
26	Sea & Air Transport	0.00	0.00	0.00	0.00	0.00	0.00	0.00	0.00	0.00
27	Other Transport	0.00	0.00	0.00	0.00	0.00	0.00	0.00	0.00	0.00
28	Communications	0.00	0.02	0.05	0.22	0.58	1.13	2.29	4.47	6.26
29	Bank., Finance & Ins.	0.00	0.00	0.00	0.00	0.00	0.06	0.12	0.22	0.21
30	Other Market Serv.	0.00	0.00	0.00	0.00	0.03	0.06	0.17	0.30	0.63
31	Non-market Services	0.00	0.00	0.00	-0.01	0.00	0.00	0.00	-0.01	0.00
32	Unallocated	0.00	0.00	0.00	0.00	0.00	0.00	0.00	0.00	0.00
	Total	0.00	0.00	0.00	0.00	0.00	0.00	0.00	0.00	0.00

Source: Cambridge Econometrics Forecast C51F1Y-C51F1R/da1/F1RA.

Table D2i.   Model simulation 1: detailed unit labour cost results – Netherlands (%)

		1985	1986	1987	1988	1989	1990	1991	1992	1993
1	Agriculture etc.	0.00	0.00	0.00	-0.13	-0.41	-0.90	-1.85	-3.07	-3.82
2	Coal & Coke	0.00	0.00	0.00	0.50	3.12	2.38	3.17	1.92	0.00
3	Oil & Gas Extraction	0.00	0.00	0.00	0.00	0.72	0.00	0.52	0.57	1.31
4	Gas Distribution	0.00	0.00	0.00	0.00	0.00	0.00	-0.21	-0.20	-0.21
5	Refined Oil	0.00	0.00	-0.15	-0.33	-0.89	-1.78	-4.56	-1.09	162.61
6	Electricity etc.	0.00	0.00	0.00	0.00	0.00	0.00	0.00	0.00	0.00
7	Water Supply	0.00	0.00	0.00	0.00	0.00	-0.15	-0.14	0.00	0.47
8	Ferrous & Non-f. Metals	0.00	0.00	0.05	0.10	0.34	0.20	-0.17	-0.77	-1.64
9	Non-metallic Min.Pr.	0.00	0.00	-0.16	-0.39	-1.01	-1.79	-3.90	-5.88	-7.28
10	Chemicals	0.00	0.00	0.00	-0.05	-0.11	-0.26	-0.43	-0.88	-1.38
11	Metal Products	0.00	0.00	0.03	0.00	-0.08	-0.16	-0.75	-3.31	-5.70
12	Agri. & Indust. Mach.	0.00	0.00	0.09	0.42	1.19	3.94	11.10	35.73	49.54
13	Office Machines	0.00	-0.05	-0.16	-0.14	-0.42	-0.56	-1.41	-1.65	-1.49
14	Electrical Goods	0.00	0.04	0.25	0.98	2.98	5.83	12.30	20.96	25.01
15	Transport Equipment	0.00	0.00	-0.08	-0.13	-0.28	-0.73	-1.74	-3.21	-4.68
16	Food, Drink & Tobacco	0.00	0.00	0.00	0.21	0.57	0.85	1.64	2.92	3.54
17	Tex., Cloth. & Footw.	0.00	0.00	-0.04	-0.11	-0.26	-0.73	-1.63	-3.14	-5.03
18	Paper & Printing Pr.	0.00	0.00	0.00	0.00	0.04	-0.07	-0.17	-0.82	-1.19
19	Rubber & Plastic Pr.	0.00	-0.04	-0.04	-0.04	-0.18	-0.34	-0.91	-2.11	-3.35
21	Other Manufactures	0.00	0.00	0.00	-0.12	-0.29	-0.57	-1.30	-2.43	-2.95
22	Construction	0.00	0.00	0.00	-0.04	-0.04	-0.07	-0.13	-0.26	-0.37
23	Distribution etc.	0.00	0.00	0.00	0.03	0.03	0.04	0.02	-0.02	-0.11
24	Lodging & Catering	0.00	0.00	0.00	-0.04	-0.08	-0.29	-0.61	-0.57	-2.04
25	Inland Transport	0.00	0.00	0.00	0.02	0.02	0.07	0.14	0.29	0.60
26	Sea & Air Transport	0.00	0.00	0.00	0.00	-0.06	-0.21	-0.61	-1.67	-3.69
27	Other Transport	0.00	0.00	-0.03	-0.05	-0.15	-0.25	-0.53	-1.59	-2.57
28	Communications	0.00	0.00	0.00	-0.02	-0.10	-0.23	-0.48	-0.85	-1.16
29	Bank., Finance & Ins.	0.00	0.00	0.00	0.00	0.00	0.05	0.05	0.00	-0.07
30	Other Market Serv.	0.00	0.00	0.00	0.00	0.00	0.03	0.05	0.10	0.14
31	Non-market Services	0.00	0.00	0.00	0.00	0.00	0.00	0.00	-0.02	-0.05
32	Unallocated	0.00	0.00	0.00	0.00	0.00	0.00	0.00	0.00	0.00
	Total	0.00	0.00	0.00	0.00	0.00	0.83	3.19	13.91	25.51

Source: Cambridge Econometrics Forecast C51F1Y-C51F1R/da1/F1RA.

Table D2j. Model simulation 1: detailed unit labour cost results – Portugal (%)

		1985	1986	1987	1988	1989	1990	1991	1992	1993
1	Agriculture etc.	0.00	0.00	0.00	0.00	0.00	0.15	0.07	0.25	0.60
2	Coal & Coke	0.00	0.00	0.00	-0.78	-3.36	-7.89	-10.00	-19.15	-37.07
3	Oil & Gas Extraction	0.00	0.00	0.00	0.00	0.00	0.00	0.00	0.00	0.00
4	Gas Distribution	0.00	0.00	0.00	0.00	0.00	0.00	0.00	0.00	0.00
5	Refined Oil	0.00	0.00	0.00	-0.30	-0.94	-2.23	-10.21	-17.47	-19.90
6	Electricity etc.	0.00	0.00	0.00	0.00	0.00	-0.19	-0.26	-0.54	-0.43
7	Water Supply	0.00	0.00	0.00	0.00	0.10	0.07	0.00	-0.24	0.86
8	Ferrous & Non-f. Metals	0.00	0.00	-0.05	-0.23	-0.66	-1.18	-2.39	-4.47	-9.07
9	Non-metallic Min.Pr.	0.00	0.00	0.00	-0.07	-0.15	-0.38	-1.24	-2.39	-4.83
10	Chemicals	0.00	0.10	0.09	0.52	1.42	3.28	6.70	12.06	17.34
11	Metal Products	0.00	-0.04	-0.04	-0.24	-0.63	-1.51	-3.09	-5.24	-8.80
12	Agri. & Indust. Mach.	0.00	0.00	0.04	0.04	0.00	-0.04	-0.55	-2.09	-8.40
13	Office Machines	0.00	0.00	0.00	0.00	0.00	0.00	0.00	0.00	0.00
14	Electrical Goods	0.00	0.00	-0.04	-0.12	-0.41	-0.96	-2.23	-4.16	-20.05
15	Transport Equipment	0.00	0.00	-0.11	-0.34	-0.99	-2.64	-5.68	-8.98	-13.67
16	Food, Drink & Tobacco	0.00	0.00	0.00	0.10	0.10	0.16	0.20	0.31	1.24
17	Tex., Cloth. & Footw.	0.00	0.00	0.00	-0.07	-0.11	-0.36	-1.01	-2.02	-2.34
18	Paper & Printing Pr.	0.00	0.00	0.00	-0.15	-0.40	-0.84	-1.78	-3.06	-3.34
19	Rubber & Plastic Pr.	0.00	0.00	0.00	-0.05	-0.18	-0.43	-0.98	-1.62	-2.48
21	Other Manufactures	0.00	0.00	-0.07	-0.17	-0.39	-0.93	-1.91	-2.91	-3.77
22	Construction	0.00	0.00	0.00	0.00	-0.03	-0.05	-0.30	-0.80	-2.84
23	Distribution etc.	0.00	0.00	0.00	-0.03	-0.03	-0.11	-0.40	-0.88	-2.90
24	Lodging & Catering	0.00	0.00	0.00	0.00	0.05	0.09	-0.04	-0.20	0.48
25	Inland Transport	0.00	0.00	0.00	0.04	0.08	0.16	0.04	-0.10	-0.17
26	Sea & Air Transport	0.00	0.00	0.00	0.00	-0.08	-0.18	-0.66	-1.27	-2.73
27	Other Transport	0.00	0.00	0.02	0.07	0.11	0.19	0.21	0.25	0.27
28	Communications	0.00	0.00	0.00	0.00	0.00	0.02	-0.12	-0.28	-0.26
29	Bank., Finance & Ins.	0.00	0.00	0.00	-0.02	-0.02	-0.04	-0.25	-0.44	-0.42
30	Other Market Serv.	0.00	0.00	0.00	0.04	0.09	0.13	0.12	0.20	-0.50
31	Non-market Services	0.00	0.00	0.00	0.00	0.01	0.02	-0.05	-0.15	0.01
32	Unallocated	0.00	0.00	0.00	0.00	0.00	0.00	0.00	0.00	0.00
	Total	0.00	0.00	-0.07	0.00	0.00	0.00	0.00	-1.61	-2.80

Source: Cambridge Econometrics Forecast C51F1Y-C51F1R/da1/F1RA.

Table D2k.   Model simulation 1: detailed unit labour cost results – United Kingdom (%)

		1985	1986	1987	1988	1989	1990	1991	1992	1993
1	Agriculture etc.	0.00	0.00	0.00	-0.06	-0.10	-0.23	-0.54	-0.67	-0.45
2	Coal & Coke	0.00	0.00	0.02	0.02	0.06	0.17	0.29	0.52	0.73
3	Oil & Gas Extraction	0.00	0.00	0.00	0.00	0.00	-0.32	-0.58	-0.74	-1.28
4	Gas Distribution	0.00	0.00	0.00	0.00	0.07	0.13	0.19	0.37	0.41
5	Refined Oil	0.00	0.00	0.00	0.00	-0.18	-0.36	-0.74	-0.96	3.40
6	Electricity etc.	0.00	0.00	0.00	0.00	0.08	0.07	0.07	0.07	0.06
7	Water Supply	0.00	0.00	0.00	0.03	0.06	0.12	0.20	0.30	0.37
8	Ferrous & Non-f. Metals	0.00	0.00	0.05	0.35	1.51	4.94	13.55	33.09	74.91
9	Non-metallic Min.Pr.	0.00	0.00	0.07	0.21	0.52	1.07	2.52	3.71	4.76
10	Chemicals	0.00	0.00	0.00	-0.06	-0.06	-0.16	-0.34	-0.86	-2.45
11	Metal Products	0.00	0.00	0.06	0.23	0.65	1.49	2.70	4.14	5.58
12	Agri. & Indust. Mach.	0.00	0.00	0.03	0.09	0.28	0.54	0.98	1.71	1.63
13	Office Machines	0.00	0.00	-0.06	-0.35	-0.90	-1.88	-3.65	-7.49	-11.32
14	Electrical Goods	0.00	-0.03	-0.09	-0.29	-0.68	-1.49	-2.81	-5.07	-6.61
15	Transport Equipment	0.00	0.00	-0.03	-0.03	-0.16	-0.42	-1.46	-2.64	-3.14
16	Food, Drink & Tobacco	0.00	0.00	0.00	-0.08	-0.21	-0.39	-0.86	-1.57	-2.26
17	Tex., Cloth. & Footw.	0.00	0.00	0.03	0.00	0.00	0.02	-0.11	-0.33	-0.26
18	Paper & Printing Pr.	0.00	0.00	-0.03	-0.20	-0.50	-1.11	-2.03	-3.60	-4.55
19	Rubber & Plastic Pr.	0.00	0.00	0.00	0.03	0.12	0.17	0.17	0.28	0.52
21	Other Manufactures	0.00	0.00	-0.03	-0.11	-0.22	-0.44	-1.06	-1.86	-1.95
22	Construction	0.00	0.00	0.00	0.00	0.05	0.11	0.19	0.23	-0.34
23	Distribution etc.	0.00	0.00	0.00	-0.02	0.00	-0.02	-0.05	-0.09	-0.17
24	Lodging & Catering	0.00	0.00	0.00	0.05	0.10	0.19	0.42	0.85	0.74
25	Inland Transport	0.00	0.00	0.00	0.02	0.09	0.16	0.36	0.62	0.88
26	Sea & Air Transport	0.00	0.00	-0.11	-0.39	-0.95	-2.19	-4.11	-6.87	-7.24
27	Other Transport	0.00	0.03	0.03	0.07	0.14	0.29	0.64	1.40	1.20
28	Communications	0.00	0.00	0.03	0.00	0.02	0.04	0.06	0.10	0.02
29	Bank., Finance & Ins.	0.00	0.00	0.00	0.00	0.03	0.03	0.02	0.00	-0.04
30	Other Market Serv.	0.00	0.00	0.03	0.08	0.23	0.50	0.99	1.70	1.98
31	Non-market Services	0.00	0.00	0.00	-0.01	-0.01	-0.05	-0.13	-0.22	-0.25
32	Unallocated	0.00	0.00	0.00	0.00	0.00	0.00	0.00	0.00	0.00
	Total	0.00	0.00	0.00	0.00	0.03	0.09	0.17	0.41	0.74

Source: Cambridge Econometrics Forecast C51F1Y-C51F1R/da1/F1.

Table D2l.   Model simulation 1: detailed unit labour cost results – EUR-11 (%)

		1985	1986	1987	1988	1989	1990	1991	1992	1993
1	Agriculture etc.	0.00	0.00	-0.02	-0.05	-0.11	-0.22	-0.57	-1.05	-1.58
2	Coal & Coke	0.00	0.00	0.00	-10.91	-6.14	-5.48	-0.18	-0.19	-0.23
3	Oil & Gas Extraction	0.00	-17.39	-0.06	-0.18	-0.18	-0.65	-2.35	-7.21	-3.56
4	Gas Distribution	0.00	0.00	-0.01	-0.03	-0.04	-0.11	-0.23	-0.32	-0.33
5	Refined Oil	0.00	0.00	-0.02	-0.06	-0.17	-0.32	-2.79	-5.62	-62.29
6	Electricity etc.	0.00	0.00	-0.03	-0.08	-0.51	-1.12	-7.43	-21.65	-10.64
7	Water Supply	0.00	0.00	0.00	0.01	0.02	-0.17	-0.23	-7.25	0.00
8	Ferrous & Non-f. Metals	0.00	0.00	-0.02	-0.17	-0.68	-1.36	-2.34	-3.57	-3.31
9	Non-metallic Min.Pr.	0.00	0.00	-0.01	-0.01	-0.01	0.01	-0.27	-0.42	-0.69
10	Chemicals	0.00	0.02	0.06	0.40	0.35	0.75	1.34	2.68	2.39
11	Metal Products	0.00	0.00	-0.02	-0.02	0.04	-0.04	-0.41	-0.60	0.11
12	Agri. & Indust. Mach.	0.00	0.00	0.00	0.02	0.00	-0.11	1.99	15.01	30.06
13	Office Machines	0.00	-0.01	0.00	0.07	0.15	0.42	0.62	0.16	-1.39
14	Electrical Goods	0.00	0.01	0.08	0.29	1.01	1.26	1.39	1.32	0.19
15	Transport Equipment	0.00	0.01	0.14	0.08	-0.03	-0.12	-0.74	-1.29	-3.18
16	Food, Drink & Tobacco	0.00	0.00	0.00	0.01	0.03	0.01	0.05	0.01	0.10
17	Tex., Cloth. & Footw.	0.00	-0.03	-0.36	-0.88	-1.22	-1.69	-4.46	-7.86	-5.70
18	Paper & Printing Pr.	0.00	0.00	-0.01	-0.03	-0.12	-0.34	-0.59	-1.25	-1.85
19	Rubber & Plastic Pr.	0.00	0.00	0.01	0.01	0.06	0.50	0.24	-0.04	-0.69
21	Other Manufactures	0.00	0.00	0.00	-0.03	-0.05	-0.02	-0.06	-0.04	-0.33
22	Construction	0.00	0.00	0.00	0.00	0.00	0.00	-0.01	-0.03	-0.22
23	Distribution etc.	0.00	0.00	0.00	-0.01	-0.01	-0.06	-0.11	-0.20	-0.51
24	Lodging & Catering	0.00	0.00	0.00	0.03	0.04	0.12	0.37	0.90	1.64
25	Inland Transport	0.00	0.00	-0.01	0.00	-0.01	-0.03	-0.03	-0.02	-0.03
26	Sea & Air Transport	0.00	0.00	0.03	0.09	0.31	0.69	0.04	0.12	0.14
27	Other Transport	0.00	0.00	-0.01	-0.01	-0.01	0.05	0.05	0.06	0.26
28	Communications	0.00	0.00	0.01	0.02	0.05	0.11	0.12	0.28	0.58
29	Bank., Finance & Ins.	0.00	0.00	0.00	-0.01	-0.01	-0.01	-0.03	-0.03	0.06
30	Other Market Serv.	0.00	0.00	0.00	-0.01	0.01	0.01	0.00	-0.04	0.11
31	Non-market Services	0.00	0.00	0.00	0.00	-0.01	-0.02	-0.04	-0.09	-0.05
32	Unallocated	0.00	0.00	0.00	0.00	0.00	0.00	0.00	0.00	0.00
	Total	0.00	0.00	0.00	0.00	0.00	0.00	0.00	0.07	0.06

Source: Cambridge Econometrics Forecast C51F1Y-C51F1R/da1/F1RA.

Table D3a. **Model simulation 1: detailed labour cost per employee results – Belgium (%)**

		1985	1986	1987	1988	1989	1990	1991	1992	1993
1	Agriculture etc.	0.00	-0.03	0.19	0.00	-0.45	-0.16	-0.43	-1.05	-0.81
2	Coal & Coke	0.00	-0.01	0.07	0.02	-0.18	0.08	0.33	0.51	0.26
3	Oil & Gas Extraction	0.00	0.00	0.00	0.00	0.00	0.00	0.00	0.00	0.00
4	Gas Distribution	0.00	0.00	-0.02	-0.07	-0.13	-0.31	-0.71	-1.13	-1.44
5	Refined Oil	0.00	-0.02	-0.09	-0.32	-1.00	-2.24	-4.99	-8.07	-5.75
6	Electricity etc.	0.00	0.00	-0.02	0.00	-0.06	-0.06	-0.02	-0.10	-0.03
7	Water Supply	0.00	0.00	0.01	0.02	0.06	-0.14	-0.08	-0.80	-0.45
8	Ferrous & Non-f. Metals	0.00	-0.03	-0.03	-0.10	-0.07	-0.34	-0.64	-0.80	-1.12
9	Non-metallic Min.Pr.	0.00	0.00	0.02	-0.02	-0.02	0.00	0.01	0.01	-0.01
10	Chemicals	0.00	0.00	-0.02	0.00	-0.02	0.02	-0.02	-0.03	-0.01
11	Metal Products	0.00	0.01	0.00	-0.04	-0.16	-0.43	-0.68	-1.16	-1.56
12	Agri. & Indust. Mach.	0.00	0.00	0.00	0.02	-0.01	0.00	0.01	0.01	0.02
13	Office Machines	0.00	0.01	0.05	-0.02	-0.01	-0.11	-0.20	-0.39	-0.64
14	Electrical Goods	0.00	0.01	0.07	0.08	0.49	-1.70	-2.81	-4.72	-4.88
15	Transport Equipment	0.00	0.01	0.02	-0.02	-0.09	-0.34	-0.50	-0.85	-1.06
16	Food, Drink & Tobacco	0.00	0.00	-0.02	-0.03	-0.07	-0.20	-0.41	-0.68	-0.86
17	Tex., Cloth. & Footw.	0.00	0.01	0.02	-0.02	-0.05	-0.23	-0.40	-0.66	-0.86
18	Paper & Printing Pr.	0.00	0.01	-0.01	-0.03	-0.08	-0.24	-0.45	-0.78	-1.05
19	Rubber & Plastic Pr.	0.00	0.01	0.01	-0.02	-0.08	-0.24	-0.53	-0.83	-0.96
21	Other Manufactures	0.00	0.02	0.06	-0.04	-0.01	-0.22	-0.58	1.54	0.47
22	Construction	0.00	0.00	-0.03	-0.01	-0.03	-0.13	-0.28	-0.40	-0.50
23	Distribution etc.	0.00	0.00	0.00	-0.01	-0.03	-0.18	-0.33	-0.49	-0.63
24	Lodging & Catering	0.00	0.01	0.05	0.08	0.28	0.60	1.06	1.86	0.48
25	Inland Transport	0.00	0.00	-0.01	-0.04	-0.12	-0.31	-0.64	-1.22	-1.93
26	Sea & Air Transport	0.00	-0.01	-0.03	-0.04	-0.09	-0.32	-0.51	-0.62	-0.41
27	Other Transport	0.00	0.01	0.01	0.04	0.06	0.15	0.20	0.50	0.58
28	Communications	0.00	0.00	0.01	-0.01	-0.05	-0.19	-0.39	-0.65	-0.87
29	Bank., Finance & Ins.	0.00	0.00	0.00	0.00	-0.04	-0.22	-0.47	-0.75	-1.08
30	Other Market Serv.	0.00	-0.03	0.01	-0.03	-0.01	-0.21	-0.40	-0.66	-0.82
31	Non-market Services	0.00	0.00	0.00	0.00	0.01	0.00	0.00	0.00	0.01
32	Unallocated	0.00	0.00	0.00	0.00	0.00	0.00	0.00	0.00	0.00
	Total	0.00	0.00	0.00	-0.02	-0.06	-0.28	-0.52	-0.83	-0.83

Source: Cambridge Econometrics Forecast C51F1Y-C51F1R/da1/F1RA.

Table D3b.   Model simulation 1: detailed labour cost per employee results – Denmark (%)

		1985	1986	1987	1988	1989	1990	1991	1992	1993
1	Agriculture etc.	0.00	0.00	-0.10	-0.05	0.03	-0.14	-0.27	-0.41	-0.28
2	Coal & Coke	0.00	0.00	-0.14	-0.31	-0.38	-0.93	0.00	0.00	0.00
3	Oil & Gas Extraction	0.00	0.00	0.04	-0.16	-0.30	0.07	-0.08	0.53	-0.11
4	Gas Distribution	0.00	0.00	-0.01	-0.05	-0.10	0.09	-0.06	-0.09	0.00
5	Refined Oil	0.00	0.00	-0.06	0.05	0.33	1.40	5.53	40.13	-88.76
6	Electricity etc.	0.00	0.00	-0.01	-0.14	-0.32	-0.72	-1.18	-1.94	-2.46
7	Water Supply	0.00	0.00	0.00	0.01	-0.09	-0.09	-0.16	-0.35	-0.47
8	Ferrous & Non-f. Metals	0.00	0.03	0.07	0.16	0.13	0.14	0.33	1.44	-3.47
9	Non-metallic Min.Pr.	0.00	0.03	-0.02	0.03	0.12	0.31	0.93	1.46	2.95
10	Chemicals	0.00	-0.01	-0.02	-0.06	-0.20	-0.39	-0.63	-0.95	-1.14
11	Metal Products	0.00	-0.01	-0.03	-0.03	-0.08	-0.15	-0.20	-0.28	-0.49
12	Agri. & Indust. Mach.	0.00	0.00	0.01	-0.01	-0.01	0.00	0.01	-0.02	0.00
13	Office Machines	0.00	0.01	0.00	0.01	0.02	0.08	0.34	1.26	2.76
14	Electrical Goods	0.00	0.00	-0.03	-0.06	-0.21	-0.25	-0.16	-0.77	-1.90
15	Transport Equipment	0.00	0.02	-0.04	-0.15	-0.31	-0.66	-1.51	-4.45	-3.94
16	Food, Drink & Tobacco	0.00	0.00	-0.04	-0.09	-0.20	-0.50	-0.84	-1.24	-1.58
17	Tex., Cloth. & Footw.	0.00	0.00	-0.02	-0.01	-0.05	-0.08	-0.07	0.31	0.24
18	Paper & Printing Pr.	0.00	-0.03	-0.13	-0.40	-0.95	-1.79	-2.88	-4.10	-3.14
19	Rubber & Plastic Pr.	0.00	-0.01	0.00	-0.01	-0.04	-0.02	-0.12	-0.13	0.20
21	Other Manufactures	0.00	0.00	-0.01	-0.02	-0.04	-0.02	-0.04	0.24	0.54
22	Construction	0.00	0.00	0.00	-0.01	-0.04	-0.06	-0.02	-0.01	0.09
23	Distribution etc.	0.00	0.00	-0.01	-0.02	-0.06	-0.11	-0.08	-0.15	-0.12
24	Lodging & Catering	0.00	0.00	-0.04	-0.05	-0.18	-0.37	-0.64	-1.28	-1.96
25	Inland Transport	0.00	0.00	-0.01	-0.01	-0.05	-0.11	-0.18	-0.30	-0.36
26	Sea & Air Transport	0.00	0.00	-0.02	0.02	0.11	0.16	0.40	0.81	1.28
27	Other Transport	0.00	0.01	0.02	0.05	0.01	-0.03	0.14	0.15	-0.40
28	Communications	0.00	0.00	-0.01	-0.06	-0.13	-0.25	-0.44	-0.69	-0.84
29	Bank., Finance & Ins.	0.00	0.00	-0.02	-0.05	-0.14	-0.28	-0.43	-0.71	-0.92
30	Other Market Serv.	0.00	0.00	-0.01	-0.01	-0.09	-0.17	-0.25	-0.32	-0.08
31	Non-market Services	0.00	0.00	0.00	-0.02	-0.07	-0.15	-0.22	-0.39	-0.61
32	Unallocated	0.00	0.00	0.00	0.00	0.00	0.00	0.00	0.00	0.00
	Total	0.00	0.00	-0.02	-0.05	-0.13	-0.23	-0.23	-0.15	-6.97

Source: Cambridge Econometrics Forecast C51F1Y-C51F1R/da1/F1RA.

Table D3c.   Model simulation 1: detailed labour cost per employee results – West Germany (%)

		1985	1986	1987	1988	1989	1990	1991	1992	1993
1	Agriculture etc.	0.00	-0.01	0.00	0.06	0.02	0.02	-0.06	0.08	0.00
2	Coal & Coke	0.00	0.00	-0.02	-0.06	-0.16	0.10	-0.41	-0.40	1.11
3	Oil & Gas Extraction	0.00	0.00	-0.04	-0.03	0.06	-0.15	-0.08	0.00	-0.02
4	Gas Distribution	0.00	0.00	0.00	0.00	0.00	0.07	-0.20	-0.38	0.05
5	Refined Oil	0.00	0.00	-0.02	-0.14	-0.34	-0.40	-0.99	1.75	111.96
6	Electricity etc.	0.00	0.00	-0.01	-0.02	0.01	0.01	0.02	0.04	0.00
7	Water Supply	0.00	0.00	0.00	0.01	-0.01	0.01	-0.02	0.00	0.00
8	Ferrous & Non-f. Metals	0.00	0.00	0.00	-0.06	-0.06	-0.25	-0.36	-0.56	-1.20
9	Non-metallic Min.Pr.	0.00	-0.01	-0.04	-0.14	-0.29	-0.56	-0.94	-1.60	-1.09
10	Chemicals	0.00	0.00	0.03	0.02	-0.02	-0.05	0.03	0.20	0.14
11	Metal Products	0.00	0.00	-0.02	-0.04	-0.07	0.07	-0.32	-0.78	-0.55
12	Agri. & Indust. Mach.	0.00	0.00	-0.01	-0.02	-0.03	0.37	-0.26	-0.21	0.02
13	Office Machines	0.00	0.00	0.00	-0.02	0.01	-0.01	-0.15	-0.07	0.28
14	Electrical Goods	0.00	0.00	0.02	-0.01	0.09	0.39	-0.53	-1.02	-0.05
15	Transport Equipment	0.00	0.02	-0.05	-0.09	-0.25	0.17	-1.30	-1.88	-1.11
16	Food, Drink & Tobacco	0.00	0.00	-0.01	-0.01	-0.02	0.00	0.00	-0.02	0.02
17	Tex., Cloth. & Footw.	0.00	0.00	0.00	0.02	0.10	0.23	0.40	0.74	1.84
18	Paper & Printing Pr.	0.00	0.00	0.00	-0.04	-0.11	-0.46	-0.33	-0.44	-0.93
19	Rubber & Plastic Pr.	0.00	0.00	-0.03	-0.05	-0.14	-0.06	-0.79	-1.54	-1.64
21	Other Manufactures	0.00	-0.01	-0.06	-0.18	-0.40	-0.40	-1.15	-2.17	-2.34
22	Construction	0.00	0.00	0.00	0.00	0.03	-0.05	0.01	0.13	-0.07
23	Distribution etc.	0.00	0.00	0.02	-0.02	0.01	-0.12	-0.02	0.15	-0.22
24	Lodging & Catering	0.00	0.00	0.04	0.01	0.03	0.20	-0.14	-0.39	-0.81
25	Inland Transport	0.00	0.00	-0.02	-0.03	-0.03	0.52	0.88	1.63	4.68
26	Sea & Air Transport	0.00	0.03	-0.01	-0.02	-0.02	0.00	0.00	0.00	0.00
27	Other Transport	0.00	0.00	0.00	0.00	0.00	0.42	0.18	0.17	0.35
28	Communications	0.00	0.00	0.00	0.01	0.02	-0.06	-0.02	0.16	0.03
29	Bank., Finance & Ins.	0.00	0.00	-0.03	-0.03	-0.02	-0.01	-0.18	-0.07	0.27
30	Other Market Serv.	0.00	0.00	0.00	0.01	0.02	0.00	0.00	-0.02	-0.01
31	Non-market Services	0.00	0.00	0.00	0.00	0.00	0.00	0.00	0.00	0.00
32	Unallocated	0.00	0.00	0.00	0.00	0.00	0.00	0.00	0.00	0.00
	Total	0.00	0.00	-0.01	-0.03	-0.06	0.01	-0.24	-0.18	2.79

Source: Cambridge Econometrics Forecast C51F1Y-C51F1R/da1/F1RA.

Table D3d. **Model simulation 1: detailed labour cost per employee results – Spain (%)**

		1985	1986	1987	1988	1989	1990	1991	1992	1993
1	Agriculture etc.	0.00	0.00	-0.04	-0.01	0.04	0.03	0.11	-0.01	-0.38
2	Coal & Coke	0.00	0.07	0.36	1.85	4.37	10.94	24.43	51.87	76.96
3	Oil & Gas Extraction	0.00	0.01	0.03	-0.31	1.17	0.74	-0.28	-0.14	0.43
4	Gas Distribution	0.00	0.00	0.03	0.01	0.04	0.10	0.23	0.23	0.22
5	Refined Oil	0.00	0.01	-0.01	0.01	0.08	0.13	0.28	0.40	0.50
6	Electricity etc.	0.00	0.01	0.01	0.01	0.05	0.14	0.30	0.56	1.06
7	Water Supply	0.00	0.00	0.01	0.02	0.10	0.23	0.47	0.64	0.83
8	Ferrous & Non-f. Metals	0.00	-0.02	0.12	0.31	0.85	2.07	3.64	6.32	7.64
9	Non-metallic Min.Pr.	0.00	0.00	0.00	0.01	0.00	0.03	0.02	0.19	0.06
10	Chemicals	0.00	0.00	0.03	0.03	0.06	0.12	0.49	0.13	-0.04
11	Metal Products	0.00	0.00	-0.01	0.03	0.03	0.04	0.07	0.02	0.42
12	Agri. & Indust. Mach.	0.00	0.00	-0.03	0.00	0.04	0.09	0.15	0.10	0.66
13	Office Machines	0.00	0.00	-0.03	-0.01	0.01	0.00	-0.02	0.00	0.00
14	Electrical Goods	0.00	0.01	0.00	0.00	0.00	0.00	0.01	0.01	0.01
15	Transport Equipment	0.00	0.00	-0.01	0.01	0.03	0.05	0.17	0.33	0.40
16	Food, Drink & Tobacco	0.00	0.00	0.00	-0.01	0.03	-0.03	-0.10	-0.24	-0.09
17	Tex., Cloth. & Footw.	0.00	0.00	0.02	0.01	0.07	0.19	0.27	0.45	1.39
18	Paper & Printing Pr.	0.00	0.00	0.00	0.01	0.01	0.01	0.18	-0.18	-0.28
19	Rubber & Plastic Pr.	0.00	0.00	0.01	-0.02	0.00	0.01	-0.02	0.02	0.01
21	Other Manufactures	0.00	0.00	0.02	-0.01	0.02	-0.01	-0.01	-0.01	0.00
22	Construction	0.00	0.00	0.00	0.02	0.01	0.01	0.01	0.00	0.00
23	Distribution etc.	0.00	0.00	0.01	0.03	0.00	-0.02	0.02	0.00	0.00
24	Lodging & Catering	0.00	0.00	0.00	0.02	0.03	0.09	0.10	0.06	0.04
25	Inland Transport	0.00	0.00	0.02	0.02	0.07	0.19	0.34	0.59	0.85
26	Sea & Air Transport	0.00	-0.01	0.01	-0.03	-0.02	0.00	-0.01	4.13	4.23
27	Other Transport	0.00	0.00	-0.04	-0.10	-0.22	-0.38	-0.62	-1.17	-1.76
28	Communications	0.00	0.00	-0.02	0.03	0.04	0.10	0.15	0.19	0.54
29	Bank., Finance & Ins.	0.00	0.00	0.00	0.02	0.03	0.07	0.10	0.08	-0.08
30	Other Market Serv.	0.00	0.00	0.00	0.01	0.03	0.07	0.12	0.13	0.14
31	Non-market Services	0.00	0.00	0.00	0.00	0.00	0.00	0.00	0.00	0.00
32	Unallocated	0.00	0.00	0.00	0.00	0.00	0.00	0.00	0.00	0.00
	Total	0.00	0.00	0.02	0.07	0.19	0.43	0.74	1.28	1.57

Source: Cambridge Econometrics Forecast C51F1Y-C51F1R/da1/F1RA.

Table D3e.   Model simulation 1: detailed labour cost per employee results — France (%)

		1985	1986	1987	1988	1989	1990	1991	1992	1993
1	Agriculture etc.	0.00	0.01	0.07	-0.01	-0.17	-0.63	-1.47	-3.06	-5.64
2	Coal & Coke	0.00	0.03	0.29	1.30	3.77	11.02	26.40	60.41	100.10
3	Oil & Gas Extraction	0.00	0.00	-0.01	-0.05	0.32	0.06	-0.11	0.27	-0.45
4	Gas Distribution	0.00	0.00	0.00	0.00	0.01	0.02	0.03	0.06	0.19
5	Refined Oil	0.00	0.00	0.00	-0.01	-0.10	-0.14	-0.15	-0.19	6.68
6	Electricity etc.	0.00	0.00	0.01	0.05	0.06	0.13	0.27	0.39	0.65
7	Water Supply	0.00	0.00	0.00	-0.01	-0.02	-0.04	-0.04	-0.13	-0.11
8	Ferrous & Non-f. Metals	0.00	0.01	0.02	0.01	0.08	0.14	0.20	0.11	0.77
9	Non-metallic Min.Pr.	0.00	0.00	-0.01	0.03	0.05	0.12	0.30	0.31	0.93
10	Chemicals	0.00	0.00	-0.01	0.01	0.03	0.12	0.16	0.20	0.57
11	Metal Products	0.00	0.03	-0.01	0.01	0.06	0.07	0.15	0.23	0.63
12	Agri. & Indust. Mach.	0.00	0.01	0.03	0.01	0.08	0.14	0.15	0.20	0.50
13	Office Machines	0.00	-0.01	-0.02	-0.04	-0.10	-0.09	-0.03	0.03	0.89
14	Electrical Goods	0.00	0.00	0.04	0.17	0.54	1.09	1.68	2.25	2.59
15	Transport Equipment	0.00	0.00	0.00	0.06	0.05	0.19	0.35	0.51	0.84
16	Food, Drink & Tobacco	0.00	0.00	0.01	0.00	0.03	0.11	0.10	0.02	0.35
17	Tex., Cloth. & Footw.	0.00	0.00	0.00	0.01	0.04	0.07	0.02	-0.06	0.14
18	Paper & Printing Pr.	0.00	0.00	0.01	-0.01	0.03	0.04	0.02	-0.05	0.38
19	Rubber & Plastic Pr.	0.00	0.00	0.01	0.03	0.05	0.06	0.06	0.02	0.38
21	Other Manufactures	0.00	0.00	-0.01	0.02	0.04	0.10	0.16	0.20	0.57
22	Construction	0.00	0.00	0.00	0.01	0.04	0.11	0.18	0.19	0.62
23	Distribution etc.	0.00	0.00	0.00	0.00	0.06	0.13	0.25	0.30	0.71
24	Lodging & Catering	0.00	0.00	0.00	0.00	-0.01	0.00	-0.01	0.01	-0.02
25	Inland Transport	0.00	0.00	0.00	0.00	0.03	0.08	0.13	0.19	1.14
26	Sea & Air Transport	0.00	-0.01	-0.02	-0.07	-0.18	-0.38	-0.69	-1.19	-1.77
27	Other Transport	0.00	0.00	-0.03	-0.06	-0.17	-0.35	-0.74	-1.51	-2.12
28	Communications	0.00	0.00	-0.01	0.02	0.06	0.10	0.19	0.20	0.64
29	Bank., Finance & Ins.	0.00	0.00	0.02	0.02	0.09	0.18	0.32	0.42	0.79
30	Other Market Serv.	0.00	0.00	-0.01	0.01	0.04	0.08	0.06	0.08	0.42
31	Non-market Services	0.00	0.00	0.00	0.01	0.04	0.10	0.13	0.18	0.70
32	Unallocated	0.00	0.00	0.00	0.00	0.00	0.00	0.00	0.00	0.00
	Total	0.00	0.00	0.01	0.06	0.17	0.47	0.84	1.53	2.78

Source: Cambridge Econometrics Forecast C51F1Y-C51F1R/da1/F1RA.

Table D3f. Model simulation 1: detailed labour cost per employee results – Ireland (%)

		1985	1986	1987	1988	1989	1990	1991	1992	1993
1	Agriculture etc.	0.00	0.01	0.08	0.02	-0.04	0.02	-0.09	-0.11	-0.13
2	Coal & Coke	0.00	-0.02	-0.04	0.09	-0.09	0.02	0.03	-0.09	0.05
3	Oil & Gas Extraction	0.00	0.00	0.02	0.04	0.10	0.01	0.17	0.04	0.06
4	Gas Distribution	0.00	0.00	0.00	0.00	0.00	0.00	0.00	0.00	0.00
5	Refined Oil	0.00	0.00	-0.02	-0.07	-0.14	0.11	-0.02	-0.06	-0.05
6	Electricity etc.	0.00	0.00	-0.01	-0.03	-0.05	-0.13	-0.20	-0.34	-0.49
7	Water Supply	0.00	0.00	0.00	0.00	0.00	0.00	-0.05	-0.08	-0.07
8	Ferrous & Non-f. Metals	0.00	-0.03	-0.15	-0.26	-0.69	-0.94	-0.39	-5.34	-8.65
9	Non-metallic Min.Pr.	0.00	-0.03	0.00	0.00	-0.03	-0.09	0.03	-0.21	-0.09
10	Chemicals	0.00	0.04	-0.02	-0.02	-0.01	-0.03	-0.04	-0.27	-0.40
11	Metal Products	0.00	-0.02	-0.04	-0.12	-0.47	-1.16	-2.34	-4.30	-6.86
12	Agri. & Indust. Mach.	0.00	0.01	0.02	0.00	0.01	0.15	0.15	0.07	-0.79
13	Office Machines	0.00	-0.03	0.00	-0.03	0.03	-0.11	-0.25	-0.41	-0.55
14	Electrical Goods	0.00	-0.02	0.02	0.01	0.07	0.06	-0.13	-0.37	-0.97
15	Transport Equipment	0.00	0.00	-0.01	-0.01	-0.02	-0.02	-0.02	-0.08	-0.12
16	Food, Drink & Tobacco	0.00	0.00	0.03	0.01	-0.05	-0.17	-0.40	-0.70	-0.90
17	Tex., Cloth. & Footw.	0.00	0.00	0.00	-0.02	0.01	0.00	-0.01	0.00	0.00
18	Paper & Printing Pr.	0.00	0.01	0.00	-0.03	-0.07	-0.14	-0.30	-0.54	-0.67
19	Rubber & Plastic Pr.	0.00	0.00	-0.01	-0.05	-0.07	-0.19	-0.34	-0.66	-0.86
21	Other Manufactures	0.00	-0.01	-0.01	-0.03	0.00	0.00	0.00	-0.01	0.00
22	Construction	0.00	0.00	0.00	0.00	-0.01	-0.04	-0.10	-0.15	-0.12
23	Distribution etc.	0.00	0.00	0.01	-0.03	-0.01	-0.07	-0.14	-0.24	-0.29
24	Lodging & Catering	0.00	0.00	0.00	-0.01	0.00	-0.01	-0.14	-0.64	-1.70
25	Inland Transport	0.00	0.00	0.00	-0.03	-0.07	-0.16	-0.28	-0.63	-0.84
26	Sea & Air Transport	0.00	-0.01	0.00	-0.03	-0.12	-0.28	-0.53	-0.99	-1.21
27	Other Transport	0.00	-0.01	-0.03	-0.14	-0.27	-0.64	-1.15	-1.84	-2.25
28	Communications	0.00	0.00	0.00	0.00	-0.04	-0.08	-0.16	-0.29	-0.53
29	Bank., Finance & Ins.	0.00	0.00	0.00	-0.01	-0.03	-0.01	-0.15	-0.14	-0.09
30	Other Market Serv.	0.00	0.00	0.00	0.02	-0.02	-0.11	-0.20	-0.35	-0.53
31	Non-market Services	0.00	0.00	0.00	-0.02	-0.07	-0.17	-0.30	-0.55	-0.62
32	Unallocated	0.00	0.00	0.00	0.00	0.00	0.00	0.00	0.00	0.00
	Total	0.00	0.00	-0.01	-0.03	-0.08	-0.13	-0.22	-0.66	-1.02

Source: Cambridge Econometrics Forecast C51F1Y-C51F1R/da1/F1RA.

Table D3g.   Model simulation 1: detailed labour cost per employee results – Italy (%)

		1985	1986	1987	1988	1989	1990	1991	1992	1993
1	Agriculture etc.	0.00	0.00	0.00	0.03	-0.01	-0.14	-0.35	-0.78	-1.24
2	Coal & Coke	0.00	0.00	0.00	-0.01	-0.02	0.04	0.01	0.11	-0.13
3	Oil & Gas Extraction	0.00	-0.02	0.03	0.05	-0.01	-0.05	0.04	0.16	0.16
4	Gas Distribution	0.00	0.00	0.00	0.00	0.00	0.01	-0.11	0.01	-0.08
5	Refined Oil	0.00	0.00	-0.01	-0.08	-0.09	-0.21	-0.43	-0.10	12.51
6	Electricity etc.	0.00	0.00	0.01	0.05	0.12	0.17	0.34	0.56	0.99
7	Water Supply	0.00	0.00	0.00	0.02	0.04	0.09	0.19	0.28	0.00
8	Ferrous & Non-f. Metals	0.00	0.00	-0.01	-0.01	-0.04	-0.08	-0.09	-0.20	-0.18
9	Non-metallic Min.Pr.	0.00	0.01	0.05	0.09	0.24	0.47	0.79	1.17	1.63
10	Chemicals	0.00	-0.01	0.02	-0.02	-0.01	-0.01	0.00	0.02	0.01
11	Metal Products	0.00	0.01	0.00	0.03	-0.01	0.00	0.01	0.02	0.02
12	Agri. & Indust. Mach.	0.00	0.00	0.00	0.00	0.00	-0.04	-0.12	-0.16	0.11
13	Office Machines	0.00	0.00	-0.03	0.00	0.00	-0.06	-0.08	-0.17	0.04
14	Electrical Goods	0.00	0.03	0.00	0.01	0.00	-0.02	-0.22	-0.46	-0.04
15	Transport Equipment	0.00	0.00	0.00	-0.02	-0.03	-0.06	-0.08	-0.15	-0.06
16	Food, Drink & Tobacco	0.00	0.00	0.00	0.07	0.03	-0.03	0.00	-0.03	-0.02
17	Tex., Cloth. & Footw.	0.00	-0.01	0.02	0.00	-0.01	0.01	0.02	0.07	0.56
18	Paper & Printing Pr.	0.00	0.00	0.02	0.04	0.10	0.03	0.17	-0.16	0.32
19	Rubber & Plastic Pr.	0.00	0.00	0.01	-0.04	-0.03	-0.05	-0.12	-0.24	-0.09
21	Other Manufactures	0.00	0.00	0.02	0.02	-0.10	-0.09	-0.28	-0.33	-0.60
22	Construction	0.00	0.00	0.00	0.00	0.00	-0.01	-0.02	-0.04	0.09
23	Distribution etc.	0.00	0.00	0.00	0.01	-0.04	-0.01	0.00	-0.01	-0.02
24	Lodging & Catering	0.00	0.00	0.00	0.00	0.07	0.08	-0.04	-0.12	-0.57
25	Inland Transport	0.00	0.00	0.00	0.01	-0.01	-0.07	-0.06	-0.16	-0.03
26	Sea & Air Transport	0.00	0.00	-0.06	-0.11	-0.08	-0.67	-1.38	-2.44	-2.24
27	Other Transport	0.00	0.00	0.00	0.00	-0.03	-0.05	-0.10	-0.14	-0.10
28	Communications	0.00	0.00	0.00	0.01	0.01	-0.02	-0.08	-0.11	0.26
29	Bank., Finance & Ins.	0.00	0.00	0.00	-0.05	-0.11	-0.28	-0.52	-0.73	-0.73
30	Other Market Serv.	0.00	0.00	0.00	0.00	0.02	-0.03	-0.03	0.00	0.00
31	Non-market Services	0.00	0.00	0.00	0.00	0.00	-0.01	-0.02	-0.01	0.20
32	Unallocated	0.00	0.00	0.00	0.00	0.00	0.00	0.00	0.00	0.00
	Total	0.00	0.00	0.00	-0.01	-0.01	-0.07	-0.14	-0.19	0.66

Source: Cambridge Econometrics Forecast C51F1Y-C51F1R/da1/F1RA.

Table D3h. **Model simulation 1: detailed labour cost per employee results –**
Luxembourg (%)

		1985	1986	1987	1988	1989	1990	1991	1992	1993
1	Agriculture etc.	0.00	0.01	-0.09	0.09	-0.03	0.05	-0.01	0.01	-0.02
2	Coal & Coke	0.00	0.00	0.00	0.00	0.00	0.00	0.00	0.00	0.00
3	Oil & Gas Extraction	0.00	0.00	0.00	0.00	0.00	0.00	0.00	0.00	0.00
4	Gas Distribution	0.00	0.00	-0.03	-0.03	0.00	-0.01	0.00	0.01	0.00
5	Refined Oil	0.00	0.00	0.00	0.00	0.00	0.00	0.00	0.00	0.00
6	Electricity etc.	0.00	0.01	-0.02	-0.02	0.01	0.00	0.00	-0.01	0.01
7	Water Supply	0.00	0.00	0.00	-0.01	-0.01	0.01	-0.01	-0.02	0.01
8	Ferrous & Non-f. Metals	0.00	0.01	-0.04	-0.01	-0.01	-0.03	0.06	0.01	0.05
9	Non-metallic Min.Pr.	0.00	0.01	0.00	0.00	0.03	-0.02	-0.01	-0.03	0.03
10	Chemicals	0.00	0.02	0.01	-0.02	0.01	0.02	0.06	-0.04	0.02
11	Metal Products	0.00	0.01	0.01	-0.03	-0.01	0.02	0.02	-0.02	0.01
12	Agri. & Indust. Mach.	0.00	0.01	0.02	-0.01	-0.02	0.01	0.02	0.00	0.00
13	Office Machines	0.00	0.00	0.00	0.00	0.00	0.00	0.00	0.00	0.00
14	Electrical Goods	0.00	0.00	0.00	0.00	0.00	0.00	0.00	0.00	0.00
15	Transport Equipment	0.00	-0.01	0.00	-0.02	-0.05	-0.08	-0.01	0.01	0.01
16	Food, Drink & Tobacco	0.00	0.00	0.00	0.03	0.02	0.00	0.02	-0.02	0.00
17	Tex., Cloth. & Footw.	0.00	0.01	-0.01	0.02	0.04	0.04	-0.03	0.00	-0.02
18	Paper & Printing Pr.	0.00	0.00	-0.02	0.00	0.01	-0.01	0.00	0.00	-0.01
19	Rubber & Plastic Pr.	0.00	0.03	0.02	-0.02	0.03	0.00	0.00	-0.02	0.00
21	Other Manufactures	0.00	0.00	-0.01	-0.03	0.04	-0.04	0.04	0.06	-0.10
22	Construction	0.00	0.00	0.00	0.01	0.01	0.00	-0.01	-0.01	-0.01
23	Distribution etc.	0.00	0.00	0.01	0.02	-0.09	-0.05	0.01	0.07	0.07
24	Lodging & Catering	0.00	0.00	0.00	0.00	0.02	-0.15	-0.05	-0.10	-0.03
25	Inland Transport	0.00	-0.01	0.00	-0.02	0.00	0.00	0.00	0.00	0.00
26	Sea & Air Transport	0.00	0.00	0.00	0.00	0.00	0.00	0.00	0.00	0.00
27	Other Transport	0.00	0.00	0.00	0.00	0.00	0.00	0.00	0.00	0.00
28	Communications	0.00	0.01	0.00	0.01	0.02	-0.01	0.00	-0.01	0.00
29	Bank., Finance & Ins.	0.00	0.00	0.00	-0.01	-0.02	0.01	0.00	0.01	0.01
30	Other Market Serv.	0.00	0.00	0.00	0.00	0.01	-0.01	0.00	-0.01	-0.01
31	Non-market Services	0.00	0.00	0.00	-0.01	0.00	0.00	0.00	-0.01	0.00
32	Unallocated	0.00	0.00	0.00	0.00	0.00	0.00	0.00	0.00	0.00
	Total	0.00	0.01	0.00	-0.01	0.00	-0.01	0.01	-0.01	0.00

Source: Cambridge Econometrics Forecast C51F1Y-C51F1R/da1/F1RA.

**Table D3i.   Model simulation 1: detailed labour cost per employee results –
Netherlands (%)**

		1985	1986	1987	1988	1989	1990	1991	1992	1993
1	Agriculture etc.	0.00	0.01	0.04	0.00	-0.03	-0.04	-0.20	-0.53	-0.70
2	Coal & Coke	0.00	0.00	0.00	-0.21	0.49	-0.47	1.22	0.23	-0.16
3	Oil & Gas Extraction	0.00	0.00	-0.01	-0.05	0.63	-0.19	0.19	-0.08	0.13
4	Gas Distribution	0.00	0.00	0.00	0.00	0.01	0.02	-0.19	-0.20	-0.26
5	Refined Oil	0.00	0.00	-0.14	-0.28	-0.76	-1.62	-3.87	-2.36	115.17
6	Electricity etc.	0.00	0.00	0.00	-0.02	-0.05	-0.07	-0.11	-0.17	-0.19
7	Water Supply	0.00	0.00	-0.01	-0.05	-0.15	-0.55	-1.05	-1.83	-2.12
8	Ferrous & Non-f. Metals	0.00	-0.01	-0.01	-0.03	-0.07	-0.23	-0.44	-0.90	-1.34
9	Non-metallic Min.Pr.	0.00	0.02	-0.03	-0.01	-0.02	0.01	-0.01	0.01	0.04
10	Chemicals	0.00	0.00	0.02	0.00	0.03	0.12	0.33	0.61	1.13
11	Metal Products	0.00	-0.01	-0.04	-0.12	-0.24	-0.32	-0.52	-0.94	0.49
12	Agri. & Indust. Mach.	0.00	-0.01	0.01	0.00	0.00	0.00	0.00	-0.01	-0.02
13	Office Machines	0.00	0.00	-0.07	0.05	0.00	0.10	-0.01	0.09	-0.10
14	Electrical Goods	0.00	0.00	0.05	0.02	0.20	-0.58	-1.20	-2.11	-2.05
15	Transport Equipment	0.00	0.00	-0.04	-0.03	-0.01	0.00	0.00	0.00	0.03
16	Food, Drink & Tobacco	0.00	-0.01	-0.06	-0.01	-0.01	-0.35	-0.62	-1.10	-1.55
17	Tex., Cloth. & Footw.	0.00	0.00	0.02	0.08	0.28	0.52	1.27	2.21	3.34
18	Paper & Printing Pr.	0.00	0.00	0.02	0.05	0.20	0.39	0.88	1.10	1.50
19	Rubber & Plastic Pr.	0.00	-0.04	-0.02	0.03	-0.01	-0.02	0.00	-0.01	0.01
21	Other Manufactures	0.00	0.01	0.04	-0.01	-0.01	-0.09	-0.25	-0.70	-1.54
22	Construction	0.00	0.00	0.00	-0.03	-0.02	0.00	0.00	-0.09	-0.24
23	Distribution etc.	0.00	0.00	0.00	0.01	-0.02	-0.06	-0.14	-0.30	-0.41
24	Lodging & Catering	0.00	0.00	0.03	0.03	0.11	0.18	-0.07	-0.91	-5.30
25	Inland Transport	0.00	0.00	-0.02	-0.07	-0.22	-0.25	-0.48	-1.32	-3.59
26	Sea & Air Transport	0.00	0.00	-0.02	0.00	0.01	0.24	1.24	3.43	8.24
27	Other Transport	0.00	0.01	0.02	0.14	0.31	0.64	0.99	2.79	3.55
28	Communications	0.00	0.00	0.03	0.05	0.08	0.18	0.44	0.73	0.82
29	Bank., Finance & Ins.	0.00	0.00	0.00	-0.01	-0.03	-0.03	-0.09	-0.17	-0.19
30	Other Market Serv.	0.00	0.00	0.00	-0.01	-0.03	-0.05	-0.08	-0.12	-0.11
31	Non-market Services	0.00	0.00	0.00	0.00	0.00	0.00	0.01	0.01	0.00
32	Unallocated	0.00	0.00	0.00	0.00	0.00	0.00	0.00	0.00	0.00
	Total	0.00	0.00	-0.02	-0.01	-0.01	-0.09	-0.17	-0.14	2.22

Source: Cambridge Econometrics Forecast C51F1Y-C51F1R/da1/F1RA.

Table D3j.   **Model simulation 1: detailed labour cost per employee results – Portugal (%)**

		1985	1986	1987	1988	1989	1990	1991	1992	1993
1	Agriculture etc.	0.00	0.00	0.01	0.01	-0.02	0.06	-0.10	-0.15	0.18
2	Coal & Coke	0.00	0.02	0.19	0.07	-0.35	0.62	-0.38	-0.34	-0.73
3	Oil & Gas Extraction	0.00	0.08	0.00	0.00	0.00	0.00	0.00	0.00	0.00
4	Gas Distribution	0.00	0.00	0.00	0.00	0.00	0.00	0.00	0.00	0.00
5	Refined Oil	0.00	0.01	0.06	-0.06	0.04	-0.12	-2.14	-3.56	0.00
6	Electricity etc.	0.00	0.00	0.02	0.06	0.16	0.17	0.15	0.23	1.80
7	Water Supply	0.00	0.00	0.00	0.00	0.10	0.07	0.00	-0.24	0.86
8	Ferrous & Non-f. Metals	0.00	0.01	-0.01	0.02	0.06	0.07	-0.07	-0.19	0.82
9	Non-metallic Min.Pr.	0.00	0.00	0.03	0.00	0.04	0.08	0.00	-0.07	1.00
10	Chemicals	0.00	0.09	-0.03	0.02	-0.02	0.01	0.02	0.02	0.04
11	Metal Products	0.00	-0.03	0.04	0.10	0.19	0.40	0.57	0.61	2.23
12	Agri. & Indust. Mach.	0.00	0.00	0.04	0.05	0.06	0.15	0.35	0.83	4.52
13	Office Machines	0.00	0.00	0.00	0.00	0.00	0.00	0.00	0.00	0.00
14	Electrical Goods	0.00	0.01	0.01	0.00	0.07	0.10	-0.15	0.00	1.90
15	Transport Equipment	0.00	0.01	-0.01	-0.01	0.01	0.02	-0.02	-0.03	0.01
16	Food, Drink & Tobacco	0.00	0.00	-0.01	0.07	-0.01	-0.10	-0.26	-0.65	0.07
17	Tex., Cloth. & Footw.	0.00	0.00	0.02	0.02	0.13	0.23	0.21	0.20	1.13
18	Paper & Printing Pr.	0.00	0.01	0.04	0.02	0.04	0.06	-0.04	-0.14	0.60
19	Rubber & Plastic Pr.	0.00	0.00	0.02	0.03	0.02	0.03	-0.11	-0.15	0.28
21	Other Manufactures	0.00	0.01	-0.02	0.03	0.11	0.14	0.02	-0.06	1.15
22	Construction	0.00	0.00	0.00	0.01	0.02	0.07	0.02	0.09	1.71
23	Distribution etc.	0.00	0.00	0.01	0.00	0.04	0.06	-0.04	-0.02	1.47
24	Lodging & Catering	0.00	0.00	0.00	0.00	0.04	0.08	-0.04	-0.18	0.77
25	Inland Transport	0.00	0.00	-0.01	0.02	0.03	0.07	-0.07	-0.17	0.98
26	Sea & Air Transport	0.00	0.00	0.01	0.05	0.06	0.20	0.43	1.52	3.63
27	Other Transport	0.00	0.00	-0.01	0.00	-0.03	-0.03	-0.08	0.05	1.66
28	Communications	0.00	0.00	0.00	0.01	0.02	0.07	0.02	0.00	1.00
29	Bank., Finance & Ins.	0.00	0.00	0.01	0.01	0.04	0.08	-0.03	-0.10	0.85
30	Other Market Serv.	0.00	0.00	0.00	0.03	0.06	0.06	-0.05	-0.14	0.85
31	Non-market Services	0.00	0.00	0.00	0.00	0.01	0.03	-0.04	-0.13	0.19
32	Unallocated	0.00	0.00	0.00	0.00	0.00	0.00	0.00	0.00	0.00
	Total	0.00	0.01	0.01	0.02	0.04	0.09	-0.07	-0.08	1.32

Source: Cambridge Econometrics Forecast C51F1Y-C51F1R/da1/F1RA.

Table D3k.   Model simulation 1: detailed labour cost per employee results – United Kingdom (%)

		1985	1986	1987	1988	1989	1990	1991	1992	1993
1	Agriculture etc.	0.00	0.00	0.03	-0.01	0.00	-0.02	-0.02	-0.08	-0.05
2	Coal & Coke	0.00	0.00	0.00	-0.06	-0.17	-0.43	-0.84	-1.60	-2.93
3	Oil & Gas Extraction	0.00	0.00	0.01	0.02	0.08	-0.11	-0.13	0.04	0.04
4	Gas Distribution	0.00	0.00	0.00	-0.02	0.02	0.03	0.00	0.03	-0.03
5	Refined Oil	0.00	0.00	0.02	0.08	0.03	0.13	0.26	0.51	3.05
6	Electricity etc.	0.00	0.00	0.00	-0.01	0.05	0.00	-0.04	-0.11	-0.18
7	Water Supply	0.00	0.00	0.00	0.01	0.00	0.01	0.01	0.01	-0.04
8	Ferrous & Non-f. Metals	0.00	0.00	-0.01	-0.11	-0.39	-1.11	-2.62	-6.42	-12.72
9	Non-metallic Min.Pr.	0.00	-0.01	-0.04	-0.09	-0.24	-0.40	-0.96	-1.22	-1.26
10	Chemicals	0.00	0.00	0.01	-0.03	0.00	-0.01	-0.02	-0.12	-0.10
11	Metal Products	0.00	-0.01	-0.07	-0.26	-0.74	-1.59	-2.72	-3.93	-4.36
12	Agri. & Indust. Mach.	0.00	0.00	0.00	-0.01	0.01	0.00	0.03	0.01	0.01
13	Office Machines	0.00	0.01	0.03	-0.04	-0.07	-0.09	-0.09	-0.18	-0.18
14	Electrical Goods	0.00	-0.02	0.02	0.05	0.16	0.32	0.44	0.66	0.47
15	Transport Equipment	0.00	0.00	-0.01	0.01	-0.02	-0.01	-0.08	-0.14	-0.14
16	Food, Drink & Tobacco	0.00	0.00	0.02	-0.01	-0.01	0.02	0.05	0.13	0.27
17	Tex., Cloth. & Footw.	0.00	0.00	0.03	-0.01	-0.02	0.00	0.02	0.00	0.17
18	Paper & Printing Pr.	0.00	0.01	0.04	0.02	0.08	0.14	0.23	0.42	0.45
19	Rubber & Plastic Pr.	0.00	0.00	0.00	-0.01	0.02	-0.03	-0.02	-0.05	-0.08
21	Other Manufactures	0.00	0.00	0.00	0.00	0.03	0.01	-0.01	-0.04	0.17
22	Construction	0.00	0.00	-0.01	-0.03	-0.02	-0.05	-0.08	-0.14	0.01
23	Distribution etc.	0.00	0.00	0.00	-0.02	0.00	-0.03	-0.05	-0.07	-0.05
24	Lodging & Catering	0.00	0.00	-0.01	0.01	0.01	0.00	-0.03	0.03	0.00
25	Inland Transport	0.00	0.00	-0.01	-0.01	0.01	-0.02	-0.04	-0.06	-0.07
26	Sea & Air Transport	0.00	0.02	0.04	0.10	0.25	0.41	0.87	1.33	2.32
27	Other Transport	0.00	0.03	0.01	-0.01	-0.04	-0.02	-0.09	-0.04	-0.02
28	Communications	0.00	0.00	0.03	-0.01	-0.02	-0.04	-0.09	-0.11	-0.07
29	Bank., Finance & Ins.	0.00	0.00	0.00	-0.01	0.01	0.00	-0.04	-0.07	-0.08
30	Other Market Serv.	0.00	0.00	0.00	-0.01	-0.01	0.00	-0.01	-0.04	-0.06
31	Non-market Services	0.00	0.00	0.00	0.00	0.01	-0.01	-0.03	-0.03	-0.02
32	Unallocated	0.00	0.00	0.00	0.00	0.00	0.00	0.00	0.00	0.00
	Total	0.00	0.00	0.00	-0.01	-0.03	-0.10	-0.20	-0.40	-0.58

Source: Cambridge Econometrics Forecast C51F1Y-C51F1R/da1/F1RA.

Table D31.   Model simulation 1: detailed labour cost per employee results – EUR-11 (%)

		1985	1986	1987	1988	1989	1990	1991	1992	1993
1	Agriculture etc.	0.00	0.00	0.01	0.01	-0.02	-0.15	-0.35	-0.75	-1.20
2	Coal & Coke	0.00	0.00	0.00	-0.03	-0.08	-0.16	-0.43	-0.62	-1.33
3	Oil & Gas Extraction	0.00	0.00	0.00	0.00	0.13	-0.11	-0.09	0.04	0.04
4	Gas Distribution	0.00	0.00	0.00	-0.01	0.01	0.03	-0.08	-0.10	0.02
5	Refined Oil	0.00	0.00	-0.02	-0.06	-0.16	-0.22	-0.61	0.20	21.34
6	Electricity etc.	0.00	0.00	0.00	0.00	0.04	0.05	0.11	0.17	0.35
7	Water Supply	0.00	0.00	0.00	0.01	0.01	0.00	0.02	-0.03	0.08
8	Ferr. & Non-f. Metals	0.00	0.00	0.01	-0.02	-0.01	-0.12	-0.33	-0.92	-1.73
9	Non-metallic Min.Pr.	0.00	0.00	-0.01	-0.03	-0.05	-0.08	-0.18	-0.27	0.19
10	Chemicals	0.00	0.00	0.01	0.00	0.00	0.03	0.11	0.23	0.31
11	Metal Products	0.00	0.00	-0.02	-0.06	-0.16	-0.28	-0.63	-1.03	-0.65
12	Agri. & Indust. Mach.	0.00	0.00	0.00	-0.01	0.00	0.17	-0.09	-0.08	0.24
13	Office Machines	0.00	0.00	0.00	-0.02	-0.02	-0.03	-0.04	0.05	0.26
14	Electrical Goods	0.00	0.00	0.02	0.04	0.17	0.42	-0.03	-0.22	0.23
15	Transport Equipment	0.00	0.01	-0.02	-0.02	-0.09	0.08	-0.48	-0.75	-0.30
16	Food, Drink & Tobacco	0.00	0.00	0.00	0.00	0.00	-0.02	-0.07	-0.14	-0.01
17	Tex., Cloth. & Footw.	0.00	0.00	0.01	0.01	0.02	0.05	0.03	0.02	0.52
18	Paper & Printing Pr.	0.00	0.00	0.01	-0.01	-0.01	-0.11	-0.01	-0.09	0.11
19	Rubber & Plastic Pr.	0.00	0.00	-0.01	-0.02	-0.04	-0.03	-0.36	-0.72	-0.71
21	Other Manufactures	0.00	0.00	-0.01	-0.05	-0.12	-0.14	-0.38	-0.63	-0.52
22	Construction	0.00	0.00	0.00	-0.01	0.01	-0.01	-0.01	-0.03	-0.05
23	Distribution etc.	0.00	0.00	0.01	-0.01	0.00	-0.04	-0.02	-0.01	-0.04
24	Lodging & Catering	0.00	0.00	0.01	0.01	0.03	0.07	0.01	-0.05	-0.50
25	Inland Transport	0.00	0.00	-0.01	-0.01	-0.02	0.02	0.02	0.00	0.53
26	Sea & Air Transport	0.00	0.01	-0.01	-0.02	0.02	-0.05	0.02	0.23	0.48
27	Other Transport	0.00	0.01	0.00	0.00	-0.03	0.03	-0.10	-0.19	-0.41
28	Communications	0.00	0.00	0.00	0.01	0.01	-0.01	0.01	0.04	0.16
29	Bank., Finance & Ins.	0.00	0.00	0.00	-0.01	-0.01	-0.03	-0.10	-0.13	0.01
30	Other Market Serv.	0.00	0.00	0.00	0.01	0.03	0.02	0.01	0.00	-0.03
31	Non-market Services	0.00	0.00	0.00	0.00	0.01	0.01	0.01	0.01	0.12
32	Unallocated	0.00	0.00	0.00	0.00	0.00	0.00	0.00	0.00	0.00
	Total	0.00	0.00	0.00	0.00	0.02	0.07	0.04	0.16	1.14

Source: Cambridge Econometrics Forecast C51F1Y-C51F1R/da1/F1RA.

Table D4a.   Model simulation 2: detailed employment results – Belgium (%)

		1985	1986	1987	1988	1989	1990	1991	1992	1993
1	Agriculture etc.	0.00	0.00	-0.01	-0.04	-0.10	-0.19	-0.23	-0.49	-0.76
2	Coal & Coke	0.00	0.00	0.00	0.00	0.00	0.00	0.00	0.00	0.00
3	Oil & Gas Extraction	0.00	0.00	0.00	0.00	0.00	0.00	0.00	0.00	0.00
4	Gas Distribution	0.00	0.00	0.00	0.00	0.00	0.00	0.00	0.00	0.00
5	Refined Oil	0.00	0.00	0.02	0.03	0.11	0.35	0.18	0.52	1.04
6	Electricity etc.	0.00	0.00	0.00	0.00	-0.01	-0.01	-0.03	-0.09	-0.18
7	Water Supply	0.00	0.00	0.00	-0.02	-0.05	-0.05	-0.14	-0.34	-0.55
8	Ferrous & Non-f. Metals	0.00	0.01	0.06	0.23	0.67	1.43	2.72	4.72	6.61
9	Non-metallic Min.Pr.	0.00	0.00	0.00	-0.01	-1.39	-2.82	-4.78	-7.95	-12.02
10	Chemicals	0.00	0.00	0.00	0.00	0.00	0.01	0.02	0.04	0.09
11	Metal Products	0.00	-0.02	-0.13	-0.49	-1.14	-1.95	-3.26	-5.76	-5.83
12	Agri. & Indust. Mach.	0.00	0.00	0.00	0.00	0.00	0.00	0.00	0.00	0.00
13	Office Machines	0.00	0.00	0.00	0.00	0.00	0.00	0.00	0.00	0.00
14	Electrical Goods	0.00	0.00	0.02	0.12	0.29	0.57	-1.45	-5.25	-11.75
15	Transport Equipment	0.00	0.00	0.03	0.18	0.46	1.03	2.18	3.53	5.11
16	Food, Drink & Tobacco	0.00	0.00	0.00	-0.01	-0.02	-0.03	-0.05	-0.08	-0.09
17	Tex., Cloth. & Footw.	0.00	0.00	0.04	0.15	0.31	0.83	1.75	2.96	4.18
18	Paper & Printing Pr.	0.00	0.00	0.03	0.15	0.38	0.70	1.56	1.92	2.83
19	Rubber & Plastic Pr.	0.00	0.00	0.04	0.14	0.25	0.68	0.94	1.33	1.55
21	Other Manufactures	0.00	0.01	0.12	0.16	0.33	0.83	1.44	10.04	16.76
22	Construction	0.00	0.00	0.02	0.11	0.37	0.89	1.70	3.06	5.68
23	Distribution etc.	0.00	0.00	0.00	-0.01	-0.02	-0.05	-0.08	-0.16	-0.32
24	Lodging & Catering	0.00	0.01	0.12	0.60	1.38	2.69	5.40	10.15	12.89
25	Inland Transport	0.00	0.00	-0.01	-0.05	-0.15	-0.28	-0.55	-1.09	-1.65
26	Sea & Air Transport	0.00	0.04	0.31	1.16	2.22	4.34	6.37	8.46	8.29
27	Other Transport	0.00	0.00	0.02	0.13	0.31	0.70	1.24	1.90	2.92
28	Communications	0.00	0.01	0.11	0.59	1.74	4.11	8.13	14.22	20.87
29	Bank., Finance & Ins.	0.00	0.00	0.00	0.00	0.00	0.00	0.00	0.00	0.00
30	Other Market Serv.	0.00	0.00	0.03	0.27	1.38	3.85	8.10	14.72	23.58
31	Non-market Services	0.00	0.00	0.00	0.00	0.00	0.00	0.01	0.01	0.02
32	Unallocated	0.00	0.00	0.00	0.00	0.00	0.00	0.00	0.00	0.00
	Total	0.00	0.00	0.02	0.10	0.37	1.02	2.12	3.87	6.02

Source: Cambridge Econometrics Forecast C51F1V-C51F1W/da1/F1WA.

Table D4b. Model simulation 2: detailed employment results – Denmark (%)

		1985	1986	1987	1988	1989	1990	1991	1992	1993
1	Agriculture etc.	0.00	0.00	0.00	0.01	0.02	0.02	-0.03	0.70	0.99
2	Coal & Coke	0.00	0.00	0.00	0.00	0.00	0.00	0.00	0.00	0.00
3	Oil & Gas Extraction	0.00	0.00	0.00	0.00	0.00	0.00	0.00	0.00	0.00
4	Gas Distribution	0.00	0.00	0.00	0.00	0.00	0.00	0.00	0.00	0.00
5	Refined Oil	0.00	0.00	0.00	0.00	0.02	0.00	0.23	0.61	2.42
6	Electricity etc.	0.00	0.00	0.00	0.00	-0.01	-0.04	-0.11	-0.05	-0.16
7	Water Supply	0.00	0.00	-0.01	-0.04	-0.10	-0.25	-0.41	-0.95	-1.19
8	Ferrous & Non-f. Metals	0.00	-0.02	-0.28	-1.17	-2.66	-5.68	-10.26	-24.76	-21.28
9	Non-metallic Min.Pr.	0.00	0.00	0.02	0.08	0.22	0.50	0.88	1.21	1.51
10	Chemicals	0.00	0.00	0.00	0.02	-0.12	-0.44	-0.95	-0.26	-0.53
11	Metal Products	0.00	0.00	-0.02	0.06	0.25	0.39	1.95	7.11	11.68
12	Agri. & Indust. Mach.	0.00	-0.05	-0.40	-1.24	-2.91	-5.69	-9.92	-11.93	-10.92
13	Office Machines	0.00	0.01	0.09	0.29	0.73	1.81	4.15	6.19	9.20
14	Electrical Goods	0.00	0.00	-0.03	-0.12	-0.27	-1.88	-2.71	-3.16	-3.47
15	Transport Equipment	0.00	-0.01	-0.10	-0.34	-0.68	-1.84	-3.35	-7.99	-8.55
16	Food, Drink & Tobacco	0.00	0.00	0.00	0.00	0.02	0.03	0.09	0.06	0.04
17	Tex., Cloth. & Footw.	0.00	0.00	0.04	0.08	0.23	0.45	1.27	-2.09	0.55
18	Paper & Printing Pr.	0.00	0.00	0.00	0.02	0.07	0.20	0.37	0.62	0.81
19	Rubber & Plastic Pr.	0.00	-0.01	-0.10	-0.29	-0.69	-1.20	-2.30	-3.62	-3.29
21	Other Manufactures	0.00	0.01	0.09	0.28	0.69	1.33	2.28	1.38	1.21
22	Construction	0.00	0.00	0.00	0.02	0.04	0.10	0.21	0.59	0.88
23	Distribution etc.	0.00	0.00	0.00	-0.01	-0.02	-0.04	-0.10	-0.06	-0.04
24	Lodging & Catering	0.00	0.00	0.01	0.01	0.07	0.03	0.26	0.70	1.20
25	Inland Transport	0.00	0.00	0.01	0.06	0.14	0.31	0.60	1.13	1.42
26	Sea & Air Transport	0.00	0.00	0.03	0.10	0.25	0.61	1.09	1.75	2.49
27	Other Transport	0.00	0.00	0.01	0.06	0.27	0.66	1.47	3.09	5.63
28	Communications	0.00	0.00	0.00	0.00	-0.01	-0.01	-0.01	-0.04	-0.07
29	Bank., Finance & Ins.	0.00	0.00	0.03	0.13	0.22	0.57	0.86	1.12	0.46
30	Other Market Serv.	0.00	0.00	-0.01	-0.02	-0.03	-0.04	-0.03	0.01	0.30
31	Non-market Services	0.00	0.00	0.00	0.00	0.00	-0.02	-0.04	-0.03	-0.07
32	Unallocated	0.00	0.00	0.00	0.00	0.00	0.00	0.00	0.00	0.00
	Total	0.00	0.00	-0.01	-0.03	-0.07	-0.12	-0.16	-0.12	0.14

Source: Cambridge Econometrics Forecast C51F1V-C51F1W/da1/F1WA.

Table D4c.   Model simulation 2: detailed employment results – West Germany (%)

		1985	1986	1987	1988	1989	1990	1991	1992	1993
1	Agriculture etc.	0.00	0.00	-0.02	-0.01	0.14	-3.05	-3.51	-3.09	-3.49
2	Coal & Coke	0.00	0.00	0.00	0.00	0.00	0.00	0.00	0.00	0.00
3	Oil & Gas Extraction	0.00	0.00	0.00	0.00	0.00	0.00	0.00	0.00	0.00
4	Gas Distribution	0.00	0.00	-0.01	-0.04	-0.07	0.13	-0.26	-0.10	-0.05
5	Refined Oil	0.00	0.00	0.00	0.00	-0.02	0.00	-0.03	0.09	0.12
6	Electricity etc.	0.00	0.00	0.00	0.00	0.00	0.00	0.00	0.00	0.00
7	Water Supply	0.00	0.00	0.00	0.00	0.00	0.00	0.00	0.00	0.00
8	Ferrous & Non-f. Metals	0.00	0.00	-0.05	-0.17	-0.42	-0.93	-1.84	-2.90	-3.55
9	Non-metallic Min.Pr.	0.00	0.00	0.00	0.00	-0.03	-0.11	0.07	-0.06	-0.49
10	Chemicals	0.00	0.00	0.00	0.00	0.01	-0.35	-0.31	-0.17	0.11
11	Metal Products	0.00	0.00	0.00	0.02	0.16	0.65	0.64	1.20	2.07
12	Agri. & Indust. Mach.	0.00	0.00	-0.04	-0.16	-0.40	-0.95	-6.44	-7.20	-6.38
13	Office Machines	0.00	0.00	-0.02	-0.06	-0.14	-0.27	-3.05	-3.65	-3.34
14	Electrical Goods	0.00	0.00	-0.02	-0.03	-0.01	-10.06	-6.55	-2.83	-1.72
15	Transport Equipment	0.00	0.00	0.00	0.00	0.00	0.00	0.00	0.00	0.00
16	Food, Drink & Tobacco	0.00	0.00	-0.01	-0.03	-0.08	0.88	-4.30	-2.24	-3.48
17	Tex., Cloth. & Footw.	0.00	0.00	0.00	-0.01	-0.06	-0.38	-0.25	-1.25	-3.72
18	Paper & Printing Pr.	0.00	0.00	0.00	0.01	-0.01	-0.74	1.57	0.33	-1.85
19	Rubber & Plastic Pr.	0.00	0.00	0.02	0.02	0.01	-3.46	-6.14	-7.64	-10.62
21	Other Manufactures	0.00	0.00	0.00	0.00	0.00	0.03	-0.71	-1.31	-1.55
22	Construction	0.00	0.00	0.03	0.13	0.50	1.34	2.28	4.59	7.20
23	Distribution etc.	0.00	0.00	0.00	-0.01	0.00	-0.16	-0.45	-0.18	-0.55
24	Lodging & Catering	0.00	0.00	-0.01	-0.06	-0.21	-0.66	-1.07	-3.02	-4.91
25	Inland Transport	0.00	-0.01	-0.11	-0.53	-1.82	0.76	-5.97	-17.95	-32.42
26	Sea & Air Transport	0.00	0.00	0.01	0.02	0.05	0.20	0.05	0.14	0.17
27	Other Transport	0.00	0.00	-0.01	-0.05	-0.22	-0.84	-1.09	-2.88	-4.91
28	Communications	0.00	0.00	0.01	0.07	0.30	-3.17	3.17	0.09	8.91
29	Bank., Finance & Ins.	0.00	0.00	0.02	0.07	0.18	0.78	0.77	1.26	-0.26
30	Other Market Serv.	0.00	0.02	0.14	0.51	1.45	3.38	7.52	12.84	18.45
31	Non-market Services	0.00	0.00	0.00	0.00	0.00	0.00	0.02	-0.01	-0.01
32	Unallocated	0.00	0.00	0.00	0.00	0.00	0.00	0.00	0.00	0.00
	Total	0.00	0.00	0.01	0.03	0.11	-0.33	-0.07	0.72	1.16

Source: Cambridge Econometrics Forecast C51F1V-C51F1W/da1/F1WA.

Table D4d. Model simulation 2: detailed employment results – Spain (%)

		1985	1986	1987	1988	1989	1990	1991	1992	1993
1	Agriculture etc.	0.00	0.00	0.00	0.00	0.00	0.00	0.00	0.00	0.00
2	Coal & Coke	0.00	0.00	0.00	0.00	0.00	0.00	0.00	0.00	0.00
3	Oil & Gas Extraction	0.00	0.00	0.00	0.00	0.00	0.00	0.00	0.00	0.00
4	Gas Distribution	0.00	0.00	0.01	0.03	0.05	0.12	0.20	0.29	0.24
5	Refined Oil	0.00	0.00	-0.04	-0.17	-0.33	-0.88	-1.58	-2.51	-2.47
6	Electricity etc.	0.00	0.00	0.00	0.01	0.06	0.11	0.24	0.42	0.34
7	Water Supply	0.00	0.00	0.00	0.00	0.00	0.00	0.00	0.00	0.00
8	Ferrous & Non-f. Metals	0.00	0.00	-0.01	-0.05	-0.15	-0.45	-0.79	-1.70	-3.43
9	Non-metallic Min.Pr.	0.00	0.00	0.00	-0.01	-0.08	-0.30	-0.72	-1.66	-2.86
10	Chemicals	0.00	0.00	0.00	0.02	0.07	0.15	0.56	0.99	0.90
11	Metal Products	0.00	0.01	0.08	0.25	0.52	1.41	2.28	2.71	2.22
12	Agri. & Indust. Mach.	0.00	0.00	0.01	0.01	0.03	0.14	0.28	0.63	0.20
13	Office Machines	0.00	0.00	0.00	0.00	-0.03	-0.15	-0.27	-0.47	-0.88
14	Electrical Goods	0.00	0.00	0.01	0.03	0.08	0.01	0.05	0.28	0.59
15	Transport Equipment	0.00	0.00	0.00	0.00	-0.01	-0.13	-0.17	0.01	0.63
16	Food, Drink & Tobacco	0.00	0.00	0.02	0.08	0.26	0.57	1.27	2.43	3.69
17	Tex., Cloth. & Footw.	0.00	0.00	0.04	0.20	0.69	1.94	4.78	8.22	11.48
18	Paper & Printing Pr.	0.00	0.00	0.00	0.00	0.00	0.00	0.00	0.00	0.00
19	Rubber & Plastic Pr.	0.00	0.00	0.00	0.01	0.04	0.05	0.29	0.10	-0.84
21	Other Manufactures	0.00	0.00	0.02	0.08	0.13	0.08	0.24	-0.27	-0.90
22	Construction	0.00	0.00	0.00	0.00	-0.02	-0.02	0.52	0.98	1.98
23	Distribution etc.	0.00	0.00	0.00	0.00	0.01	0.04	0.10	0.19	0.27
24	Lodging & Catering	0.00	0.00	-0.01	-0.03	-0.09	-0.17	-0.31	-0.49	-0.79
25	Inland Transport	0.00	0.00	0.00	0.00	-0.02	0.00	0.00	-0.06	-0.11
26	Sea & Air Transport	0.00	0.00	0.00	0.00	0.00	0.00	0.00	0.00	0.00
27	Other Transport	0.00	0.00	0.00	0.00	0.00	0.00	0.00	0.00	0.00
28	Communications	0.00	0.00	0.00	-0.02	-0.04	-0.09	-0.16	-0.21	-0.34
29	Bank., Finance & Ins.	0.00	0.00	0.00	-0.01	-0.06	-0.13	-0.33	-0.83	-1.69
30	Other Market Serv.	0.00	0.00	0.00	0.00	0.00	0.00	0.00	0.00	0.00
31	Non-market Services	0.00	0.00	0.00	0.00	0.00	0.00	0.00	0.00	0.00
32	Unallocated	0.00	0.00	0.00	0.00	0.00	0.00	0.00	0.00	0.00
	Total	0.00	0.00	0.00	0.02	0.04	0.10	0.27	0.42	0.53

Source: Cambridge Econometrics Forecast C51F1V-C51F1W/da1/F1WA.

Table D4e.   Model simulation 2: detailed employment results – France (%)

		1985	1986	1987	1988	1989	1990	1991	1992	1993
1	Agriculture etc.	0.00	0.00	0.00	0.00	-0.01	-0.01	-0.05	-0.15	-0.44
2	Coal & Coke	0.00	0.00	0.00	0.00	0.00	0.00	0.00	0.00	0.00
3	Oil & Gas Extraction	0.00	0.00	0.00	0.00	0.00	0.00	0.00	0.00	0.00
4	Gas Distribution	0.00	0.00	0.00	0.00	0.00	0.00	-0.01	-0.04	-0.17
5	Refined Oil	0.00	0.00	0.00	0.01	0.02	0.02	0.04	0.18	0.65
6	Electricity etc.	0.00	0.00	0.00	-0.01	-0.01	-0.05	-0.18	-0.49	-1.34
7	Water Supply	0.00	0.00	-0.01	-0.03	-0.09	-0.23	-0.51	-1.02	-1.64
8	Ferrous & Non-f. Metals	0.00	0.00	0.00	0.04	0.16	0.38	0.80	1.54	2.58
9	Non-metallic Min.Pr.	0.00	0.00	0.00	0.00	0.00	-0.01	-0.02	-0.04	-0.05
10	Chemicals	0.00	0.00	0.01	0.03	0.12	0.33	0.76	1.54	2.74
11	Metal Products	0.00	0.00	0.03	0.09	0.20	0.51	0.90	1.53	2.56
12	Agri. & Indust. Mach.	0.00	0.00	-0.03	-0.11	-0.33	-0.69	-1.40	-2.67	-4.10
13	Office Machines	0.00	0.00	0.04	0.17	0.53	1.59	3.24	6.24	10.05
14	Electrical Goods	0.00	0.00	0.00	0.00	-0.02	0.15	-0.10	-0.15	-0.46
15	Transport Equipment	0.00	0.00	0.00	0.00	0.00	-0.01	-0.05	-0.13	-0.40
16	Food, Drink & Tobacco	0.00	0.00	0.00	-0.01	-0.03	-0.05	-0.16	-0.23	-0.31
17	Tex., Cloth. & Footw.	0.00	0.00	0.02	0.06	0.18	0.41	0.85	1.55	2.41
18	Paper & Printing Pr.	0.00	0.00	0.00	0.00	-0.02	0.12	-0.02	0.36	0.14
19	Rubber & Plastic Pr.	0.00	0.00	-0.03	-0.09	-0.22	-0.41	-0.71	-1.25	-1.35
21	Other Manufactures	0.00	-0.01	-0.05	-0.19	-0.53	-1.11	-2.09	-3.89	-5.43
22	Construction	0.00	0.00	0.00	0.01	0.03	0.11	0.16	0.23	-0.13
23	Distribution etc.	0.00	0.00	0.00	0.02	0.05	0.14	0.29	0.52	0.64
24	Lodging & Catering	0.00	0.00	0.00	0.00	0.00	0.00	0.01	0.03	0.06
25	Inland Transport	0.00	0.00	0.00	0.02	0.05	0.10	0.03	-0.17	-1.11
26	Sea & Air Transport	0.00	0.00	0.00	0.00	-0.01	-0.03	-0.08	-0.20	-0.37
27	Other Transport	0.00	0.00	0.02	0.03	0.05	0.07	0.22	0.08	-0.29
28	Communications	0.00	0.00	0.00	0.00	0.00	-0.01	-0.01	-0.02	-0.05
29	Bank., Finance & Ins.	0.00	0.00	0.00	0.00	0.00	-0.02	-0.21	-0.59	-1.39
30	Other Market Serv.	0.00	0.00	-0.01	-0.03	-0.08	-0.16	-0.40	-0.79	-1.45
31	Non-market Services	0.00	0.00	0.00	0.00	0.00	0.01	0.01	0.01	-0.05
32	Unallocated	0.00	0.00	0.00	0.00	0.00	0.00	0.00	0.00	0.00
	Total	0.00	0.00	0.00	0.00	0.00	0.02	-0.01	-0.05	-0.25

Source: Cambridge Econometrics Forecast C51F1V-C51F1W/da1/F1WA

Table D4f.   Model simulation 2: detailed employment results – Ireland (%)

		1985	1986	1987	1988	1989	1990	1991	1992	1993
1	Agriculture etc.	0.00	0.00	0.00	0.00	0.01	-0.01	-0.05	-0.11	-0.14
2	Coal & Coke	0.00	0.00	0.00	0.00	0.00	0.00	0.00	0.00	0.00
3	Oil & Gas Extraction	0.00	0.00	0.00	0.00	0.00	0.00	0.00	0.00	0.00
4	Gas Distribution	0.00	0.00	0.00	0.00	0.00	0.00	0.00	0.00	0.00
5	Refined Oil	0.00	0.00	0.00	0.00	0.00	0.00	0.00	0.00	0.00
6	Electricity etc.	0.00	0.00	-0.03	-0.10	-0.32	-0.45	-0.59	-0.76	-0.74
7	Water Supply	0.00	0.00	0.00	0.00	0.00	0.00	0.00	0.00	0.00
8	Ferrous & Non-f. Metals	0.00	-0.02	-0.16	-0.37	-0.74	-0.63	-1.99	-2.34	1.77
9	Non-metallic Min.Pr.	0.00	0.01	0.06	0.26	0.08	-0.16	-0.50	-1.17	-4.37
10	Chemicals	0.00	-0.01	-0.13	-0.46	-0.82	-1.95	-2.99	-3.75	-3.46
11	Metal Products	0.00	0.01	0.16	0.84	2.85	7.68	18.89	43.47	91.78
12	Agri. & Indust. Mach.	0.00	0.00	0.00	0.00	0.00	0.00	0.00	0.00	0.00
13	Office Machines	0.00	0.00	0.00	0.00	0.00	0.00	0.00	0.00	0.00
14	Electrical Goods	0.00	0.01	0.07	0.27	0.56	0.74	2.21	3.26	3.29
15	Transport Equipment	0.00	0.00	0.01	0.03	0.05	0.18	1.15	2.03	2.06
16	Food, Drink & Tobacco	0.00	0.00	0.00	0.03	0.08	0.07	0.14	0.52	0.97
17	Tex., Cloth. & Footw.	0.00	0.02	0.25	1.22	2.71	7.39	16.05	28.98	32.25
18	Paper & Printing Pr.	0.00	0.00	0.00	-0.02	-0.17	-0.37	-0.87	-1.71	-1.76
19	Rubber & Plastic Pr.	0.00	0.00	-0.01	-0.05	-0.15	-0.36	-0.63	-0.98	-1.32
21	Other Manufactures	0.00	-0.02	-0.14	-0.48	-1.21	-2.31	-3.53	-4.83	-4.45
22	Construction	0.00	0.00	0.00	-0.02	-0.02	-0.09	-0.12	-0.20	-0.29
23	Distribution etc.	0.00	0.00	0.00	0.00	0.00	0.00	0.00	-0.01	-0.05
24	Lodging & Catering	0.00	0.00	0.01	0.02	0.03	-0.10	-0.17	-0.33	-0.30
25	Inland Transport	0.00	0.00	0.00	-0.01	-0.02	-0.04	-0.09	-0.07	0.05
26	Sea & Air Transport	0.00	0.00	-0.01	-0.05	-0.15	-0.43	-0.89	-1.53	-2.63
27	Other Transport	0.00	-0.01	-0.05	-0.16	-0.27	-0.15	-0.73	-1.24	-0.21
28	Communications	0.00	0.00	0.00	0.00	0.00	-0.01	0.01	0.05	0.15
29	Bank., Finance & Ins.	0.00	0.00	0.00	0.00	0.00	0.01	0.01	0.02	0.05
30	Other Market Serv.	0.00	0.00	0.00	0.00	-0.01	-0.05	-0.02	-0.05	-0.05
31	Non-market Services	0.00	0.00	0.00	0.01	0.02	0.03	0.05	0.08	0.09
32	Unallocated	0.00	0.00	0.00	0.00	0.00	0.00	0.00	0.00	0.00
	Total	0.00	0.00	0.01	0.03	0.08	0.19	0.44	0.77	0.98

Source: Cambridge Econometrics Forecast C51F1V-C51F1W/da1/F1WA.

Table D4g.   Model simulation 2: detailed employment results – Italy (%)

		1985	1986	1987	1988	1989	1990	1991	1992	1993
1	Agriculture etc.	0.00	0.01	0.06	0.26	0.72	1.67	3.39	6.01	8.75
2	Coal & Coke	0.00	0.00	0.00	0.00	0.00	0.00	0.00	0.00	0.00
3	Oil & Gas Extraction	0.00	0.00	0.00	0.00	0.00	0.00	0.00	0.00	0.00
4	Gas Distribution	0.00	0.00	0.00	0.00	0.04	-0.01	0.03	-0.03	0.05
5	Refined Oil	0.00	0.01	0.06	0.21	0.54	1.13	1.68	2.08	-2.86
6	Electricity etc.	0.00	0.00	0.00	0.00	-0.03	-0.07	-0.03	-0.14	-0.22
7	Water Supply	0.00	0.00	-0.01	-0.04	-0.24	-0.50	-0.78	-1.53	-0.13
8	Ferrous & Non-f. Metals	0.00	0.00	0.00	0.00	-0.01	-0.04	-0.09	-0.18	-0.30
9	Non-metallic Min.Pr.	0.00	0.00	0.01	0.03	0.09	0.23	0.35	0.62	0.87
10	Chemicals	0.00	-0.01	-0.09	-0.29	-0.63	-0.99	-2.35	-4.10	-7.27
11	Metal Products	0.00	0.00	0.00	-0.01	0.00	0.00	0.16	0.28	0.78
12	Agri. & Indust. Mach.	0.00	0.00	-0.02	-0.08	-0.21	-0.47	-1.05	-1.82	-2.92
13	Office Machines	0.00	-0.01	-0.06	-0.23	-0.60	-0.83	-1.61	-3.09	-3.15
14	Electrical Goods	0.00	0.00	0.00	0.03	0.07	0.10	0.03	-0.05	-1.19
15	Transport Equipment	0.00	-0.01	-0.07	-0.24	-0.61	-1.34	-2.57	-4.35	-5.80
16	Food, Drink & Tobacco	0.00	0.00	0.01	0.04	0.11	0.29	0.23	0.54	0.96
17	Tex., Cloth. & Footw.	0.00	0.00	0.00	0.00	0.00	0.00	0.00	-0.01	-0.04
18	Paper & Printing Pr.	0.00	0.00	0.01	0.05	0.13	0.58	0.72	0.92	0.40
19	Rubber & Plastic Pr.	0.00	0.00	0.00	0.01	0.02	0.05	0.05	0.05	-0.46
21	Other Manufactures	0.00	0.00	0.00	0.00	0.00	0.01	0.07	0.04	-0.07
22	Construction	0.00	0.00	0.00	-0.01	0.01	-0.12	-0.21	-0.23	-0.10
23	Distribution etc.	0.00	0.00	0.00	0.00	-0.01	-0.02	-0.05	-0.12	-0.18
24	Lodging & Catering	0.00	0.00	-0.01	-0.04	-0.11	-0.18	-0.30	-0.47	-0.51
25	Inland Transport	0.00	0.00	0.00	0.02	0.07	0.21	0.36	0.57	0.38
26	Sea & Air Transport	0.00	-0.03	-0.32	-1.42	-3.93	-4.81	-8.54	-13.34	-15.16
27	Other Transport	0.00	0.00	0.00	0.00	0.00	0.01	0.02	0.03	0.05
28	Communications	0.00	0.00	-0.01	-0.03	-0.07	-0.17	-0.28	-0.42	-0.55
29	Bank., Finance & Ins.	0.00	0.00	0.00	0.00	0.01	0.04	0.12	0.27	0.57
30	Other Market Serv.	0.00	0.00	0.00	-0.01	-0.03	-0.09	-0.16	-0.29	-0.32
31	Non-market Services	0.00	0.00	0.00	0.00	0.00	0.02	0.00	-0.04	-0.34
32	Unallocated	0.00	0.00	0.00	0.00	0.00	0.00	0.00	0.00	0.00
	Total	0.00	0.00	0.00	0.01	0.02	0.10	0.18	0.29	0.30

Source: Cambridge Econometrics Forecast C51F1V-C51F1W/da1/F1WA.

Table D4h.   Model simulation 2: detailed employment results – Luxembourg (%)

		1985	1986	1987	1988	1989	1990	1991	1992	1993
1	Agriculture etc.	0.00	0.00	-0.01	-0.04	0.05	0.26	0.48	1.45	3.87
2	Coal & Coke	0.00	0.00	0.00	0.00	0.00	0.00	0.00	0.00	0.00
3	Oil & Gas Extraction	0.00	0.00	0.00	0.00	0.00	0.00	0.00	0.00	0.00
4	Gas Distribution	0.00	0.00	0.00	0.00	0.00	0.00	0.00	0.00	0.00
5	Refined Oil	0.00	0.00	0.00	0.00	0.00	0.00	0.00	0.00	0.00
6	Electricity etc.	0.00	0.00	0.00	0.00	0.00	0.00	0.00	0.00	0.00
7	Water Supply	0.00	0.00	0.00	0.00	0.00	0.00	0.00	0.00	0.00
8	Ferrous & Non-f. Metals	0.00	0.00	-0.04	-0.19	-0.39	-0.87	-1.83	-2.94	-3.69
9	Non-metallic Min.Pr.	0.00	0.00	0.04	0.08	0.23	0.39	1.60	2.07	2.34
10	Chemicals	0.00	0.00	0.00	0.00	0.00	0.00	0.00	0.00	0.00
11	Metal Products	0.00	0.00	0.02	0.06	0.16	0.25	0.35	-0.11	-0.37
12	Agri. & Indust. Mach.	0.00	0.01	0.13	0.22	0.65	0.76	1.55	-1.34	-1.51
13	Office Machines	0.00	0.00	0.00	0.00	0.00	0.00	0.00	0.00	0.00
14	Electrical Goods	0.00	0.00	0.00	0.00	0.00	0.00	0.00	0.00	0.00
15	Transport Equipment	0.00	0.00	0.00	0.00	0.00	0.00	0.00	0.00	0.00
16	Food, Drink & Tobacco	0.00	0.00	-0.01	-0.07	-0.13	-0.57	-1.96	0.78	0.54
17	Tex., Cloth. & Footw.	0.00	0.00	0.00	0.00	0.00	0.00	0.00	0.00	0.00
18	Paper & Printing Pr.	0.00	0.00	0.00	-0.03	-0.07	-0.02	0.04	-0.23	-1.79
19	Rubber & Plastic Pr.	0.00	-0.01	-0.13	-0.64	-1.11	2.77	-5.06	-7.89	-11.28
21	Other Manufactures	0.00	0.00	0.00	0.00	0.00	0.00	0.00	0.00	0.00
22	Construction	0.00	0.00	0.00	-0.01	-0.03	-0.05	-0.10	-0.26	-0.28
23	Distribution etc.	0.00	0.00	0.00	0.00	0.00	-0.01	0.00	0.00	0.01
24	Lodging & Catering	0.00	0.00	0.00	0.00	0.00	0.02	0.05	0.03	-0.10
25	Inland Transport	0.00	0.00	0.00	0.01	-0.01	-0.02	-0.07	-0.24	-0.43
26	Sea & Air Transport	0.00	0.00	0.00	0.00	0.00	0.00	0.00	0.00	0.00
27	Other Transport	0.00	0.00	0.00	0.00	0.00	0.00	0.00	0.00	0.00
28	Communications	0.00	0.00	0.00	0.00	-0.01	-0.05	-0.11	-0.22	-0.47
29	Bank., Finance & Ins.	0.00	0.00	0.00	0.00	0.00	0.00	-0.01	-0.01	-0.02
30	Other Market Serv.	0.00	0.00	-0.03	-0.19	-0.63	-1.57	-2.83	-6.28	-10.04
31	Non-market Services	0.00	0.00	0.00	0.00	0.00	0.00	0.00	0.00	0.00
32	Unallocated	0.00	0.00	0.00	0.00	0.00	0.00	0.00	0.00	0.00
	Total	0.00	0.00	-0.01	-0.06	-0.14	-0.21	-0.71	-1.47	-2.18

Source: Cambridge Econometrics Forecast C51F1V-C51F1W/da1/F1WA.

Table D4i.   Model simulation 2: detailed employment results – Netherlands (%)

		1985	1986	1987	1988	1989	1990	1991	1992	1993
1	Agriculture etc.	0.00	0.00	0.00	-0.02	-0.05	-0.12	-0.24	-0.49	-0.94
2	Coal & Coke	0.00	0.00	0.00	0.00	0.00	0.00	0.00	0.00	0.00
3	Oil & Gas Extraction	0.00	0.00	0.00	0.00	0.00	0.00	0.00	0.00	0.00
4	Gas Distribution	0.00	0.00	0.00	0.01	0.01	0.05	0.12	0.30	0.41
5	Refined Oil	0.00	0.01	-0.02	0.21	-0.19	0.02	0.62	6.92	51.14
6	Electricity etc.	0.00	0.00	0.00	0.00	0.00	0.00	0.00	0.00	0.00
7	Water Supply	0.00	-0.01	-0.08	-0.44	-1.34	-3.08	-6.68	-13.77	-26.44
8	Ferrous & Non-f. Metals	0.00	0.00	-0.02	-0.06	-0.09	-0.41	-1.20	-2.38	-3.74
9	Non-metallic Min.Pr.	0.00	-0.01	-0.06	-0.16	-0.36	-0.07	-0.05	0.45	0.82
10	Chemicals	0.00	0.00	0.00	0.01	0.01	-0.21	-0.39	-0.54	-0.83
11	Metal Products	0.00	0.00	-0.06	-0.27	-0.74	-1.35	-2.54	-5.12	-8.09
12	Agri. & Indust. Mach.	0.00	0.00	0.00	0.00	0.00	0.00	0.00	0.00	0.00
13	Office Machines	0.00	0.00	0.00	0.00	0.00	0.00	0.00	0.00	0.00
14	Electrical Goods	0.00	-0.01	-0.18	-0.54	-1.08	-0.50	0.66	1.13	2.97
15	Transport Equipment	0.00	0.00	-0.01	0.00	-0.01	-0.12	-1.53	-4.04	-7.70
16	Food, Drink & Tobacco	0.00	0.00	-0.01	-0.06	-0.19	-0.34	-1.02	-1.65	-2.75
17	Tex., Cloth. & Footw.	0.00	0.00	0.00	0.00	-0.01	-0.03	-0.07	-0.14	-0.24
18	Paper & Printing Pr.	0.00	0.00	-0.01	-0.03	-0.08	0.03	-0.25	-0.72	-1.88
19	Rubber & Plastic Pr.	0.00	0.00	0.04	0.14	0.39	0.75	1.11	1.15	1.17
21	Other Manufactures	0.00	0.00	-0.01	-0.03	-0.11	-0.19	-0.34	-0.74	-1.75
22	Construction	0.00	0.00	0.00	0.00	0.00	0.00	0.00	0.00	0.00
23	Distribution etc.	0.00	0.00	0.00	-0.02	-0.06	-0.15	-0.31	-0.63	-1.26
24	Lodging & Catering	0.00	0.00	-0.05	-0.22	-0.71	-1.84	-3.21	-4.75	-7.93
25	Inland Transport	0.00	0.01	0.00	0.04	-0.15	0.39	0.75	2.06	6.28
26	Sea & Air Transport	0.00	-0.03	-0.38	-1.63	-4.07	-8.82	-16.31	-27.31	-38.79
27	Other Transport	0.00	0.00	0.00	0.00	0.00	0.00	0.00	0.00	0.00
28	Communications	0.00	0.00	-0.02	-0.09	-0.22	-0.53	-0.70	-1.23	-0.99
29	Bank., Finance & Ins.	0.00	0.00	0.00	-0.02	-0.05	-0.09	-0.15	-0.28	-0.54
30	Other Market Serv.	0.00	0.00	0.01	0.04	0.08	0.20	0.41	0.79	1.16
31	Non-market Services	0.00	0.00	0.00	0.00	0.00	0.00	0.00	-0.01	-0.02
32	Unallocated	0.00	0.00	0.00	0.00	0.00	0.00	0.00	0.00	0.00
	Total	0.00	0.00	-0.01	-0.03	-0.10	-0.13	-0.23	-0.35	-0.32

Source: Cambridge Econometrics Forecast C51F1V-C51F1W/da1/F1WA.

Table D4j. Model simulation 2: detailed employment results – Portugal (%)

		1985	1986	1987	1988	1989	1990	1991	1992	1993
1	Agriculture etc.	0.00	0.00	0.00	0.00	-0.01	-0.01	-0.03	-0.10	-0.24
2	Coal & Coke	0.00	0.00	0.00	0.00	0.00	0.00	0.00	0.00	0.00
3	Oil & Gas Extraction	0.00	0.00	0.00	0.00	0.00	0.00	0.00	0.00	0.00
4	Gas Distribution	0.00	0.00	0.00	0.00	0.00	0.00	0.00	0.00	0.00
5	Refined Oil	0.00	0.00	0.00	0.00	0.00	0.00	0.00	0.00	0.00
6	Electricity etc.	0.00	0.00	0.00	0.01	0.02	0.04	0.03	0.05	0.17
7	Water Supply	0.00	0.00	0.00	0.00	0.00	0.00	0.00	0.00	0.00
8	Ferrous & Non-f. Metals	0.00	0.01	0.02	-0.01	0.00	0.47	1.55	3.54	7.08
9	Non-metallic Min.Pr.	0.00	0.00	0.02	0.07	0.18	0.43	1.23	2.49	5.85
10	Chemicals	0.00	0.00	0.00	0.00	-0.01	-0.03	-0.07	-0.39	-0.88
11	Metal Products	0.00	0.00	0.00	0.00	0.01	0.05	0.20	0.55	0.72
12	Agri. & Indust. Mach.	0.00	0.00	0.00	0.00	0.00	0.02	0.20	0.76	3.33
13	Office Machines	0.00	0.00	0.00	0.00	0.00	0.00	0.00	0.00	0.00
14	Electrical Goods	0.00	0.00	0.01	0.01	0.44	-0.06	-0.92	0.11	5.77
15	Transport Equipment	0.00	0.00	0.00	0.00	0.01	0.03	0.08	0.10	0.08
16	Food, Drink & Tobacco	0.00	0.00	0.01	0.04	0.05	0.11	0.18	0.47	0.69
17	Tex., Cloth. & Footw.	0.00	0.00	0.01	0.02	0.05	0.10	0.18	0.31	0.45
18	Paper & Printing Pr.	0.00	0.00	0.02	0.08	0.21	0.29	1.31	1.87	3.12
19	Rubber & Plastic Pr.	0.00	0.00	0.03	0.10	0.25	0.56	1.07	1.68	2.59
21	Other Manufactures	0.00	0.00	0.01	0.03	0.07	0.14	0.26	0.29	0.33
22	Construction	0.00	0.00	0.00	0.00	0.01	0.03	0.20	0.84	4.72
23	Distribution etc.	0.00	0.00	0.00	0.00	0.01	0.03	0.08	0.23	0.47
24	Lodging & Catering	0.00	0.00	0.00	0.00	0.00	0.00	-0.02	0.00	0.01
25	Inland Transport	0.00	0.00	-0.01	-0.04	-0.09	-0.18	-0.25	-0.26	0.42
26	Sea & Air Transport	0.00	0.00	0.00	0.00	-0.01	-0.01	-0.19	-0.92	-1.77
27	Other Transport	0.00	0.00	0.00	0.03	0.17	0.52	1.20	2.36	3.98
28	Communications	0.00	0.00	0.00	0.00	-0.01	-0.01	-0.03	-0.05	-0.09
29	Bank., Finance & Ins.	0.00	0.00	0.00	-0.02	-0.05	-0.10	-0.16	-0.27	-0.55
30	Other Market Serv.	0.00	0.00	-0.01	-0.03	-0.09	-0.20	-0.34	-0.45	-0.58
31	Non-market Services	0.00	0.00	0.00	-0.01	-0.01	-0.03	-0.05	-0.07	-0.19
32	Unallocated	0.00	0.00	0.00	0.00	0.00	0.00	0.00	0.00	0.00
	Total	0.00	0.00	0.00	0.00	0.01	0.02	0.08	0.23	0.87

Source: Cambridge Econometrics Forecast C51F1V-C51F1W/da1/F1WA.

Table D4k. Model simulation 2: detailed employment results – United Kingdom (%)

		1985	1986	1987	1988	1989	1990	1991	1992	1993
1	Agriculture etc.	0.00	0.00	0.00	0.01	0.01	0.03	0.04	0.01	-0.33
2	Coal & Coke	0.00	0.00	0.00	0.00	0.00	0.00	0.00	0.00	0.00
3	Oil & Gas Extraction	0.00	0.00	0.00	0.00	0.00	0.00	0.00	0.00	0.00
4	Gas Distribution	0.00	0.00	0.00	0.00	0.00	0.02	0.05	0.15	0.18
5	Refined Oil	0.00	0.00	0.01	0.03	0.02	0.03	-0.03	0.79	3.80
6	Electricity etc.	0.00	0.00	0.00	0.00	0.00	-0.01	-0.02	-0.04	-0.08
7	Water Supply	0.00	0.00	0.00	0.00	0.01	0.01	0.01	-0.05	-0.07
8	Ferrous & Non-f. Metals	0.00	-0.01	-0.08	-0.15	0.17	2.10	7.58	23.52	64.53
9	Non-metallic Min.Pr.	0.00	0.00	0.00	0.00	0.00	0.00	-0.01	-0.11	-0.33
10	Chemicals	0.00	0.01	0.05	0.17	0.33	0.60	1.19	2.93	0.50
11	Metal Products	0.00	0.01	0.08	0.26	0.71	1.50	2.53	3.03	4.15
12	Agri. & Indust. Mach.	0.00	0.00	0.00	0.00	0.00	0.00	0.00	0.00	0.00
13	Office Machines	0.00	0.00	0.03	0.14	0.37	0.89	1.73	3.73	5.26
14	Electrical Goods	0.00	0.00	0.00	-0.02	-0.06	-0.17	-0.42	-0.94	-1.65
15	Transport Equipment	0.00	0.00	0.00	0.01	0.02	0.06	0.24	0.45	0.46
16	Food, Drink & Tobacco	0.00	0.00	0.00	-0.02	-0.04	-0.11	-0.29	-0.74	-1.61
17	Tex., Cloth. & Footw.	0.00	0.00	-0.05	-0.23	-0.65	-1.41	-2.52	-3.86	-4.84
18	Paper & Printing Pr.	0.00	0.00	-0.03	-0.10	-0.23	-0.14	-1.04	-0.04	-0.86
19	Rubber & Plastic Pr.	0.00	0.00	0.00	0.01	0.05	0.17	0.24	0.30	0.62
21	Other Manufactures	0.00	0.00	0.00	0.00	-0.01	-0.02	-0.06	-0.13	-0.25
22	Construction	0.00	0.00	-0.01	-0.03	-0.07	-0.16	-0.33	-0.79	-1.97
23	Distribution etc.	0.00	0.00	0.00	-0.01	-0.02	-0.04	-0.09	-0.18	-0.35
24	Lodging & Catering	0.00	0.00	0.00	0.00	0.00	0.00	-0.01	-0.04	-0.11
25	Inland Transport	0.00	0.00	0.00	-0.01	-0.03	-0.06	-0.11	-0.19	-0.26
26	Sea & Air Transport	0.00	0.00	0.02	0.08	0.22	0.50	1.01	1.98	2.57
27	Other Transport	0.00	0.00	0.00	-0.01	-0.06	-0.16	-0.34	-0.77	-2.03
28	Communications	0.00	0.00	0.00	0.00	0.00	0.00	-0.01	-0.06	-0.25
29	Bank., Finance & Ins.	0.00	0.00	0.00	0.00	0.00	-0.01	-0.03	-0.07	-0.14
30	Other Market Serv.	0.00	0.00	-0.04	-0.13	-0.36	-0.76	-1.48	-2.56	-3.26
31	Non-market Services	0.00	0.00	0.00	0.00	-0.01	-0.01	-0.08	-0.10	-0.10
32	Unallocated	0.00	0.00	0.00	0.00	0.00	0.00	0.00	0.00	0.00
	Total	0.00	0.00	-0.01	-0.03	-0.07	-0.14	-0.29	-0.45	-0.67

Source: Cambridge Econometrics Forecast C51F1V-C51F1W/da1/F1WA.

Table D4l.   Model simulation 2: detailed employment results – EUR-11 (%)

		1985	1986	1987	1988	1989	1990	1991	1992	1993
1	Agriculture etc.	0.00	0.00	0.01	0.06	0.19	0.06	0.44	1.12	1.70
2	Coal & Coke	0.00	0.00	0.00	0.00	0.00	0.00	0.00	0.00	0.00
3	Oil & Gas Extraction	0.00	0.00	0.00	0.00	0.00	0.00	0.00	0.00	0.00
4	Gas Distribution	0.00	0.00	0.00	-0.01	-0.01	0.05	-0.04	0.03	0.03
5	Refined Oil	0.00	0.00	0.01	0.02	0.05	0.10	0.12	0.39	1.07
6	Electricity etc.	0.00	0.00	0.00	0.00	0.00	-0.01	-0.02	-0.06	-0.18
7	Water Supply	0.00	0.00	0.00	-0.01	-0.04	-0.12	-0.24	-0.48	-0.77
8	Ferrous & Non-f. Metals	0.00	0.00	-0.02	-0.06	-0.06	0.13	0.72	2.61	6.72
9	Non-metallic Min.Pr.	0.00	0.00	0.00	0.01	-0.03	-0.06	-0.07	-0.22	-0.48
10	Chemicals	0.00	0.00	0.00	-0.01	-0.02	-0.12	-0.14	0.04	-0.64
11	Metal Products	0.00	0.00	0.02	0.07	0.20	0.58	0.86	1.26	1.84
12	Agri. & Indust. Mach.	0.00	0.00	-0.03	-0.11	-0.28	-0.60	-3.03	-3.45	-3.24
13	Office Machines	0.00	0.00	0.00	0.02	0.06	0.25	-0.52	0.23	0.64
14	Electrical Goods	0.00	0.00	-0.01	-0.03	-0.04	-4.37	-2.84	-1.30	-1.01
15	Transport Equipment	0.00	0.00	-0.01	-0.03	-0.07	-0.16	-0.31	-0.57	-0.86
16	Food, Drink & Tobacco	0.00	0.00	0.00	0.00	0.01	0.26	-0.96	-0.35	-0.61
17	Tex., Cloth. & Footw.	0.00	0.00	0.00	0.00	0.02	0.10	0.42	0.62	0.78
18	Paper & Printing Pr.	0.00	0.00	0.00	-0.01	-0.04	-0.15	0.39	0.30	-0.63
19	Rubber & Plastic Pr.	0.00	0.00	0.00	0.00	-0.01	-1.23	-2.42	-3.07	-4.16
21	Other Manufactures	0.00	0.00	0.00	-0.01	-0.04	-0.10	-0.33	-0.62	-0.91
22	Construction	0.00	0.00	0.00	0.03	0.10	0.26	0.52	1.10	1.81
23	Distribution etc.	0.00	0.00	0.00	0.00	0.00	-0.03	-0.08	-0.03	-0.16
24	Lodging & Catering	0.00	0.00	0.00	-0.01	-0.05	-0.14	-0.20	-0.44	-0.75
25	Inland Transport	0.00	0.00	-0.03	-0.12	-0.44	0.17	-0.68	-2.33	-5.07
26	Sea & Air Transport	0.00	0.00	-0.06	-0.27	-1.08	-0.80	-1.06	-1.65	-1.39
27	Other Transport	0.00	0.00	0.00	0.00	-0.03	-0.17	-0.21	-0.70	-1.43
28	Communications	0.00	0.00	0.00	0.03	0.11	-0.63	0.95	0.40	2.63
29	Bank., Finance & Ins.	0.00	0.00	0.01	0.02	0.04	0.17	0.15	0.21	-0.33
30	Other Market Serv.	0.00	0.00	0.02	0.07	0.22	0.56	1.32	2.34	3.39
31	Non-market Services	0.00	0.00	0.00	0.00	0.00	0.00	-0.01	-0.03	-0.09
32	Unallocated	0.00	0.00	0.00	0.00	0.00	0.00	0.00	0.00	0.00
	Total	0.00	0.00	0.00	0.01	0.02	-0.04	0.04	0.26	0.38

Source: Cambridge Econometrics Forecast C51F1V-C51F1W/da1/F1WA.

Table D5a.   Model simulation 2: detailed unit labour cost results – Belgium (%)

		1985	1986	1987	1988	1989	1990	1991	1992	1993
1	Agriculture etc.	0.00	0.00	0.48	0.46	0.43	2.40	5.20	4.00	3.12
2	Coal & Coke	0.00	0.09	1.73	6.45	18.43	42.04	91.03	142.42	160.00
3	Oil & Gas Extraction	0.00	0.00	0.00	0.00	0.00	0.00	0.00	0.00	0.00
4	Gas Distribution	0.00	0.00	0.00	0.20	0.57	0.83	1.42	2.85	2.87
5	Refined Oil	0.00	0.00	0.36	1.24	3.24	6.71	12.12	24.28	40.33
6	Electricity etc.	0.00	0.00	-0.07	-0.15	-0.16	-0.35	-0.43	-0.71	-0.89
7	Water Supply	0.00	0.00	0.00	-0.18	-0.19	-0.52	-1.00	-0.43	-0.54
8	Ferrous & Non-f. Metals	0.00	0.07	0.55	1.90	5.17	10.23	19.18	32.98	39.56
9	Non-metallic Min.Pr.	0.00	0.03	0.50	2.36	4.04	5.89	7.78	8.22	2.75
10	Chemicals	0.00	0.00	0.00	-0.05	-0.10	-0.29	-0.53	-0.87	-0.89
11	Metal Products	0.00	0.03	0.23	0.98	2.29	3.50	4.62	4.70	2.13
12	Agri. & Indust. Mach.	0.00	0.00	0.05	0.19	0.67	1.51	2.51	3.69	3.73
13	Office Machines	0.00	0.00	0.08	0.48	1.10	0.38	2.39	3.76	4.26
14	Electrical Goods	0.00	0.09	0.82	3.37	11.70	14.20	16.53	17.74	19.09
15	Transport Equipment	0.00	0.00	-0.11	-0.51	-1.52	-3.75	-8.29	-14.80	-22.38
16	Food, Drink & Tobacco	0.00	0.00	0.00	0.08	0.15	0.00	1.58	-1.72	-1.39
17	Tex., Cloth. & Footw.	0.00	0.00	-0.09	-0.15	-0.68	-1.57	-2.74	-4.27	-6.42
18	Paper & Printing Pr.	0.00	0.00	0.03	0.13	0.30	0.63	0.60	2.10	1.90
19	Rubber & Plastic Pr.	0.00	0.00	0.00	0.09	0.18	0.00	0.42	0.96	0.89
21	Other Manufactures	0.00	0.00	0.00	0.17	0.38	0.30	1.07	15.26	24.96
22	Construction	0.00	0.00	0.00	-0.15	-0.30	-0.41	-0.55	-1.09	-1.67
23	Distribution etc.	0.00	0.00	0.02	0.07	0.08	-0.12	-0.13	-0.31	-1.65
24	Lodging & Catering	0.00	0.00	0.00	-0.34	-0.44	-1.49	-3.42	-6.86	-8.42
25	Inland Transport	0.00	0.00	0.02	0.07	0.22	0.31	0.81	1.55	1.96
26	Sea & Air Transport	0.00	0.00	-0.21	-0.76	-1.59	-3.22	-5.00	-7.10	-8.35
27	Other Transport	0.00	0.00	0.00	-0.14	-0.12	-0.52	-0.60	-0.07	-1.03
28	Communications	0.00	-0.03	-0.20	-0.82	-1.61	-2.69	-3.74	-4.11	-1.64
29	Bank., Finance & Ins.	0.00	0.00	0.03	0.06	0.15	0.09	0.31	1.18	1.20
30	Other Market Serv.	0.00	-0.08	-0.75	-3.09	-6.36	-10.64	-15.32	-18.57	-16.72
31	Non-market Services	0.00	0.00	0.00	0.00	0.00	-0.01	-0.02	-0.04	-0.05
32	Unallocated	0.00	0.00	0.00	0.00	0.00	0.00	0.00	0.00	0.00
	Total	0.00	0.00	0.06	0.18	0.52	0.30	0.17	0.04	-0.55

Source: Cambridge Econometrics Forecast C51F1V-C51F1W/da1/F1WA.

Table D5b.   Model simulation 2: detailed unit labour cost results – Denmark (%)

		1985	1986	1987	1988	1989	1990	1991	1992	1993
1	Agriculture etc.	0.00	0.00	0.00	0.00	0.00	0.12	0.11	-1.60	-1.43
2	Coal & Coke	0.00	0.00	0.00	0.00	0.00	0.00	0.00	0.00	0.00
3	Oil & Gas Extraction	0.00	0.00	0.00	0.00	-0.74	-1.32	-2.25	-3.89	-3.65
4	Gas Distribution	0.00	0.00	0.00	0.00	0.00	0.26	0.22	0.24	0.26
5	Refined Oil	0.00	0.00	-0.13	-0.13	-0.88	-2.76	-5.30	-20.95	-94.20
6	Electricity etc.	0.00	0.00	0.00	0.00	-0.07	0.10	0.00	0.38	1.33
7	Water Supply	0.00	0.00	0.00	0.00	0.00	0.11	0.10	0.43	1.01
8	Ferrous & Non-f. Metals	0.00	-0.06	-0.10	-0.52	-1.32	-2.71	-4.91	-10.51	-7.69
9	Non-metallic Min.Pr.	0.00	0.00	0.09	0.36	0.94	1.82	3.24	3.72	3.53
10	Chemicals	0.00	0.00	0.00	0.00	0.08	0.51	0.54	0.50	0.80
11	Metal Products	0.00	0.00	0.00	0.06	0.20	0.37	1.65	3.05	4.47
12	Agri. & Indust. Mach.	0.00	-0.03	-0.05	-0.15	-0.30	-0.41	-0.50	0.32	1.88
13	Office Machines	0.00	-0.02	-0.17	-0.69	-1.54	-3.93	-7.71	-7.09	-10.60
14	Electrical Goods	0.00	0.03	0.07	0.34	0.67	3.24	2.01	4.34	6.20
15	Transport Equipment	0.00	0.03	0.18	0.59	1.21	1.86	6.36	17.70	16.58
16	Food, Drink & Tobacco	0.00	0.00	0.00	0.13	0.45	0.82	2.64	2.08	3.63
17	Tex., Cloth. & Footw.	0.00	0.00	0.11	0.33	0.91	2.07	4.19	-0.73	1.08
18	Paper & Printing Pr.	0.00	0.00	0.12	0.49	1.30	3.36	4.35	8.05	9.27
19	Rubber & Plastic Pr.	0.00	0.00	0.03	0.12	0.27	-0.03	1.33	1.96	2.00
21	Other Manufactures	0.00	0.00	-0.10	-0.32	-0.82	-1.92	-4.33	-6.47	-9.20
22	Construction	0.00	0.00	0.00	0.03	0.06	0.17	0.13	0.42	0.29
23	Distribution etc.	0.00	0.00	0.00	0.05	0.09	0.45	0.28	0.58	-0.02
24	Lodging & Catering	0.00	0.00	0.00	0.08	0.10	0.57	1.34	3.99	3.79
25	Inland Transport	0.00	0.00	-0.02	-0.07	-0.14	-0.24	-0.70	-1.21	-1.86
26	Sea & Air Transport	0.00	0.00	-0.13	-0.49	-0.96	-1.92	-3.59	-5.21	-6.94
27	Other Transport	0.00	-0.04	-0.29	-1.01	-2.09	-4.48	-8.73	-13.35	-14.89
28	Communications	0.00	0.00	0.00	0.00	0.03	0.03	0.16	0.00	0.06
29	Bank., Finance & Ins.	0.00	0.00	-0.07	-0.24	-0.57	-1.33	-3.07	-4.56	-6.71
30	Other Market Serv.	0.00	0.00	-0.05	0.00	-0.04	-0.04	-0.16	-0.31	-0.43
31	Non-market Services	0.00	0.00	0.00	0.03	0.04	0.11	-0.02	0.36	0.26
32	Unallocated	0.00	0.00	0.00	0.00	0.00	0.00	0.00	0.00	0.00
	Total	0.00	0.00	-0.02	0.00	0.00	-0.06	-0.15	-0.24	-9.34

Source: Cambridge Econometrics Forecast C51F1V-C51F1W/da1/F1WA.

Table D5c.   Model simulation 2: detailed unit labour cost results – West Germany (%)

		1985	1986	1987	1988	1989	1990	1991	1992	1993
1	Agriculture etc.	0.00	0.00	0.11	0.35	0.97	-1.10	0.34	3.79	1.57
2	Coal & Coke	0.00	0.00	0.00	0.20	0.50	-3.23	-0.16	1.86	5.37
3	Oil & Gas Extraction	0.00	0.09	0.00	0.35	0.75	0.95	1.10	2.22	1.44
4	Gas Distribution	0.00	0.00	0.00	0.00	0.10	-1.03	-0.10	0.78	1.95
5	Refined Oil	0.00	0.00	0.00	0.09	-0.08	-2.92	0.19	9.85	143.01
6	Electricity etc.	0.00	0.00	0.00	0.00	0.06	0.00	-0.12	-0.25	-0.66
7	Water Supply	0.00	0.00	-0.05	-0.20	-0.74	-11.95	-44.83	-99.99	-99.97
8	Ferrous & Non-f. Metals	0.00	0.00	0.07	0.30	0.49	1.57	2.17	2.65	1.51
9	Non-metallic Min.Pr.	0.00	0.00	0.06	0.16	0.27	-0.97	0.74	1.01	0.91
10	Chemicals	0.00	0.00	0.00	-0.04	-0.12	0.44	0.76	0.56	-0.07
11	Metal Products	0.00	0.00	-0.03	-0.05	-0.11	-3.45	-2.53	-2.59	-1.74
12	Agri. & Indust. Mach.	0.00	0.00	0.07	0.18	0.41	-3.05	-1.21	-0.54	0.48
13	Office Machines	0.00	0.00	0.05	0.14	0.27	-1.19	-0.22	0.83	1.58
14	Electrical Goods	0.00	0.00	0.02	0.00	0.02	-17.07	-8.36	-4.39	-2.45
15	Transport Equipment	0.00	0.00	0.19	0.49	1.08	-5.14	3.75	4.33	2.14
16	Food, Drink & Tobacco	0.00	0.00	0.00	0.07	0.13	1.81	-5.61	-0.49	-2.26
17	Tex., Cloth. & Footw.	0.00	0.00	-0.04	-0.16	-0.47	-2.45	-2.48	-4.45	-6.70
18	Paper & Printing Pr.	0.00	0.00	0.04	0.03	-0.03	3.15	1.57	1.27	-3.37
19	Rubber & Plastic Pr.	0.00	0.00	0.03	0.06	0.06	-7.35	-6.47	-7.33	-8.45
21	Other Manufactures	0.00	0.00	0.00	-0.09	-0.37	-3.64	-2.07	-5.27	-8.20
22	Construction	0.00	0.00	-0.03	-0.03	-0.22	0.56	0.36	-1.00	-2.32
23	Distribution etc.	0.00	0.00	0.00	0.05	0.00	1.93	1.58	1.00	-1.68
24	Lodging & Catering	0.00	0.04	0.09	0.38	1.12	-0.66	4.07	6.94	13.38
25	Inland Transport	0.00	0.00	-0.04	-0.16	-0.52	-2.39	-5.86	-11.16	-17.41
26	Sea & Air Transport	0.00	0.00	0.06	0.23	0.46	2.15	0.05	0.14	0.17
27	Other Transport	0.00	0.04	0.36	1.29	2.84	-1.43	3.23	12.03	19.94
28	Communications	0.00	0.02	0.00	0.05	0.05	-2.96	2.99	-3.27	1.31
29	Bank., Finance & Ins.	0.00	-0.03	-0.06	-0.25	-0.68	-2.13	-3.05	-6.35	-10.15
30	Other Market Serv.	0.00	0.00	0.00	0.07	-0.14	-0.62	-0.07	-1.39	-0.83
31	Non-market Services	0.00	0.00	0.00	0.00	0.00	0.00	0.02	-0.02	-0.01
32	Unallocated	0.00	0.00	0.00	0.00	0.00	0.00	0.00	0.00	0.00
	Total	0.00	0.00	0.00	0.00	0.00	-3.49	0.05	-1.33	-0.53

Source: Cambridge Econometrics Forecast C51F1V-C51F1W/da1/F1WA.

Table D5d. **Model simulation 2: detailed unit labour cost results – Spain (%)**

		1985	1986	1987	1988	1989	1990	1991	1992	1993
1	Agriculture etc.	0.00	0.00	0.00	0.07	0.18	0.53	1.21	2.59	3.36
2	Coal & Coke	0.00	-0.08	-0.20	-0.73	-1.47	-3.91	-7.46	-13.33	-13.86
3	Oil & Gas Extraction	0.00	-0.64	-1.06	-7.02	-16.00	-47.41	-100.00	-100.00	-100.00
4	Gas Distribution	0.00	0.00	0.00	-0.07	-0.08	-0.26	-0.54	-0.79	-0.97
5	Refined Oil	0.00	0.00	-0.15	-0.77	-1.69	-4.11	-7.37	-11.40	-12.42
6	Electricity etc	0.00	0.00	-0.05	-0.04	-0.08	-0.32	-0.53	-0.73	-0.38
7	Water Supply	0.00	0.00	0.00	0.00	0.04	0.12	0.42	1.34	2.78
8	Ferrous & Non-f. Metals	0.00	0.00	-0.21	-1.00	-2.95	-7.11	-12.52	-22.30	-29.21
9	Non-metallic Min.Pr.	0.00	0.00	0.03	0.17	0.44	0.77	0.82	1.33	1.03
10	Chemicals	0.00	0.00	0.00	0.00	0.00	0.00	0.84	0.44	-0.04
11	Metal Products	0.00	0.00	0.03	0.03	0.05	0.05	0.10	-0.24	-0.29
12	Agri. & Indust. Mach.	0.00	0.00	0.04	0.04	0.23	0.57	0.92	2.10	2.41
13	Office Machines	0.00	0.00	-0.06	-0.06	-0.19	-0.93	-0.36	-1.28	-1.92
14	Electrical Goods	0.00	0.00	0.00	-0.03	-0.06	-0.11	-0.54	-0.34	0.10
15	Transport Equipment	0.00	0.00	0.00	0.04	0.37	0.04	-0.27	-1.01	-1.59
16	Food, Drink & Tobacco	0.00	0.00	0.07	0.29	0.96	1.97	4.11	8.25	12.10
17	Tex., Cloth. & Footw.	0.00	0.00	0.13	0.67	2.32	6.66	16.70	35.79	62.55
18	Paper & Printing Pr.	0.00	0.00	0.07	0.30	0.89	2.20	3.89	6.25	8.42
19	Rubber & Plastic Pr.	0.00	0.00	0.00	0.19	0.55	0.80	1.74	3.09	3.54
21	Other Manufactures	0.00	0.00	0.04	0.04	0.09	0.09	0.25	0.07	-0.04
22	Construction	0.00	0.00	0.00	0.00	0.00	0.02	0.12	0.45	0.77
23	Distribution etc.	0.00	0.00	0.00	-0.04	-0.03	-0.05	-0.06	-0.02	0.12
24	Lodging & Catering	0.00	0.00	0.00	0.00	0.13	0.23	0.40	0.69	1.30
25	Inland Transport	0.00	0.00	0.00	0.00	0.04	0.02	0.11	0.40	0.70
26	Sea & Air Transport	0.00	0.00	0.03	0.16	0.34	0.47	-6.55	-13.39	-71.15
27	Other Transport	0.00	0.00	0.03	0.06	0.23	0.56	1.55	2.90	4.83
28	Communications	0.00	0.00	0.00	0.04	0.09	0.21	0.40	0.72	1.20
29	Bank., Finance & Ins.	0.00	0.00	0.00	0.00	0.01	0.00	0.03	0.08	0.06
30	Other Market Serv.	0.00	0.00	0.00	0.00	0.00	-0.03	-0.05	-0.04	-0.02
31	Non-market Services	0.00	0.00	0.00	0.00	0.00	0.00	0.00	0.00	0.00
32	Unallocated	0.00	0.00	0.00	0.00	0.00	0.00	0.00	0.00	0.00
	Total	0.00	0.00	0.00	0.00	0.00	0.19	-0.02	-97.97	-96.38

Source: Cambridge Econometrics Forecast C51F1V-C51F1W/da1/F1WA.

Table D5e.   Model simulation 2: detailed unit labour cost results – France (%)

		1985	1986	1987	1988	1989	1990	1991	1992	1993
1	Agriculture etc.	0.00	0.00	-0.13	-0.13	-0.13	-0.27	-0.79	-1.84	-2.56
2	Coal & Coke	0.00	0.00	0.00	0.00	-0.28	-0.43	-0.81	-1.14	-0.99
3	Oil & Gas Extraction	0.00	0.00	0.00	0.00	-0.39	-2.02	-0.51	1.22	6.45
4	Gas Distribution	0.00	0.00	0.00	0.00	0.00	0.00	0.00	-0.17	0.00
5	Refined Oil	0.00	0.00	0.00	-0.08	-0.24	-0.72	-0.09	1.30	11.28
6	Electricity etc.	0.00	0.00	0.00	0.00	0.05	0.08	0.15	0.23	0.23
7	Water Supply	0.00	0.00	0.00	0.06	0.06	0.21	0.37	0.83	1.62
8	Ferrous & Non-f. Metals	0.00	-0.06	-0.06	-0.20	-0.67	-1.22	-1.92	-2.70	-2.15
9	Non-metallic Min.Pr.	0.00	0.00	0.00	0.07	0.23	0.49	1.01	2.20	3.72
10	Chemicals	0.00	0.00	0.10	0.40	1.01	2.19	4.32	7.85	12.04
11	Metal Products	0.00	0.00	0.12	0.40	0.95	2.01	3.67	6.65	10.15
12	Agri. & Indust. Mach.	0.00	0.00	0.06	0.10	0.33	0.50	1.18	2.23	2.43
13	Office Machines	0.00	0.03	0.16	0.69	2.01	6.02	10.91	20.82	32.18
14	Electrical Goods	0.00	0.00	0.00	0.00	0.00	-0.29	-1.20	-1.78	-2.55
15	Transport Equipment	0.00	0.00	0.04	0.08	0.25	0.29	0.62	1.27	0.74
16	Food, Drink & Tobacco	0.00	0.00	0.00	0.06	0.13	0.06	0.69	0.31	0.90
17	Tex., Cloth. & Footw.	0.00	0.00	-0.03	-0.03	-0.10	-0.26	-0.54	-1.03	-1.37
18	Paper & Printing Pr.	0.00	0.00	0.00	0.11	0.36	0.72	1.84	3.07	5.20
19	Rubber & Plastic Pr.	0.00	0.00	0.00	0.07	0.17	0.10	0.70	1.44	2.11
21	Other Manufactures	0.00	0.00	0.03	0.17	0.45	0.90	1.47	2.81	4.09
22	Construction	0.00	0.00	0.00	0.03	0.13	0.18	0.47	1.04	1.86
23	Distribution etc.	0.00	0.00	0.00	0.02	0.02	0.00	0.07	0.18	0.45
24	Lodging & Catering	0.00	0.00	0.00	0.00	0.02	0.05	0.12	0.24	0.40
25	Inland Transport	0.00	0.00	-0.02	-0.02	-0.07	-0.17	-0.30	-0.45	-0.32
26	Sea & Air Transport	0.00	0.00	0.12	0.53	1.38	3.25	7.56	16.01	26.12
27	Other Transport	0.00	0.00	0.00	0.05	0.10	0.27	0.59	1.15	1.80
28	Communications	0.00	0.00	0.00	0.03	0.06	0.03	0.18	0.35	0.80
29	Bank., Finance & Ins.	0.00	-0.03	-0.04	-0.15	-0.45	-1.34	-2.01	-3.39	-4.51
30	Other Market Serv.	0.00	0.00	0.04	0.04	0.08	0.12	0.25	0.43	0.32
31	Non-market Services	0.00	0.00	-0.01	-0.01	-0.03	-0.12	-0.19	-0.31	-0.06
32	Unallocated	0.00	0.00	0.00	0.00	0.00	0.00	0.00	0.00	0.00
	Total	0.00	0.00	0.00	0.00	0.21	0.39	0.85	1.58	2.49

Source: Cambridge Econometrics Forecast C51F1V-C51F1W/da1/F1WA.

Table D5f.   Model simulation 2: detailed unit labour cost results – Ireland (%)

		1985	1986	1987	1988	1989	1990	1991	1992	1993
1	Agriculture etc.	0.00	0.00	0.00	-0.26	-0.47	-0.76	-0.56	-1.75	-2.07
2	Coal & Coke	0.00	0.07	0.33	1.27	3.63	8.11	12.92	19.01	21.59
3	Oil & Gas Extraction	0.00	0.10	0.17	0.50	0.91	1.83	3.18	3.70	3.60
4	Gas Distribution	0.00	0.00	0.00	0.00	0.00	0.00	0.00	0.00	0.00
5	Refined Oil	0.00	0.00	0.00	0.00	0.22	0.17	0.32	0.58	0.43
6	Electricity etc.	0.00	0.00	0.07	0.17	0.47	0.72	0.99	1.22	1.10
7	Water Supply	0.00	0.00	0.00	0.05	0.16	0.19	0.42	0.85	1.60
8	Ferrous & Non-f. Metals	0.00	0.02	0.19	0.53	-1.92	-1.59	2.17	2.84	-0.31
9	Non-metallic Min.Pr.	0.00	0.03	0.19	0.96	1.37	2.15	3.48	6.35	4.39
10	Chemicals	0.00	0.06	0.48	2.04	4.66	11.39	19.64	29.28	35.47
11	Metal Products	0.00	-0.03	-0.11	-0.26	0.03	1.13	4.27	14.25	41.15
12	Agri. & Indust. Mach.	0.00	0.00	-0.11	-0.65	-1.88	-6.30	-12.62	-20.75	-17.88
13	Office Machines	0.00	0.00	0.68	2.42	5.64	17.26	29.87	52.97	66.87
14	Electrical Goods	0.00	0.00	-0.14	-0.42	-0.97	-0.48	-3.31	-4.15	-4.01
15	Transport Equipment	0.00	0.00	-0.02	-0.02	-0.06	-0.32	-0.92	-1.59	-1.91
16	Food, Drink & Tobacco	0.00	0.00	0.00	0.09	0.18	0.00	1.38	0.93	1.14
17	Tex., Cloth. & Footw.	0.00	-0.03	-0.40	-1.35	-3.42	-5.12	-8.87	-13.99	-13.62
18	Paper & Printing Pr.	0.00	0.02	0.14	0.48	0.86	1.32	1.55	-0.95	-3.58
19	Rubber & Plastic Pr.	0.00	0.00	-0.06	-0.32	-0.76	-2.28	-4.11	-6.69	-8.96
21	Other Manufactures	0.00	0.05	0.19	0.71	1.85	3.90	6.75	10.42	12.57
22	Construction	0.00	0.00	0.00	0.00	0.00	0.00	0.03	0.04	0.03
23	Distribution etc.	0.00	0.00	0.02	0.05	0.14	0.30	0.76	0.88	1.02
24	Lodging & Catering	0.00	0.00	-0.02	-0.09	-0.24	-0.61	-1.02	-1.20	-1.12
25	Inland Transport	0.00	0.00	0.00	0.00	-0.04	-0.13	-0.29	-0.61	-0.86
26	Sea & Air Transport	0.00	-0.02	-0.06	-0.33	-0.83	-2.34	-4.60	-7.92	-13.20
27	Other Transport	0.00	0.04	0.16	0.64	1.56	2.98	6.19	11.92	16.20
28	Communications	0.00	0.00	0.03	0.00	0.00	-0.05	-0.06	-0.30	-0.55
29	Bank., Finance & Ins.	0.00	0.00	0.00	0.00	-0.07	-0.29	-0.30	-0.62	-0.95
30	Other Market Serv.	0.00	0.00	0.00	0.03	-0.03	-0.11	-0.13	-0.39	-0.85
31	Non-market Services	0.00	0.00	0.00	-0.02	-0.05	-0.14	-0.28	-0.49	-0.42
32	Unallocated	0.00	0.00	0.00	0.00	0.00	0.00	0.00	0.00	0.00
	Total	0.00	0.00	0.00	0.00	0.00	0.00	0.00	-1.67	0.00

Source: Cambridge Econometrics Forecast C51F1V-C51F1W/da1/F1WA.

Table D5g.   Model simulation 2: detailed unit labour cost results – Italy (%)

		1985	1986	1987	1988	1989	1990	1991	1992	1993
1	Agriculture etc.	0.00	0.00	0.17	0.54	1.35	3.03	6.12	10.39	14.32
2	Coal & Coke	0.00	0.00	0.00	18.38	25.38	33.97	69.63	72.41	41.74
3	Oil & Gas Extraction	0.00	-0.05	-0.27	-1.06	-2.34	-7.05	-10.32	-16.45	-21.11
4	Gas Distribution	0.00	0.00	0.00	0.00	0.00	0.00	0.22	0.11	0.36
5	Refined Oil	0.00	0.00	0.00	0.00	-0.06	-0.35	-0.27	0.32	8.07
6	Electricity etc.	0.00	0.00	0.00	0.07	0.28	0.27	0.68	0.82	1.54
7	Water Supply	0.00	0.00	0.00	0.04	0.20	0.00	0.48	0.47	1.03
8	Ferrous & Non-f. Metals	0.00	0.00	0.07	0.23	0.31	0.89	1.26	1.96	1.76
9	Non-metallic Min.Pr.	0.00	0.00	-0.04	-0.08	-0.22	-0.60	-0.39	-1.08	-0.99
10	Chemicals	0.00	0.00	-0.06	-0.19	-0.30	-0.44	-1.28	-2.41	-5.31
11	Metal Products	0.00	0.00	0.00	-0.04	-0.08	-0.15	-0.03	0.37	0.89
12	Agri. & Indust. Mach.	0.00	0.00	0.08	0.27	0.60	1.35	2.73	4.66	5.27
13	Office Machines	0.00	0.00	0.08	0.24	0.66	0.89	1.60	3.13	3.27
14	Electrical Goods	0.00	0.00	0.08	0.21	0.45	0.47	0.50	1.17	0.16
15	Transport Equipment	0.00	0.00	0.04	0.07	0.07	0.18	-0.14	0.10	-0.13
16	Food, Drink & Tobacco	0.00	0.00	0.00	0.00	0.08	0.21	0.33	0.35	0.98
17	Tex., Cloth. & Footw.	0.00	0.00	0.09	0.28	0.68	1.26	2.42	3.98	4.98
18	Paper & Printing Pr.	0.00	0.00	0.00	0.00	0.07	1.18	-1.22	1.32	-0.61
19	Rubber & Plastic Pr.	0.00	0.00	0.00	0.04	0.04	-0.12	-0.20	-0.04	-0.64
21	Other Manufactures	0.00	0.00	0.00	-0.05	-0.16	-0.25	0.29	0.14	-0.48
22	Construction	0.00	0.00	0.00	0.00	0.00	0.07	0.12	0.22	0.31
23	Distribution etc.	0.00	0.00	0.00	0.00	0.05	0.04	0.12	0.15	0.06
24	Lodging & Catering	0.00	0.00	0.07	0.14	0.45	0.29	1.30	1.51	3.65
25	Inland Transport	0.00	0.00	0.00	-0.06	-0.16	-0.23	-0.51	-0.87	-1.56
26	Sea & Air Transport	0.00	-0.05	-0.49	-2.13	-6.20	-5.59	-8.13	-11.56	-9.93
27	Other Transport	0.00	-0.04	-0.04	-0.08	-0.23	-0.63	-1.06	-1.10	-0.54
28	Communications	0.00	0.00	-0.02	-0.06	-0.17	-0.38	-0.66	-0.74	-0.63
29	Bank., Finance & Ins.	0.00	0.00	0.00	-0.02	-0.06	-0.16	-0.34	-0.67	-0.98
30	Other Market Serv.	0.00	0.00	0.00	0.00	0.00	0.04	0.07	0.09	0.05
31	Non-market Services	0.00	0.00	0.00	0.00	-0.02	-0.02	-0.03	-0.04	-0.11
32	Unallocated	0.00	0.00	0.00	0.00	0.00	0.00	0.00	0.00	0.00
	Total	0.00	0.00	0.00	0.00	0.00	0.15	0.11	0.26	0.44

Source: Cambridge Econometrics Forecast C51F1V-C51F1W/da1/F1WA.

Table D5h.   Model simulation 2: detailed unit labour cost results – Luxembourg (%)

		1985	1986	1987	1988	1989	1990	1991	1992	1993
1	Agriculture etc.	0.00	-0.14	-0.27	-1.24	-2.71	-4.57	-12.03	-22.46	-29.64
2	Coal & Coke	0.00	0.00	0.00	0.00	0.00	0.00	0.00	0.00	0.00
3	Oil & Gas Extraction	0.00	0.00	0.00	0.00	0.00	0.00	0.00	0.00	0.00
4	Gas Distribution	0.00	0.00	-0.05	-0.06	-0.15	-0.28	-0.51	-0.69	-0.77
5	Refined Oil	0.00	0.00	0.00	0.00	0.00	0.00	0.00	0.00	0.00
6	Electricity etc.	0.00	0.00	-0.03	0.22	-0.83	-2.50	-10.89	-11.48	-15.01
7	Water Supply	0.00	0.00	0.00	0.12	0.18	0.41	0.92	3.00	4.46
8	Ferrous & Non-f. Metals	0.00	0.00	-0.32	-1.08	-2.47	-5.72	-11.15	-17.04	-20.03
9	Non-metallic Min.Pr.	0.00	0.00	-0.09	-0.21	-0.59	-1.00	-3.75	-5.81	-8.31
10	Chemicals	0.00	0.18	1.11	9.40	0.90	-6.32	-10.14	33.00	0.19
11	Metal Products	0.00	0.00	-0.13	-0.40	-0.82	-3.08	-6.72	-10.30	-13.81
12	Agri. & Indust. Mach.	0.00	-0.03	-0.14	-0.15	-0.50	-0.31	-0.89	0.32	-0.69
13	Office Machines	0.00	0.00	0.00	0.00	0.00	0.00	0.00	0.00	0.00
14	Electrical Goods	0.00	0.00	0.00	0.00	0.00	0.00	0.00	0.00	0.00
15	Transport Equipment	0.00	0.14	1.94	1.16	2.11	5.07	6.82	12.11	10.02
16	Food, Drink & Tobacco	0.00	0.00	0.00	-0.07	0.07	0.45	0.40	0.50	0.66
17	Tex., Cloth. & Footw.	0.00	-0.36	-3.20	-7.49	-13.08	-20.74	-46.78	-65.50	-65.79
18	Paper & Printing Pr.	0.00	0.00	0.00	0.13	0.02	0.06	-0.14	0.34	-4.36
19	Rubber & Plastic Pr.	0.00	0.05	0.27	0.96	2.76	10.55	11.12	16.53	20.63
21	Other Manufactures	0.00	0.00	0.00	-0.63	-2.18	-1.88	1.08	0.16	0.32
22	Construction	0.00	0.00	0.00	0.00	0.00	0.00	0.04	0.09	0.11
23	Distribution etc.	0.00	0.00	0.00	0.00	-0.14	0.00	0.00	0.00	-0.13
24	Lodging & Catering	0.00	0.00	0.00	0.00	0.00	0.00	0.00	0.16	0.31
25	Inland Transport	0.00	0.00	-0.01	0.05	-0.04	-0.11	-0.06	0.53	0.21
26	Sea & Air Transport	0.00	0.00	0.00	0.00	0.00	0.00	0.00	0.00	0.00
27	Other Transport	0.00	0.00	0.00	0.00	0.00	0.00	0.00	0.00	0.00
28	Communications	0.00	0.02	0.05	0.30	0.72	1.44	2.86	5.49	6.57
29	Bank., Finance & Ins.	0.00	0.00	0.00	0.00	0.00	0.06	0.18	0.22	0.21
30	Other Market Serv.	0.00	0.00	0.00	-0.04	0.03	0.09	0.20	0.34	0.73
31	Non-market Services	0.00	0.00	0.00	-0.01	0.00	0.00	0.00	-0.01	0.00
32	Unallocated	0.00	0.00	0.00	0.00	0.00	0.00	0.00	0.00	0.00
	Total	0.00	0.00	0.00	0.00	0.00	0.00	0.00	0.00	0.00

Source: Cambridge Econometrics Forecast C51F1V-C51F1W/da1/F1WA.

Table D5i.   Model simulation 2: detailed unit labour cost results – Netherlands (%)

		1985	1986	1987	1988	1989	1990	1991	1992	1993
1	Agriculture etc.	0.00	0.00	0.00	-0.13	-0.27	-0.30	-0.58	-1.57	0.15
2	Coal & Coke	0.00	0.00	0.00	2.04	6.52	6.41	8.33	10.20	9.30
3	Oil & Gas Extraction	0.00	0.00	0.00	0.42	0.00	1.09	1.65	3.05	4.26
4	Gas Distribution	0.00	0.00	0.00	0.20	0.23	0.70	1.06	2.08	4.38
5	Refined Oil	0.00	0.00	0.00	0.39	-0.35	-0.61	3.02	25.99	406.56
6	Electricity etc.	0.00	0.00	0.00	0.00	0.00	-0.16	0.00	0.14	1.08
7	Water Supply	0.00	0.00	0.00	0.12	0.14	0.15	0.28	0.41	0.60
8	Ferrous & Non-f. Metals	0.00	0.00	0.18	0.65	1.65	2.18	3.24	5.61	7.25
9	Non-metallic Min.Pr.	0.00	0.04	0.19	0.82	1.98	3.11	2.58	0.20	-3.61
10	Chemicals	0.00	0.00	0.00	0.00	0.05	0.00	-0.21	-0.27	-0.21
11	Metal Products	0.00	0.00	0.07	0.15	0.28	0.33	0.19	-2.00	-4.33
12	Agri. & Indust. Mach.	0.00	0.00	0.00	0.00	0.00	0.00	0.00	0.00	0.00
13	Office Machines	0.00	-0.05	-0.08	-0.14	-0.42	-1.38	-0.47	-1.28	-0.85
14	Electrical Goods	0.00	0.04	0.21	0.68	2.35	4.38	9.07	14.91	19.00
15	Transport Equipment	0.00	0.00	-0.04	-0.04	-0.19	-0.53	-2.44	-5.14	-8.58
16	Food, Drink & Tobacco	0.00	0.00	0.10	0.32	0.67	0.93	3.59	4.25	8.65
17	Tex., Cloth. & Footw.	0.00	0.00	0.00	0.11	0.11	-0.15	-0.50	-1.00	-3.82
18	Paper & Printing Pr.	0.00	0.00	0.00	0.00	0.00	0.81	-1.40	-0.28	-0.94
19	Rubber & Plastic Pr.	0.00	0.00	-0.04	-0.13	-0.32	-0.88	-1.80	-3.54	-4.39
21	Other Manufactures	0.00	0.00	0.00	0.04	0.12	0.27	0.11	0.51	1.31
22	Construction	0.00	0.00	0.00	0.04	0.18	0.32	0.61	1.24	2.77
23	Distribution etc.	0.00	0.00	0.05	0.16	0.41	0.66	1.29	2.55	5.00
24	Lodging & Catering	0.00	0.00	0.05	0.08	0.72	0.90	-0.26	-1.93	13.12
25	Inland Transport	0.00	0.00	0.00	0.02	-0.06	-0.02	-0.32	-0.46	-0.06
26	Sea & Air Transport	0.00	-0.05	-0.22	-0.91	-2.41	-5.80	-11.33	-20.14	-30.17
27	Other Transport	0.00	-0.05	-0.60	-2.48	-5.94	-13.27	-22.49	-35.56	-42.24
28	Communications	0.00	0.00	0.00	0.00	0.00	-0.16	0.35	-0.02	1.38
29	Bank., Finance & Ins.	0.00	0.00	0.03	0.06	0.14	0.21	0.45	1.04	2.28
30	Other Market Serv.	0.00	0.00	0.00	0.06	0.14	0.30	0.66	1.32	2.53
31	Non-market Services	0.00	0.00	0.00	0.00	0.00	0.00	0.00	-0.02	-0.05
32	Unallocated	0.00	0.00	0.00	0.00	0.00	0.00	0.00	0.00	0.00
	Total	0.00	0.00	0.00	0.00	0.00	-0.49	-1.09	-2.09	-0.37

Source: Cambridge Econometrics Forecast C51F1V-C51F1W/da1/F1WA.

Table D5j.   Model simulation 2: detailed unit labour cost results – Portugal (%)

		1985	1986	1987	1988	1989	1990	1991	1992	1993
1	Agriculture etc.	0.00	0.00	-0.09	-0.08	-0.08	0.00	-0.33	-0.75	-0.77
2	Coal & Coke	0.00	-0.37	0.00	0.74	0.72	-5.33	-8.33	-19.15	-32.20
3	Oil & Gas Extraction	0.00	0.00	0.00	0.00	0.00	0.00	0.00	0.00	0.00
4	Gas Distribution	0.00	0.00	0.00	0.00	0.00	0.00	0.00	0.00	0.00
5	Refined Oil	0.00	0.00	0.00	-0.30	-0.95	-2.03	-9.30	-15.64	-17.03
6	Electricity etc.	0.00	0.00	0.00	0.12	0.00	0.00	-0.13	-0.11	-0.09
7	Water Supply	0.00	0.00	0.00	0.10	0.30	0.44	0.54	0.73	1.71
8	Ferrous & Non-f. Metals	0.00	0.00	0.00	0.09	0.08	-0.64	-1.87	-3.87	-7.08
9	Non-metallic Min.Pr.	0.00	0.00	0.00	0.00	-0.03	-0.14	-0.78	-1.55	-3.54
10	Chemicals	0.00	0.00	0.18	0.52	1.34	3.00	6.27	11.57	16.78
11	Metal Products	0.00	0.00	0.00	-0.03	-0.10	-0.51	-1.38	-1.93	-4.37
12	Agri. & Indust. Mach.	0.00	0.00	0.00	0.09	0.14	0.20	-0.14	-1.22	-6.51
13	Office Machines	0.00	0.00	0.00	0.00	0.00	0.00	0.00	0.00	0.00
14	Electrical Goods	0.00	0.00	0.00	0.00	-0.37	-0.06	-2.78	-3.54	-10.53
15	Transport Equipment	0.00	0.00	-0.11	-0.20	-0.62	-2.23	-5.20	-7.25	-11.70
16	Food, Drink & Tobacco	0.00	0.00	0.00	0.10	0.10	0.24	0.20	0.19	0.59
17	Tex., Cloth. & Footw.	0.00	0.00	0.00	0.00	-0.07	-0.21	-0.82	-1.72	-2.64
18	Paper & Printing Pr.	0.00	0.00	-0.10	-0.40	-0.80	-1.11	-4.53	-4.47	-4.55
19	Rubber & Plastic Pr.	0.00	0.00	0.00	-0.10	-0.19	-0.52	-1.05	-1.44	-1.97
21	Other Manufactures	0.00	0.00	0.00	-0.07	-0.16	-0.41	-1.17	-1.54	-2.26
22	Construction	0.00	0.00	0.00	0.00	0.03	0.02	-0.26	-0.65	-2.44
23	Distribution etc.	0.00	0.00	0.00	0.00	0.03	0.08	-0.19	-0.43	-1.95
24	Lodging & Catering	0.00	0.00	0.00	0.11	0.15	0.30	0.30	0.40	0.79
25	Inland Transport	0.00	0.00	0.04	0.12	0.26	0.49	0.47	0.57	0.21
26	Sea & Air Transport	0.00	0.00	-0.04	-0.21	-0.51	-1.01	-2.24	-3.22	-5.44
27	Other Transport	0.00	0.00	-0.04	-0.14	-0.32	-0.47	-0.95	-1.18	-0.39
28	Communications	0.00	0.00	0.00	0.00	0.02	0.02	-0.15	-0.21	-0.23
29	Bank., Finance & Ins.	0.00	0.00	-0.02	-0.09	-0.19	-0.34	-0.89	-1.46	-1.94
30	Other Market Serv.	0.00	0.00	0.00	0.05	0.19	0.32	0.30	0.67	0.34
31	Non-market Services	0.00	0.00	0.01	0.03	0.06	0.11	0.05	0.08	0.17
32	Unallocated	0.00	0.00	0.00	0.00	0.00	0.00	0.00	0.00	0.00
	Total	0.00	0.00	-0.07	0.00	0.00	0.00	0.00	-1.32	-2.14

Source: Cambridge Econometrics Forecast C51F1V-C51F1W/da1/F1WA.

Table D5k.   Model simulation 2: detailed unit labour cost results – United Kingdom (%)

		1985	1986	1987	1988	1989	1990	1991	1992	1993
1	Agriculture etc.	0.00	-0.06	-0.06	-0.22	-0.46	-0.83	-1.60	-2.90	-3.56
2	Coal & Coke	0.00	0.00	0.02	0.02	0.04	0.07	0.22	0.42	0.68
3	Oil & Gas Extraction	0.00	0.00	0.00	0.00	0.66	1.12	1.45	3.47	3.60
4	Gas Distribution	0.00	0.00	0.00	0.07	0.21	0.39	0.64	1.12	1.37
5	Refined Oil	0.00	0.00	0.19	0.37	0.73	1.65	2.26	7.01	14.26
6	Electricity etc.	0.00	0.00	0.00	0.00	0.08	0.13	0.27	0.46	0.41
7	Water Supply	0.00	0.00	0.04	0.09	0.25	0.49	0.85	1.37	1.58
8	Ferrous & Non-f. Metals	0.00	0.00	0.00	0.34	1.54	5.11	13.79	35.49	79.61
9	Non-metallic Min.Pr.	0.00	0.00	0.07	0.21	0.58	1.14	2.83	4.06	4.63
10	Chemicals	0.00	0.00	0.31	0.91	2.06	4.17	7.68	15.38	13.15
11	Metal Products	0.00	0.00	0.12	0.44	1.21	2.63	4.82	7.73	11.02
12	Agri. & Indust. Mach.	0.00	0.00	0.06	0.19	0.47	1.11	2.19	3.44	3.39
13	Office Machines	0.00	0.00	-0.18	-0.76	-1.84	-4.14	-7.29	-15.29	-17.44
14	Electrical Goods	0.00	0.00	-0.03	-0.06	-0.23	-0.18	-1.11	-1.93	-2.09
15	Transport Equipment	0.00	0.00	0.00	-0.03	-0.10	-0.34	-1.21	-2.43	-2.71
16	Food, Drink & Tobacco	0.00	0.00	-0.08	-0.08	-0.21	-0.45	-0.78	-2.04	-2.87
17	Tex., Cloth. & Footw.	0.00	0.03	0.10	0.39	1.14	2.40	4.19	6.40	9.06
18	Paper & Printing Pr.	0.00	0.00	-0.10	-0.40	-0.95	-1.50	-3.77	-4.03	-6.61
19	Rubber & Plastic Pr.	0.00	0.00	0.07	0.26	0.70	1.48	2.79	5.20	7.10
21	Other Manufactures	0.00	0.00	0.05	0.22	0.55	1.16	2.17	4.23	4.77
22	Construction	0.00	0.00	0.03	0.06	0.10	0.23	0.40	0.50	0.02
23	Distribution etc.	0.00	0.00	0.00	0.00	0.02	0.03	0.06	0.09	0.10
24	Lodging & Catering	0.00	0.00	0.00	0.00	0.05	0.14	0.38	0.60	0.49
25	Inland Transport	0.00	0.03	0.02	0.05	0.18	0.35	0.82	1.45	1.98
26	Sea & Air Transport	0.00	0.00	-0.06	-0.22	-0.59	-1.42	-2.90	-5.96	-5.24
27	Other Transport	0.00	0.00	0.05	0.22	0.56	1.06	2.16	4.09	4.83
28	Communications	0.00	0.00	0.00	0.02	0.07	0.13	0.26	0.48	0.52
29	Bank. Finance & Ins.	0.00	0.00	0.00	0.03	0.06	0.11	0.15	0.29	0.27
30	Other Market Serv.	0.00	0.00	0.06	0.16	0.41	0.83	1.64	2.79	3.35
31	Non-market Services	0.00	0.00	0.02	0.01	0.03	0.05	0.00	0.03	0.15
32	Unallocated	0.00	0.00	0.00	0.00	0.00	0.00	0.00	0.00	0.00
	Total	0.00	0.00	0.02	0.00	0.25	0.56	1.04	1.95	2.62

Source: Cambridge Econometrics Forecast C51F1V-C51F1W/da1/F1WA.

Table D5l.   Model simulation 2: detailed unit labour cost results – EUR-11 (%)

		1985	1986	1987	1988	1989	1990	1991	1992	1993
1	Agriculture etc.	0.00	-0.02	0.00	-0.02	0.01	0.03	-0.31	-1.16	-1.67
2	Coal & Coke	0.00	0.00	0.00	2.86	3.04	2.53	0.16	0.10	0.07
3	Oil & Gas Extraction	0.00	-0.10	-0.11	-0.31	-0.48	-2.69	-99.98	-99.97	-99.97
4	Gas Distribution	0.00	0.00	-0.01	0.02	0.06	-0.05	0.06	0.24	0.43
5	Refined Oil	0.00	0.00	0.02	0.04	-0.05	-0.85	-2.20	-3.79	-60.94
6	Electricity etc.	0.00	0.00	-0.01	0.05	-0.14	-0.56	-5.44	-7.65	-6.92
7	Water Supply	0.00	0.00	0.00	0.03	0.07	-4.03	-11.29	-99.98	-99.96
8	Ferrous & Non-f. Metals	0.00	0.00	0.04	0.11	-0.11	-0.16	0.34	0.75	1.53
9	Non-metallic Min. Pr.	0.00	0.01	0.10	0.44	0.80	1.13	1.59	1.78	0.54
10	Chemicals	0.00	0.02	0.14	0.72	0.73	1.55	2.44	5.95	5.31
11	Metal Products	0.00	0.00	0.03	0.12	0.39	0.28	0.83	1.84	3.45
12	Agri. & Indust. Mach.	0.00	-0.01	0.00	-0.02	-0.11	-1.50	-2.74	-4.34	-1.85
13	Office Machines	0.00	-0.01	0.02	0.12	0.27	0.26	0.89	1.69	-0.09
14	Electrical Goods	0.00	0.02	0.13	0.47	1.40	-1.37	0.27	1.31	1.35
15	Transport Equipment	0.00	0.01	0.21	0.14	0.13	-0.83	-0.60	-1.15	-3.70
16	Food, Drink & Tobacco	0.00	0.00	0.01	0.08	0.25	0.58	0.55	1.13	2.00
17	Tex., Cloth. & Footw.	0.00	-0.03	-0.38	-0.77	-0.89	-1.01	-4.04	-7.38	-3.29
18	Paper & Printing Pr.	0.00	0.00	0.03	0.13	0.29	1.05	0.47	1.17	-0.04
19	Rubber & Plastic Pr.	0.00	0.00	0.02	0.08	0.22	-0.54	-0.44	-0.32	-0.47
21	Other Manufactures	0.00	0.00	0.02	0.04	0.08	-0.17	0.34	0.76	0.39
22	Construction	0.00	0.00	0.00	0.00	0.00	0.10	0.13	0.11	-0.05
23	Distribution etc.	0.00	0.00	0.01	0.04	0.08	0.33	0.37	0.46	0.06
24	Lodging & Catering	0.00	0.00	0.01	0.04	0.18	-0.02	0.53	1.18	3.95
25	Inland Transport	0.00	0.00	0.00	0.01	0.00	-0.13	-0.38	-0.65	-1.17
26	Sea & Air Transport	0.00	-0.01	-0.09	-0.42	-2.57	-0.95	0.05	0.14	0.17
27	Other Transport	0.00	-0.01	-0.05	-0.21	-0.42	-1.49	-2.04	-2.63	-2.20
28	Communications	0.00	0.00	-0.02	-0.05	-0.08	-0.35	0.09	-0.07	0.89
29	Bank., Finance & Ins.	0.00	-0.01	-0.02	-0.07	-0.16	-0.46	-0.85	-1.35	-1.92
30	Other Market Serv.	0.00	-0.01	-0.06	-0.24	-0.48	-0.88	-1.19	-1.40	-0.85
31	Non-market Services	0.00	0.00	0.00	0.00	0.00	0.01	-0.03	-0.02	0.02
32	Unallocated	0.00	0.00	0.00	0.00	0.00	0.00	0.00	0.00	0.00
	Total	0.00	0.00	0.00	0.00	0.00	0.00	0.00	-0.83	3.32

Source: Cambridge Econometrics Forecast C51F1V-C51F1W/da1/F1WA.

Table D6a. **Model simulation 2: detailed labour cost per employee results – Belgium (%)**

		1985	1986	1987	1988	1989	1990	1991	1992	1993
1	Agriculture etc.	0.00	-0.01	0.34	0.11	-0.06	0.36	1.03	2.32	3.05
2	Coal & Coke	0.00	-0.07	0.06	0.00	0.14	0.07	0.73	-1.62	1.46
3	Oil & Gas Extraction	0.00	0.00	0.00	0.00	0.00	0.00	0.00	0.00	0.00
4	Gas Distribution	0.00	0.00	-0.03	0.10	0.23	0.23	0.29	0.92	0.58
5	Refined Oil	0.00	-0.05	-0.13	-0.43	-1.16	-3.28	-6.56	-10.24	-7.33
6	Electricity etc.	0.00	0.00	-0.06	-0.04	0.08	0.06	0.24	0.39	0.80
7	Water Supply	0.00	0.00	0.05	0.06	0.39	0.56	0.90	2.68	3.55
8	Ferrous & Non-f. Metals	0.00	0.02	0.03	0.13	0.40	0.39	0.95	2.09	2.21
9	Non-metallic Min.Pr.	0.00	-0.02	-0.01	0.03	0.02	0.00	0.02	-0.01	0.02
10	Chemicals	0.00	0.00	0.02	-0.02	0.01	0.03	0.02	-0.02	0.00
11	Metal Products	0.00	0.01	0.03	0.13	0.11	-0.07	-0.04	0.36	-0.57
12	Agri. & Indust. Mach.	0.00	0.00	0.00	-0.01	-0.01	-0.03	0.00	0.00	0.01
13	Office Machines	0.00	-0.01	0.00	0.11	0.26	0.61	0.84	1.75	1.99
14	Electrical Goods	0.00	0.03	0.12	0.18	0.92	-2.03	-2.31	-3.47	-2.53
15	Transport Equipment	0.00	0.01	0.01	0.19	0.30	0.06	-0.10	0.84	-0.58
16	Food, Drink & Tobacco	0.00	0.00	0.01	0.21	0.50	0.69	1.17	2.67	2.91
17	Tex., Cloth. & Footw.	0.00	0.01	0.01	0.23	0.44	0.57	1.15	2.33	2.40
18	Paper & Printing Pr.	0.00	0.00	0.05	0.23	0.49	0.67	1.23	2.59	2.91
19	Rubber & Plastic Pr.	0.00	0.00	0.01	0.14	0.33	0.37	1.05	2.00	2.26
21	Other Manufactures	0.00	0.01	0.08	0.24	0.62	0.75	1.46	5.71	5.71
22	Construction	0.00	0.01	0.07	0.17	0.42	0.56	1.11	2.19	2.50
23	Distribution etc.	0.00	0.00	0.04	0.18	0.37	0.42	0.93	1.91	1.73
24	Lodging & Catering	0.00	0.02	0.20	0.60	1.76	2.71	4.87	8.19	11.34
25	Inland Transport	0.00	0.00	0.02	0.10	0.19	0.22	0.38	0.81	0.50
26	Sea & Air Transport	0.00	0.05	0.17	0.80	1.71	3.30	5.93	9.75	11.95
27	Other Transport	0.00	0.03	0.26	0.98	2.18	3.89	7.45	13.33	13.88
28	Communications	0.00	0.00	0.05	0.20	0.41	0.55	1.12	2.26	2.38
29	Bank., Finance & Ins.	0.00	0.00	0.07	0.24	0.54	0.82	1.45	2.86	3.21
30	Other Market Serv.	0.00	0.00	0.01	0.20	0.51	0.71	1.50	3.05	3.42
31	Non-market Services	0.00	0.00	0.00	0.00	0.00	0.00	0.00	-0.01	0.01
32	Unallocated	0.00	0.00	0.00	0.00	0.00	0.00	0.00	0.00	0.00
	Total	0.00	0.00	0.05	0.20	0.53	0.70	1.40	2.83	3.59

Source: Cambridge Econometrics Forecast C51F1Y-C51F1R/da1/F1RA.

Table D6b. Model simulation 2: detailed labour cost per employee results – Denmark (%)

		1985	1986	1987	1988	1989	1990	1991	1992	1993
1	Agriculture etc.	0.00	0.00	0.01	0.02	0.04	0.15	0.00	0.41	0.31
2	Coal & Coke	0.00	0.00	-0.14	-0.28	-0.36	-0.89	0.00	0.00	0.00
3	Oil & Gas Extraction	0.00	0.01	0.07	0.28	0.05	0.06	0.09	0.43	-0.09
4	Gas Distribution	0.00	0.00	-0.01	-0.03	-0.08	0.11	-0.03	0.07	0.16
5	Refined Oil	0.00	0.01	0.01	0.48	1.13	2.97	13.70	84.31	157.21
6	Electricity etc.	0.00	0.00	-0.01	-0.01	-0.11	0.05	-0.12	0.49	1.66
7	Water Supply	0.00	0.00	0.01	0.03	0.09	0.36	0.46	1.37	2.19
8	Ferrous & Non-f. Metals	0.00	-0.04	0.08	0.26	0.36	0.94	1.30	6.06	0.19
9	Non-metallic Min.Pr.	0.00	-0.01	-0.02	-0.06	-0.10	-0.55	-0.14	-0.59	-0.61
10	Chemicals	0.00	0.00	0.00	0.01	-0.01	0.24	0.06	0.55	0.16
11	Metal Products	0.00	0.00	0.00	0.03	0.05	0.38	0.10	0.27	-0.17
12	Agri. & Indust. Mach.	0.00	-0.02	0.01	0.00	-0.01	0.00	0.01	0.01	0.00
13	Office Machines	0.00	0.01	0.12	0.47	1.13	2.98	6.07	9.44	11.68
14	Electrical Goods	0.00	0.01	-0.05	-0.12	-0.38	1.48	-0.90	-1.86	-3.63
15	Transport Equipment	0.00	0.02	-0.03	-0.17	-0.37	-1.67	-1.79	-5.09	-5.32
16	Food, Drink & Tobacco	0.00	0.00	-0.05	-0.06	-0.06	-0.04	-0.74	-0.55	-0.49
17	Tex., Cloth. & Footw.	0.00	-0.01	-0.03	-0.11	-0.29	-0.49	-1.08	1.09	0.05
18	Paper & Printing Pr.	0.00	-0.03	-0.14	-0.39	-0.88	-1.17	-3.21	-3.05	-2.48
19	Rubber & Plastic Pr.	0.00	0.00	0.00	0.02	0.03	-0.48	0.52	0.69	0.67
21	Other Manufactures	0.00	0.01	0.02	0.13	0.33	0.59	1.34	2.56	3.29
22	Construction	0.00	0.00	0.00	0.03	0.05	0.17	0.11	0.39	0.19
23	Distribution etc.	0.00	0.00	-0.01	0.03	0.05	0.37	0.11	0.56	0.02
24	Lodging & Catering	0.00	0.00	-0.03	-0.07	-0.22	-0.03	-0.75	-0.76	-2.22
25	Inland Transport	0.00	0.00	0.00	0.03	0.08	0.22	0.20	0.44	0.05
26	Sea & Air Transport	0.00	0.02	0.02	0.02	0.12	0.26	0.24	0.47	-0.14
27	Other Transport	0.00	0.01	0.18	0.68	1.25	2.45	4.85	8.47	8.22
28	Communications	0.00	0.00	0.00	0.02	0.05	0.06	0.19	0.17	0.29
29	Bank., Finance & Ins.	0.00	0.01	0.00	0.02	0.03	0.21	0.12	0.58	0.45
30	Other Market Serv.	0.00	0.00	-0.02	0.10	0.14	0.32	0.50	0.80	0.82
31	Non-market Services	0.00	0.00	0.00	0.03	0.04	0.13	0.02	0.39	0.34
32	Unallocated	0.00	0.00	0.00	0.00	0.00	0.00	0.00	0.00	0.00
	Total	0.00	0.00	0.00	0.04	0.07	0.29	0.51	1.56	1.75

Source: Cambridge Econometrics Forecast C51F1Y-C51F1R/da1/F1RA.

Table D6c. **Model simulation 2: detailed labour cost per employee results – West Germany (%)**

		1985	1986	1987	1988	1989	1990	1991	1992	1993
1	Agriculture etc.	0.00	-0.01	0.00	-0.05	-0.12	0.72	1.86	1.29	-0.03
2	Coal & Coke	0.00	-0.01	-0.11	-0.15	-0.23	-4.23	-2.12	-1.04	3.21
3	Oil & Gas Extraction	0.00	0.09	-0.07	0.08	-0.03	-0.09	0.08	-0.01	-0.21
4	Gas Distribution	0.00	0.00	0.00	0.00	0.08	-1.26	0.00	0.61	1.93
5	Refined Oil	0.00	0.00	-0.02	0.04	-0.24	-3.12	0.03	9.22	141.62
6	Electricity etc.	0.00	0.00	-0.01	-0.03	0.02	0.03	0.02	-0.01	-0.03
7	Water Supply	0.00	0.00	-0.01	0.00	-0.03	0.02	0.01	0.00	0.00
8	Ferrous & Non-f. Metals	0.00	-0.01	-0.04	0.00	-0.18	0.51	-0.04	-0.31	-1.07
9	Non-metallic Min.Pr.	0.00	-0.01	-0.03	-0.14	-0.23	-1.76	-0.96	-0.86	0.90
10	Chemicals	0.00	0.00	0.00	-0.04	-0.14	0.98	1.05	0.43	-0.92
11	Metal Products	0.00	0.00	-0.03	-0.04	-0.02	-3.17	-2.50	-2.05	-1.00
12	Agri. & Indust. Mach.	0.00	-0.01	-0.02	-0.08	-0.19	-4.55	0.23	-1.24	-1.09
13	Office Machines	0.00	0.00	0.01	-0.01	-0.10	-0.59	0.85	0.51	0.26
14	Electrical Goods	0.00	0.00	0.03	0.03	0.18	-6.76	-1.46	-0.54	1.44
15	Transport Equipment	0.00	-0.04	-0.13	-0.45	-1.10	-9.11	-3.52	-6.20	-8.08
16	Food, Drink & Tobacco	0.00	0.00	-0.01	0.02	0.01	0.02	-0.01	0.03	-0.02
17	Tex., Cloth. & Footw.	0.00	0.00	0.01	0.04	0.11	-0.98	-0.20	0.52	2.31
18	Paper & Printing Pr.	0.00	0.00	0.01	-0.04	-0.10	3.28	0.99	0.38	-0.16
19	Rubber & Plastic Pr.	0.00	-0.01	-0.02	-0.02	0.01	-2.60	0.86	3.00	6.66
21	Other Manufactures	0.00	0.00	0.01	0.04	0.14	-2.54	1.55	2.25	2.53
22	Construction	0.00	0.00	-0.01	0.07	0.15	1.38	1.53	1.66	1.31
23	Distribution etc.	0.00	0.00	-0.01	0.03	-0.02	2.11	2.06	1.37	-0.34
24	Lodging & Catering	0.00	0.04	0.08	0.34	1.01	-0.98	2.76	4.07	9.13
25	Inland Transport	0.00	0.02	0.12	0.59	2.06	-0.87	6.08	21.55	47.35
26	Sea & Air Transport	0.00	-0.01	-0.01	0.00	-0.05	0.00	0.00	0.00	0.00
27	Other Transport	0.00	0.05	0.44	1.63	3.94	2.02	11.84	32.49	57.82
28	Communications	0.00	0.02	0.01	0.07	0.10	1.30	2.44	1.94	1.79
29	Bank., Finance & Ins.	0.00	-0.02	-0.01	-0.04	-0.08	-0.72	0.97	0.36	0.52
30	Other Market Serv.	0.00	0.00	-0.01	0.06	-0.05	0.01	0.06	0.04	-0.02
31	Non-market Services	0.00	0.00	0.00	0.00	0.00	0.00	0.00	0.00	0.00
32	Unallocated	0.00	0.00	0.00	0.00	0.00	0.00	0.00	0.00	0.00
	Total	0.00	0.00	0.00	0.03	0.07	-1.08	0.56	1.84	6.30

Source: Cambridge Econometrics Forecast C51F1Y-C51F1R/da1/F1RA.

Table D6d. Model simulation 2: detailed labour cost per employee results – Spain (%)

		1985	1986	1987	1988	1989	1990	1991	1992	1993
1	Agriculture etc.	0.00	0.00	-0.01	0.03	-0.01	0.02	0.10	0.08	-0.23
2	Coal & Coke	0.00	-0.04	0.22	1.15	2.55	6.95	14.76	30.63	43.33
3	Oil & Gas Extraction	0.00	-0.02	0.26	-0.64	-0.44	-1.01	0.16	-0.22	-0.34
4	Gas Distribution	0.00	0.00	0.02	-0.01	0.06	0.07	0.10	0.24	0.30
5	Refined Oil	0.00	0.02	0.05	0.08	0.14	0.19	0.32	0.28	0.25
6	Electricity etc.	0.00	0.00	-0.03	0.04	0.04	0.08	0.14	0.21	0.47
7	Water Supply	0.00	0.00	0.00	0.00	0.00	-0.01	-0.19	-0.85	-1.81
8	Ferrous & Non-f. Metals	0.00	0.03	0.12	0.28	0.68	1.81	2.96	5.38	6.48
9	Non-metallic Min.Pr.	0.00	0.00	0.00	0.01	-0.01	-0.06	-0.19	-0.09	-0.26
10	Chemicals	0.00	0.00	0.01	0.03	0.01	0.10	0.46	-0.13	-0.30
11	Metal Products	0.00	0.00	0.01	0.00	0.01	-0.14	-0.19	-0.23	0.05
12	Agri. & Indust. Mach.	0.00	0.00	0.02	-0.04	-0.07	-0.25	-0.20	-0.35	0.23
13	Office Machines	0.00	0.00	-0.04	0.02	0.00	0.06	0.02	0.00	0.00
14	Electrical Goods	0.00	0.00	0.01	0.00	0.00	0.01	0.00	0.00	0.01
15	Transport Equipment	0.00	0.00	0.00	-0.03	-0.02	-0.18	-0.23	0.04	-0.13
16	Food, Drink & Tobacco	0.00	-0.01	0.00	-0.01	0.04	0.08	0.07	0.13	0.25
17	Tex., Cloth. & Footw.	0.00	-0.01	0.01	0.06	0.13	0.19	0.39	0.89	1.94
18	Paper & Printing Pr.	0.00	-0.01	-0.02	-0.08	-0.24	-0.48	-1.03	-1.86	-2.49
19	Rubber & Plastic Pr.	0.00	0.00	-0.03	0.04	0.05	-0.04	-0.10	0.19	0.24
21	Other Manufactures	0.00	0.00	0.02	-0.02	0.00	0.02	0.00	0.01	-0.01
22	Construction	0.00	0.00	0.00	0.00	0.00	0.02	0.01	0.00	0.00
23	Distribution etc.	0.00	0.00	0.00	-0.03	-0.01	0.00	0.00	0.00	-0.01
24	Lodging & Catering	0.00	0.00	0.01	0.03	0.21	0.38	0.66	1.04	1.83
25	Inland Transport	0.00	0.00	0.00	0.01	0.02	-0.04	0.03	0.01	-0.04
26	Sea & Air Transport	0.00	-0.01	-0.02	0.02	-0.02	-0.05	-0.08	8.93	9.57
27	Other Transport	0.00	0.00	-0.01	-0.10	-0.23	-0.42	-0.31	-0.84	-0.44
28	Communications	0.00	0.00	0.00	0.05	0.07	0.21	0.35	0.46	0.74
29	Bank., Finance & Ins.	0.00	0.00	0.00	0.00	-0.01	-0.01	0.02	0.02	0.06
30	Other Market Serv.	0.00	0.00	0.00	0.01	0.00	-0.01	-0.02	-0.07	-0.14
31	Non-market Services	0.00	0.00	0.00	0.00	0.00	0.00	0.00	0.00	0.00
32	Unallocated	0.00	0.00	0.00	0.00	0.00	0.00	0.00	0.00	0.00
	Total	0.00	0.00	0.02	0.05	0.10	0.25	0.41	0.86	0.95

Source: Cambridge Econometrics Forecast C51F1Y-C51F1R/da1/F1RA.

Table D6e.   Model simulation 2: detailed labour cost per employee results – France (%)

		1985	1986	1987	1988	1989	1990	1991	1992	1993
1	Agriculture etc.	0.00	0.00	-0.10	-0.03	0.04	-0.07	-0.05	-0.46	-1.00
2	Coal & Coke	0.00	0.00	-0.02	0.42	1.51	3.30	6.04	9.93	13.47
3	Oil & Gas Extraction	0.00	0.01	0.00	0.20	0.12	0.13	0.14	0.06	-0.06
4	Gas Distribution	0.00	0.00	0.00	0.00	0.00	0.01	0.02	-0.12	0.18
5	Refined Oil	0.00	0.00	0.01	-0.04	-0.11	-0.40	0.27	1.53	10.53
6	Electricity etc.	0.00	0.00	-0.01	-0.04	-0.03	-0.10	-0.17	-0.23	0.16
7	Water Supply	0.00	0.00	0.00	0.07	0.08	0.28	0.52	1.13	2.11
8	Ferrous & Non-f. Metals	0.00	-0.06	-0.04	-0.02	-0.09	-0.25	-0.27	-0.40	0.29
9	Non-metallic Min.Pr.	0.00	0.00	-0.03	-0.05	-0.08	-0.29	-0.23	-0.37	0.08
10	Chemicals	0.00	-0.01	0.00	0.00	-0.10	-0.22	-0.39	-0.69	-0.55
11	Metal Products	0.00	-0.01	-0.01	-0.01	-0.04	-0.18	-0.31	-0.58	-0.65
12	Agri. & Indust. Mach.	0.00	-0.01	0.01	-0.06	-0.05	-0.16	-0.27	-0.44	-0.17
13	Office Machines	0.00	0.01	-0.04	-0.11	-0.24	-0.78	-0.14	-0.13	0.85
14	Electrical Goods	0.00	0.00	-0.02	-0.06	-0.10	-1.23	-2.02	-3.86	-6.04
15	Transport Equipment	0.00	-0.01	0.00	-0.06	-0.07	-0.13	-0.26	-0.48	-0.47
16	Food, Drink & Tobacco	0.00	0.00	-0.02	-0.01	-0.06	-0.26	-0.50	-0.90	-0.67
17	Tex., Cloth. & Footw.	0.00	0.00	-0.03	-0.02	-0.05	-0.13	-0.25	-0.49	-0.46
18	Paper & Printing Pr.	0.00	0.00	-0.04	-0.04	-0.05	-0.12	-0.27	-0.43	-0.13
19	Rubber & Plastic Pr.	0.00	0.00	-0.03	-0.02	-0.04	-0.26	-0.08	-0.05	0.39
21	Other Manufactures	0.00	-0.01	-0.02	-0.01	-0.05	-0.15	-0.30	-0.49	-0.37
22	Construction	0.00	0.00	-0.01	-0.03	-0.03	-0.22	-0.35	-0.58	-0.42
23	Distribution etc.	0.00	0.00	-0.01	-0.01	-0.06	-0.21	-0.36	-0.60	-0.57
24	Lodging & Catering	0.00	0.00	0.00	-0.01	0.00	-0.01	0.00	0.00	0.00
25	Inland Transport	0.00	0.00	-0.01	0.02	0.03	0.01	0.06	0.14	1.03
26	Sea & Air Transport	0.00	-0.01	0.02	0.11	0.24	0.59	1.03	1.54	1.43
27	Other Transport	0.00	-0.01	-0.06	-0.14	-0.38	-0.78	-1.76	-2.95	-4.07
28	Communications	0.00	0.00	-0.02	-0.03	-0.05	-0.17	-0.28	-0.44	-0.13
29	Bank., Finance & Ins.	0.00	-0.03	0.01	-0.01	0.10	0.14	0.16	0.31	0.86
30	Other Market Serv.	0.00	0.00	0.02	-0.02	-0.05	-0.13	-0.23	-0.39	-0.22
31	Non-market Services	0.00	0.00	-0.01	-0.02	-0.03	-0.13	-0.20	-0.32	0.00
32	Unallocated	0.00	0.00	0.00	0.00	0.00	0.00	0.00	0.00	0.00
	Total	0.00	-0.01	-0.01	0.00	0.02	-0.03	0.00	0.01	0.65

Source: Cambridge Econometrics Forecast C51F1Y-C51F1R/da1/F1RA.

Table D6f. **Model simulation 2: detailed labour cost per employee results – Ireland (%)**

		1985	1986	1987	1988	1989	1990	1991	1992	1993
1	Agriculture etc.	0.00	0.00	0.03	-0.13	-0.17	-0.14	-0.16	-0.46	-0.67
2	Coal & Coke	0.00	0.01	-0.05	0.02	0.01	0.00	0.00	0.00	0.07
3	Oil & Gas Extraction	0.00	0.09	0.03	0.11	0.01	0.04	-0.02	-0.08	0.14
4	Gas Distribution	0.00	0.00	0.00	0.00	0.00	0.00	0.00	0.00	0.00
5	Refined Oil	0.00	0.00	-0.02	-0.07	0.07	-0.07	-0.02	0.05	0.00
6	Electricity etc.	0.00	-0.01	0.00	-0.11	-0.46	-1.05	-1.88	-2.82	-3.47
7	Water Supply	0.00	0.00	0.00	0.05	0.16	0.19	0.42	0.85	1.60
8	Ferrous & Non-f. Metals	0.00	-0.02	-0.15	-0.42	-0.52	-0.62	-3.81	-7.12	-10.45
9	Non-metallic Min.Pr.	0.00	0.00	-0.06	-0.27	-0.24	-0.46	-0.97	-1.74	-0.13
10	Chemicals	0.00	0.01	-0.01	0.04	0.03	0.13	0.04	-0.14	-0.44
11	Metal Products	0.00	-0.03	-0.07	-0.26	-0.71	-1.89	-3.26	-5.30	-7.85
12	Agri. & Indust. Mach.	0.00	0.01	0.02	0.01	0.01	-0.03	-0.03	-0.09	-0.78
13	Office Machines	0.00	-0.07	-0.02	0.00	-0.03	-0.24	0.06	-0.02	-0.48
14	Electrical Goods	0.00	0.02	0.01	0.02	0.08	0.00	0.30	-0.28	-0.85
15	Transport Equipment	0.00	0.00	-0.01	-0.01	-0.04	-0.22	-0.12	-0.18	-0.30
16	Food, Drink & Tobacco	0.00	0.00	-0.03	0.00	-0.08	-0.19	-0.34	-0.73	-0.92
17	Tex., Cloth. & Footw.	0.00	0.00	0.00	0.02	0.01	0.02	-0.02	-0.01	0.00
18	Paper & Printing Pr.	0.00	0.00	-0.01	-0.04	-0.09	-0.28	-0.37	-0.91	-1.00
19	Rubber & Plastic Pr.	0.00	0.01	0.00	-0.03	-0.05	-0.17	-0.38	-0.72	-0.86
21	Other Manufactures	0.00	0.03	0.01	0.02	-0.01	0.00	0.00	-0.03	-0.02
22	Construction	0.00	0.00	0.00	0.01	0.03	0.10	0.11	0.26	0.43
23	Distribution etc.	0.00	0.00	0.01	-0.01	0.00	-0.02	-0.01	-0.04	0.08
24	Lodging & Catering	0.00	0.00	0.01	0.04	0.19	0.74	1.15	2.09	2.24
25	Inland Transport	0.00	0.00	0.00	-0.01	-0.05	-0.16	-0.38	-0.66	-0.88
26	Sea & Air Transport	0.00	-0.01	-0.01	-0.10	-0.17	-0.52	-0.90	-1.63	-1.62
27	Other Transport	0.00	0.04	0.16	0.63	1.38	1.91	4.44	8.62	9.53
28	Communications	0.00	0.00	0.03	0.01	0.03	0.05	0.07	0.03	-0.04
29	Bank., Finance & Ins.	0.00	0.00	0.01	0.04	0.06	0.04	0.29	0.44	0.59
30	Other Market Serv.	0.00	0.00	0.00	0.04	0.02	0.02	0.00	-0.04	-0.27
31	Non-market Services	0.00	0.00	0.00	-0.02	-0.04	-0.14	-0.30	-0.54	-0.55
32	Unallocated	0.00	0.00	0.00	0.00	0.00	0.00	0.00	0.00	0.00
	Total	0.00	0.00	-0.01	-0.04	-0.05	-0.17	-0.40	-0.75	-1.04

Source: Cambridge Econometrics Forecast C51F1Y-C51F1R/da1/F1RA.

Table D6g.   Model simulation 2: detailed labour cost per employee results – Italy (%)

		1985	1986	1987	1988	1989	1990	1991	1992	1993
1	Agriculture etc.	0.00	-0.02	0.02	-0.05	-0.20	-0.35	-0.58	-0.90	-1.13
2	Coal & Coke	0.00	0.00	0.00	-0.01	-0.05	0.01	0.14	-0.22	-0.43
3	Oil & Gas Extraction	0.00	-0.02	-0.05	-0.14	0.21	0.20	0.07	0.14	-0.10
4	Gas Distribution	0.00	0.00	0.00	-0.02	-0.08	-0.07	0.05	-0.07	0.11
5	Refined Oil	0.00	0.00	-0.02	-0.04	-0.16	-0.45	-0.30	0.77	14.82
6	Electricity etc.	0.00	0.00	-0.01	0.04	0.23	0.21	0.38	0.50	1.35
7	Water Supply	0.00	0.00	0.00	0.06	0.38	0.40	0.74	1.13	-0.02
8	Ferrous & Non-f. Metals	0.00	-0.01	0.02	0.08	-0.05	0.07	-0.08	0.02	0.03
9	Non-metallic Min.Pr.	0.00	0.00	0.00	0.06	0.12	0.24	0.38	0.73	1.29
10	Chemicals	0.00	0.01	-0.01	-0.04	0.02	0.01	0.01	-0.01	0.00
11	Metal Products	0.00	0.00	0.01	0.01	0.01	0.00	0.01	0.00	0.00
12	Agri. & Indust. Mach.	0.00	-0.01	0.01	0.05	0.05	0.15	0.21	0.46	0.91
13	Office Machines	0.00	-0.01	0.01	0.00	0.07	0.14	0.37	0.53	0.80
14	Electrical Goods	0.00	-0.01	0.00	-0.01	0.02	-0.66	-0.92	-0.87	-0.39
15	Transport Equipment	0.00	0.00	0.01	0.01	-0.05	0.02	0.02	0.12	0.30
16	Food, Drink & Tobacco	0.00	0.00	0.00	-0.03	-0.01	0.02	-0.02	-0.02	-0.02
17	Tex., Cloth. & Footw.	0.00	-0.01	0.02	0.05	0.12	0.07	0.15	0.52	1.08
18	Paper & Printing Pr.	0.00	0.00	0.01	0.04	0.15	1.09	-1.23	1.69	0.53
19	Rubber & Plastic Pr.	0.00	0.00	-0.01	0.03	0.02	-0.02	0.05	0.53	1.11
21	Other Manufactures	0.00	0.00	0.01	-0.01	-0.07	-0.37	0.21	0.69	0.77
22	Construction	0.00	0.00	0.00	0.01	0.01	0.16	0.23	0.38	0.58
23	Distribution etc.	0.00	0.00	0.00	-0.01	0.02	-0.01	-0.01	0.01	0.01
24	Lodging & Catering	0.00	0.00	0.08	0.18	0.57	0.48	1.62	2.02	4.24
25	Inland Transport	0.00	0.00	0.01	-0.02	-0.05	0.02	-0.03	0.05	0.15
26	Sea & Air Transport	0.00	0.04	0.42	1.44	2.87	10.78	20.73	36.01	45.34
27	Other Transport	0.00	-0.02	0.08	0.34	0.79	1.61	2.81	5.00	6.56
28	Communications	0.00	0.00	0.00	0.03	0.08	0.22	0.30	0.77	1.30
29	Bank., Finance & Ins.	0.00	0.00	0.02	0.09	0.21	0.52	0.84	1.57	2.04
30	Other Market Serv.	0.00	0.00	0.00	-0.01	-0.01	0.01	0.00	0.01	0.01
31	Non-market Services	0.00	0.00	0.00	0.00	-0.02	-0.03	-0.04	-0.01	0.23
32	Unallocated	0.00	0.00	0.00	0.00	0.00	0.00	0.00	0.00	0.00
	Total	0.00	0.00	0.02	0.10	0.37	0.77	1.21	2.15	3.51

Source: Cambridge Econometrics Forecast C51F1Y-C51F1R/da1/F1RA.

Table D6h. Model simulation 2: detailed labour cost per employee results – Luxembourg (%)

		1985	1986	1987	1988	1989	1990	1991	1992	1993
1	Agriculture etc.	0.00	-0.11	0.04	-0.10	-0.03	0.03	0.00	-0.01	-0.04
2	Coal & Coke	0.00	0.00	0.00	0.00	0.00	0.00	0.00	0.00	0.00
3	Oil & Gas Extraction	0.00	0.00	0.00	0.00	0.00	0.00	0.00	0.00	0.00
4	Gas Distribution	0.00	0.00	-0.04	0.00	-0.03	0.00	-0.01	-0.01	-0.03
5	Refined Oil	0.00	0.00	0.00	0.00	0.00	0.00	0.00	0.00	0.00
6	Electricity etc.	0.00	0.00	0.00	0.00	0.00	0.00	0.00	0.00	0.00
7	Water Supply	0.00	0.00	-0.01	0.00	-0.01	-0.01	-0.02	-0.01	0.02
8	Ferrous & Non-f. Metals	0.00	0.03	-0.01	0.00	-0.02	0.01	0.01	0.05	0.01
9	Non-metallic Min.Pr.	0.00	0.01	0.00	-0.01	0.00	0.03	0.03	-0.02	-0.01
10	Chemicals	0.00	0.01	-0.05	0.04	-0.07	0.02	0.00	-0.02	-0.01
11	Metal Products	0.00	0.01	0.01	0.01	0.01	-0.02	0.01	0.01	0.01
12	Agri. & Indust. Mach.	0.00	-0.02	0.00	-0.02	-0.01	-0.01	0.00	-0.02	-0.02
13	Office Machines	0.00	0.00	0.00	0.00	0.00	0.00	0.00	0.00	0.00
14	Electrical Goods	0.00	0.00	0.00	0.00	0.00	0.00	0.00	0.00	0.00
15	Transport Equipment	0.00	0.01	0.00	0.00	-0.07	-0.03	0.04	-0.06	0.16
16	Food, Drink & Tobacco	0.00	0.00	0.00	-0.01	0.02	0.07	0.04	0.01	0.00
17	Tex., Cloth. & Footw.	0.00	-0.02	0.01	0.01	0.00	0.01	0.00	0.01	0.02
18	Paper & Printing Pr.	0.00	0.00	-0.01	-0.02	0.01	-0.01	-0.01	0.00	0.00
19	Rubber & Plastic Pr.	0.00	0.03	-0.01	0.00	-0.02	0.01	0.01	0.00	0.01
21	Other Manufactures	0.00	0.00	-0.01	0.02	0.01	-0.10	0.07	0.07	-0.11
22	Construction	0.00	0.00	0.00	0.00	-0.01	-0.02	0.01	0.00	0.01
23	Distribution etc.	0.00	0.00	0.00	0.00	-0.12	0.06	0.10	0.09	0.00
24	Lodging & Catering	0.00	0.00	0.00	0.00	0.02	0.04	0.00	0.01	0.02
25	Inland Transport	0.00	0.00	0.00	0.00	0.00	0.00	-0.01	-0.01	-0.01
26	Sea & Air Transport	0.00	0.00	0.00	0.00	0.00	0.00	0.00	0.00	0.00
27	Other Transport	0.00	0.00	0.00	0.00	0.00	0.00	0.00	0.00	0.00
28	Communications	0.00	0.01	-0.01	-0.01	-0.01	0.01	0.01	-0.01	0.00
29	Bank., Finance & Ins.	0.00	0.00	0.00	-0.01	-0.03	0.01	0.04	-0.01	0.01
30	Other Market Serv.	0.00	0.00	0.00	-0.03	0.01	0.01	0.01	-0.01	0.01
31	Non-market Services	0.00	0.00	0.00	-0.01	0.00	0.00	0.01	0.00	0.01
32	Unallocated	0.00	0.00	0.00	0.00	0.00	0.00	0.00	0.00	0.00
	Total	0.00	0.00	-0.01	0.00	-0.01	0.00	0.01	0.00	0.00

Source: Cambridge Econometrics Forecast C51F1Y-C51F1R/da1/F1RA.

**Table D6i.   Model simulation 2: detailed labour cost per employee results –
Netherlands (%)**

		1985	1986	1987	1988	1989	1990	1991	1992	1993
1	Agriculture etc.	0.00	0.01	0.07	0.11	0.40	0.81	1.55	3.19	6.11
2	Coal & Coke	0.00	0.00	0.00	0.35	0.24	-0.09	0.58	0.40	0.08
3	Oil & Gas Extraction	0.00	0.00	-0.03	0.27	-0.29	0.36	0.05	-0.08	-0.57
4	Gas Distribution	0.00	0.00	0.01	0.23	0.29	0.83	1.27	2.52	4.96
5	Refined Oil	0.00	-0.02	-0.04	-0.07	-0.65	-2.04	-0.99	9.86	199.46
6	Electricity etc.	0.00	0.00	0.02	0.08	0.17	0.27	0.86	1.73	3.19
7	Water Supply	0.00	0.01	0.09	0.57	1.53	3.41	7.67	16.94	37.59
8	Ferrous & Non-f. Metals	0.00	-0.01	0.03	0.21	0.47	0.93	1.77	3.78	6.83
9	Non-metallic Min.Pr.	0.00	0.01	-0.01	0.00	-0.01	0.02	-0.02	-0.02	-0.02
10	Chemicals	0.00	0.00	0.00	0.02	0.09	0.39	0.49	0.82	1.29
11	Metal Products	0.00	-0.02	-0.06	-0.11	-0.13	-0.20	-0.11	0.68	4.69
12	Agri. & Indust. Mach.	0.00	-0.01	0.00	0.00	0.00	0.09	0.00	0.00	0.00
13	Office Machines	0.00	-0.01	0.00	0.01	-0.07	0.14	-0.04	0.10	0.03
14	Electrical Goods	0.00	0.00	0.06	0.07	0.38	-2.28	-1.35	-1.91	-1.38
15	Transport Equipment	0.00	0.00	0.00	0.00	-0.03	-0.02	0.01	0.02	-0.02
16	Food, Drink & Tobacco	0.00	0.00	0.09	0.28	0.57	1.03	1.93	3.90	7.78
17	Tex., Cloth. & Footw.	0.00	0.00	-0.04	-0.11	-0.23	-0.54	-1.07	-2.31	-3.31
18	Paper & Printing Pr.	0.00	0.01	0.07	0.28	0.71	2.09	1.98	5.62	8.44
19	Rubber & Plastic Pr.	0.00	0.01	0.00	0.01	-0.02	-0.01	0.03	-0.03	0.01
21	Other Manufactures	0.00	0.00	0.03	0.12	0.36	0.61	1.23	2.34	4.39
22	Construction	0.00	0.00	0.02	0.11	0.35	0.67	1.26	2.55	4.51
23	Distribution etc.	0.00	0.00	0.06	0.19	0.50	0.91	1.75	3.64	7.06
24	Lodging & Catering	0.00	0.03	0.34	1.43	4.12	8.73	13.94	19.05	43.62
25	Inland Transport	0.00	0.00	0.07	0.28	0.86	0.35	0.36	-0.54	-4.06
26	Sea & Air Transport	0.00	0.03	0.67	2.78	6.84	16.46	33.96	69.89	124.55
27	Other Transport	0.00	0.15	1.06	4.10	9.31	19.39	36.91	69.97	68.65
28	Communications	0.00	0.01	0.09	0.32	0.76	1.73	3.29	5.58	7.92
29	Bank., Finance & Ins.	0.00	0.00	0.05	0.14	0.33	0.63	1.23	2.48	4.47
30	Other Market Serv.	0.00	0.00	0.01	0.08	0.20	0.43	0.93	1.85	3.36
31	Non-market Services	0.00	0.00	0.00	0.00	0.00	0.01	0.01	0.01	0.00
32	Unallocated	0.00	0.00	0.00	0.00	0.00	0.00	0.00	0.00	0.00
	Total	0.00	0.00	0.08	0.39	0.88	1.66	3.29	6.74	14.51

Source: Cambridge Econometrics Forecast C51F1Y-C51F1R/da1/F1RA.

Table D6j.   Model simulation 2: detailed labour cost per employee results – Portugal (%)

		1985	1986	1987	1988	1989	1990	1991	1992	1993
1	Agriculture etc.	0.00	0.00	-0.06	0.00	0.05	0.21	0.20	0.27	0.62
2	Coal & Coke	0.00	-0.22	0.07	0.33	0.37	-0.55	0.40	-0.39	0.00
3	Oil & Gas Extraction	0.00	0.53	0.00	0.00	0.00	0.00	0.00	0.00	0.00
4	Gas Distribution	0.00	0.00	0.00	0.00	0.00	0.00	0.00	0.00	0.00
5	Refined Oil	0.00	0.01	0.06	-0.06	0.03	0.08	-1.64	-2.62	1.24
6	Electricity etc.	0.00	0.00	0.02	0.18	0.16	0.36	0.42	0.75	1.89
7	Water Supply	0.00	0.00	0.00	0.10	0.30	0.44	0.54	0.73	1.71
8	Ferrous & Non-f. Metals	0.00	0.01	0.02	0.04	0.14	0.30	0.27	0.44	1.15
9	Non-metallic Min.Pr.	0.00	0.00	0.03	0.08	0.14	0.27	0.31	0.47	1.22
10	Chemicals	0.00	-0.01	0.06	0.05	0.03	0.00	0.02	0.01	0.04
11	Metal Products	0.00	0.00	0.02	0.05	0.18	0.30	0.35	0.36	1.67
12	Agri. & Indust. Mach.	0.00	0.00	0.00	0.09	0.14	0.30	0.63	1.24	4.16
13	Office Machines	0.00	0.00	0.00	0.00	0.00	0.00	0.00	0.00	0.00
14	Electrical Goods	0.00	0.00	0.03	0.06	0.18	0.16	0.15	0.53	1.60
15	Transport Equipment	0.00	0.00	-0.05	0.06	0.02	0.01	-0.01	0.00	0.00
16	Food, Drink & Tobacco	0.00	0.00	0.02	0.17	0.17	0.41	0.49	0.96	1.64
17	Tex., Cloth. & Footw.	0.00	0.00	0.03	0.11	0.24	0.53	0.66	0.92	1.48
18	Paper & Printing Pr.	0.00	0.02	0.04	0.04	0.13	0.26	0.14	0.28	0.79
19	Rubber & Plastic Pr.	0.00	0.00	0.03	0.00	0.05	0.03	-0.02	0.21	0.66
21	Other Manufactures	0.00	0.00	0.04	0.08	0.19	0.37	0.36	0.59	1.41
22	Construction	0.00	0.00	0.01	0.03	0.11	0.21	0.25	0.53	1.81
23	Distribution etc.	0.00	0.00	0.01	0.04	0.11	0.26	0.26	0.46	1.46
24	Lodging & Catering	0.00	0.00	0.00	0.10	0.14	0.28	0.29	0.40	1.02
25	Inland Transport	0.00	0.00	0.03	0.07	0.17	0.32	0.32	0.51	1.30
26	Sea & Air Transport	0.00	0.01	0.08	0.25	0.65	1.33	2.41	5.18	7.04
27	Other Transport	0.00	0.01	0.07	0.19	0.39	0.69	0.80	1.07	1.91
28	Communications	0.00	0.00	0.02	0.06	0.15	0.26	0.35	0.55	1.25
29	Bank., Finance & Ins.	0.00	0.00	0.00	0.05	0.15	0.27	0.32	0.51	1.20
30	Other Market Serv.	0.00	0.00	0.00	0.04	0.18	0.29	0.28	0.47	1.15
31	Non-market Services	0.00	0.00	0.01	0.03	0.07	0.13	0.10	0.15	0.37
32	Unallocated	0.00	0.00	0.00	0.00	0.00	0.00	0.00	0.00	0.00
	Total	0.00	0.00	0.03	0.09	0.20	0.36	0.41	0.75	1.87

Source: Cambridge Econometrics Forecast C51F1Y-C51F1R/da1/F1RA.

Table D6k.   Model simulation 2: detailed labour cost per employee results – United Kingdom (%)

		1985	1986	1987	1988	1989	1990	1991	1992	1993
1	Agriculture etc.	0.00	-0.06	0.00	-0.02	-0.04	-0.07	-0.05	-0.10	-0.04
2	Coal & Coke	0.00	0.00	-0.03	-0.15	-0.46	-1.21	-2.05	-3.92	-6.18
3	Oil & Gas Extraction	0.00	-0.01	-0.05	-0.17	0.14	0.00	-0.01	-0.04	0.05
4	Gas Distribution	0.00	0.00	-0.01	0.02	0.06	0.05	0.05	0.02	0.13
5	Refined Oil	0.00	-0.01	0.05	-0.05	-0.22	-0.40	-0.18	-0.56	2.31
6	Electricity etc	0.00	0.00	-0.01	-0.05	-0.04	-0.13	-0.19	-0.33	-0.51
7	Water Supply	0.00	0.00	0.01	0.01	0.03	0.01	0.01	-0.11	-0.19
8	Ferrous & Non-f. Metals	0.00	0.00	-0.06	-0.11	-0.35	-0.94	-2.82	-6.58	-12.94
9	Non-metallic Min.Pr.	0.00	-0.01	-0.05	-0.09	-0.22	-0.44	-1.11	-1.33	-1.50
10	Chemicals	0.00	-0.03	0.02	-0.03	-0.04	-0.03	-0.03	-0.07	-0.05
11	Metal Products	0.00	-0.03	-0.12	-0.49	-1.31	-2.67	-4.47	-6.50	-7.56
12	Agri. & Indust. Mach.	0.00	-0.01	0.00	-0.01	-0.01	0.00	0.02	-0.01	0.01
13	Office Machines	0.00	0.02	-0.01	-0.04	-0.11	-0.15	-0.19	-0.15	-0.15
14	Electrical Goods	0.00	0.00	0.00	0.02	0.03	-0.12	-0.19	-0.38	-0.88
15	Transport Equipment	0.00	0.00	0.01	-0.01	0.00	-0.07	-0.09	-0.16	-0.15
16	Food, Drink & Tobacco	0.00	0.00	-0.05	0.02	0.03	0.09	0.07	0.21	0.39
17	Tex., Cloth. & Footw.	0.00	0.02	-0.05	-0.17	-0.40	-0.85	-1.60	-2.53	-2.31
18	Paper & Printing Pr.	0.00	0.01	0.03	0.05	0.14	0.17	0.47	0.57	0.86
19	Rubber & Plastic Pr.	0.00	-0.01	0.00	-0.01	-0.03	-0.02	-0.04	-0.08	-0.03
21	Other Manufactures	0.00	-0.01	-0.02	0.00	-0.01	-0.08	-0.07	-0.13	-0.15
22	Construction	0.00	0.00	0.01	0.01	-0.06	-0.13	-0.15	-0.18	0.01
23	Distribution etc.	0.00	0.00	-0.01	-0.02	-0.03	-0.07	-0.10	-0.13	-0.06
24	Lodging & Catering	0.00	0.00	-0.01	-0.03	-0.02	0.00	0.03	0.00	0.00
25	Inland Transport	0.00	0.02	0.00	-0.02	0.00	-0.04	0.01	0.05	0.16
26	Sea & Air Transport	0.00	0.01	0.07	0.23	0.54	1.05	1.91	3.36	4.79
27	Other Transport	0.00	0.00	0.02	0.13	0.33	0.65	1.33	2.71	3.78
28	Communications	0.00	0.00	-0.01	-0.01	-0.03	-0.08	-0.11	-0.07	0.05
29	Bank., Finance & Ins.	0.00	0.00	-0.01	0.00	-0.01	-0.03	-0.10	-0.12	-0.17
30	Other Market Serv.	0.00	-0.01	0.01	0.01	0.00	-0.02	-0.01	-0.04	-0.06
31	Non-market Services	0.00	0.00	0.01	0.01	0.02	0.04	0.05	0.09	0.20
32	Unallocated	0.00	0.00	0.00	0.00	0.00	0.00	0.00	0.00	0.00
	Total	0.00	0.00	-0.01	-0.03	-0.07	-0.18	-0.32	-0.58	-0.79

Source: Cambridge Econometrics Forecast C51F1Y-C51F1R/da1/F1RA.

Table D6l. **Model simulation 2: detailed labour cost per employee results – EUR-11 (%)**

		1985	1986	1987	1988	1989	1990	1991	1992	1993
1	Agriculture etc.	0.00	-0.02	-0.01	-0.01	-0.04	0.12	0.26	0.22	0.25
2	Coal & Coke	0.00	0.00	-0.04	-0.13	-0.38	-1.42	-1.88	-3.19	-4.69
3	Oil & Gas Extraction	0.00	0.01	-0.04	-0.06	0.08	0.04	0.01	-0.03	-0.01
4	Gas Distribution	0.00	0.00	0.00	0.02	0.05	-0.32	0.07	0.26	0.77
5	Refined Oil	0.00	0.00	0.00	0.00	-0.13	-0.93	-0.03	3.03	28.00
6	Electricity etc.	0.00	0.00	-0.01	-0.02	0.02	-0.01	-0.03	-0.06	0.06
7	Water Supply	0.00	0.00	0.01	0.04	0.11	0.22	0.39	0.70	1.19
8	Ferr. & Non-f. Metals	0.00	-0.01	-0.01	0.03	-0.03	0.18	-0.20	-0.56	-1.15
9	Non-metallic Min.Pr.	0.00	0.00	-0.02	-0.05	-0.09	-0.52	-0.42	-0.35	0.29
10	Chemicals	0.00	-0.01	0.00	-0.03	-0.08	0.28	0.26	-0.14	-0.78
11	Metal Products	0.00	-0.01	-0.03	-0.09	-0.21	-1.59	-1.62	-1.72	-1.33
12	Agri. & Indust. Mach.	0.00	-0.01	-0.01	-0.04	-0.10	-2.02	-0.15	-0.67	0.04
13	Office Machines	0.00	0.00	-0.01	-0.03	-0.08	-0.32	0.50	0.46	0.53
14	Electrical Goods	0.00	0.00	0.01	0.01	0.09	-4.05	-1.29	-1.14	-0.84
15	Transport Equipment	0.00	-0.02	-0.05	-0.18	-0.43	-3.77	-1.46	-2.61	-2.76
16	Food, Drink & Tobacco	0.00	0.00	-0.01	0.02	0.04	0.00	0.09	0.14	0.55
17	Tex., Cloth. & Footw.	0.00	0.00	-0.01	0.00	0.00	-0.27	-0.28	-0.19	0.49
18	Paper & Printing Pr.	0.00	0.00	0.00	0.00	0.03	1.14	0.06	0.56	0.62
19	Rubber & Plastic Pr.	0.00	0.00	-0.02	0.00	0.00	-1.03	0.39	1.46	3.21
21	Other Manufactures	0.00	0.00	0.00	0.01	0.04	-0.75	0.44	0.83	1.01
22	Construction	0.00	0.00	0.00	0.03	0.06	0.35	0.40	0.51	0.37
23	Distribution etc.	0.00	0.00	0.00	0.02	0.02	0.50	0.51	0.49	0.21
24	Lodging & Catering	0.00	0.01	0.04	0.15	0.46	0.34	1.42	2.51	5.22
25	Inland Transport	0.00	0.01	0.03	0.16	0.54	-0.07	0.84	2.69	5.74
26	Sea & Air Transport	0.00	0.01	0.15	0.57	1.04	3.18	4.14	7.04	6.45
27	Other Transport	0.00	0.02	0.18	0.72	1.59	2.03	5.05	10.56	13.83
28	Communications	0.00	0.01	0.00	0.04	0.07	0.42	0.53	0.76	0.26
29	Bank., Finance & Ins.	0.00	-0.01	0.00	0.01	0.04	-0.04	0.27	0.34	0.73
30	Other Market Serv.	0.00	0.00	0.01	0.04	0.05	0.09	0.09	0.08	-0.18
31	Non-market Services	0.00	0.00	0.00	0.00	0.00	-0.02	-0.04	-0.04	0.08
32	Unallocated	0.00	0.00	0.00	0.00	0.00	0.00	0.00	0.00	0.00
	Total	0.00	0.00	0.01	0.05	0.13	-0.07	0.49	1.19	2.79

Source: Cambridge Econometrics Forecast C51F1V-C51F1W/da1/F1WA.

Bibliography

Allen, C.A. and J. D. Whitley, 'Modelling Bilateral Trade', in S.G. Hall (ed.) *Applied Economic Forecasting Techniques*, Hampstead: Harvester Wheatsheaf, 1994, 144–172.

Anderton, R. *et al.*, 'Nominal Convergence in European Wage Behaviour: Achievements and Explanations', in R. Barrell (ed.) *Economic Convergence and Monetary Union in Europe*, London: Sage, 1992, 31–57.

Armah, B., 'Trade Sensitive Manufacturing Employment: Some New Insights', *Review of Black Political Economy*, 21(2), Autumn 1992, 37–54.

Assefa, A. and R.A. Wilson, *Modelling Regional Migration Using LFS Data*, presented at the European Symposium on Labour Market Developments, 20–21 May, University of Warwick, 1993.

Balassa, B., 'Tariff reduction and trade in manufactures among industrial countries', *American Economic Review*, Vol. 56, 1966.

Baldwin, R., 'The growth effects of 1992', *Economic Policy*, 1992.

Baldwin, R. and A. Venables, *Methodologies for an Aggregate Ex Post Evaluation of the Completion of the Internal Market. Feasibility Study and Literature Survey*, Final Report submitted to the European Commission DG II, 1995.

Ball, J. and E. St Cyr, 'Short-term employment functions in British manufacturing', *Review of Economic Studies*, 33, 1966, 179–207.

Barker, T.S., B. Gardiner and A. Dieppe, *E3ME Version 1.1 (E3ME11) Users' Manual*, Final Report to the European Commission DG XII, September 1995.

Barry, F., J. Bradley, A. Hannan and J. McCartan, *Trade liberalization and structural change in small-open-economy macromodels: the internal market and the EU periphery*, Economic and Social Research Institute, 1996.

Bean, C., R. Layard and S. Nickell, 'The Rise in Unemployment: A Multi-country Study', *Economica*, 1986.

Beatson, M., *Labour Market Flexibility*, Research Strategy Branch, Research Series No. 48, Sheffield: Employment Department, 1995.

Berman E., J. Bound and Z. Griliches, 'Changes in the Demand for Skilled Labour within US Manufacturing: Evidence from the Annual Survey of Manufactures', *Quarterly Journal of Economics*, November 1995, 367–397.

Borjas, G.J. and V.A. Ramey, 'Foreign Competition, Market Power and Wage Inequality', *Quarterly Journal of Economics*, November 1995, 1075–1110.

Brander, J., 'Intra-industry trade in identical commodities', *Journal of International Economics*, 11, 1981.

Briscoe, G. and R.A. Wilson, 'Explanations of the demand for labour in the United Kingdom engineering sector', *Applied Economics*, 23, 1991, 913–926.

Bruno, M., 'Aggregate Supply and Demand Factors in OECD Unemployment: An Update', *Economica*, 1986.

Buigues, P., F. Ilzkovitz and J.F. Lebrun, 'The impact of the internal market by industrial sector: the challenge for the Member States', *European Economy/Social Europe*, Special Edition, Luxembourg: Office for Official Publications of the EC 1990.

Buigues, P., F. Ilzkovitz, J.F. Lebrun and A. Sapir, 'Market Services and European Integration: the challenge for the 1990s', *European Economy/Social Europe No. 3*, Luxembourg: Office for Official Publications of the EC 1993.

Burtless, G., 'International Trade and the Rise in Earnings Inequality', *Journal of Economic Literature*, June 1995.

Cambridge Econometrics, 'A Disaggregated Analysis of UK Exports by Country and Sector', in *Industry and the British Economy*, November 1992.

Cecchini, P., *The European Challenge 1992: The Benefits of a Single Market*, Wildwood House, Aldershot, 1988.

Chan-Lee, J.H., D.T. Coe and M. Prywes, 'Microeconomic Changes and Macroeconomic Disinflation in the 1980s', *OECD Economic Studies*, 1987.

Chui, M. and J.D. Whitley, 'What Has Been Happening To European Trade', *LBS Economic Outlook*, August 1995.

Coe, D., 'Nominal Wages, the NAIRU and Wage Flexibility', *OECD Economic Studies*, 1986.

Coe, D. and R. Magnadam, 'Capital and trade as engines of growth in France: an application of Johansen's cointegration methodology', *IMF Working Paper 93/11*, 1993.

Collins, P. and M. Hutchings, 'Articles 101 and 102 of the EEC Treaty: Completing the Internal Market', *European Law Review*, 11, No. 3, 1986, 191–199.

Dauses, M.A., 'The System of the Free Movement of Goods in the EU', *American Journal of Comparative Law*, 33, No. 2 (Spring), 1985, 209–231.

DRI McGraw-Hill, 'Survey of the Trade Associations' Perception of the Effects of the Single Market', Draft Final Report to the European Commission, 1995.

Driver, C., A. Kilpatrick and B. Nesbitt, 'The Employment Effects of UK Manufacturing Trade Expansion with the EEC and the Newly Industrializing Countries', *European Economic Review*, 30(2), April 1986, 427–438.

Dunning, J. and J. Cantwell, 'Japanese direct investment in Europe', in Burgenmeier and Mucchieli (eds) *Multinationals and Europe 1992*, 1991.

Engle, R.F. and C.W.J. Granger, 'Cointegration and error correction: representation, estimation and testing', *Econometrica*, 55, 1987, 251–276.

Emerson, M., 'Regulation or Deregulation of the Labour Market: Policy Regimes for the Recruitment and Dismissal of Employees in the Industrialized Countries' *European Economic Review*, 32, 1988, 775–817.

European Commission, 'A General System for the Recognition of Higher Education Diplomas', *Bulletin of the European Communities*, Supplement 8/85, Luxembourg: Office for Official Publications of the EC, 1985.

European Commission, '*Completing the Internal Market*'. White Paper, Luxembourg: Office for Official Publications of the EC, 1985.

European Commission, 'Employment problems: views of businessmen and the workforce'/ 'COMPACT: A prototype macroeconomic model of the European Community in the World Economy', *European Economy*, No. 27, Luxembourg: Office for Official Publications of the EC, 1986.

European Commission, 'Second Report from the Commission to the Council and European Parliament on the Implementation of the Commission's White Paper on Completing the Internal Market', 1987.

European Commission, 'Developments on the labour market in the Community'/'Quest', *European Economy*, no. 47, March, Luxembourg: Office for Official Publications of the EC, 1991.

European Commission, *Employment in Europe – 1994*, Luxembourg: Office for Official Publications of the EC, 1994.

European Commission, Tableau de Bord. Brussels: DG V Employment Observatory, 1994.

European Commission, *The Single Market Review*, Subseries I: Impact on Manufacturing, No. 1: *Food, Drink and Tobacco Processing Machinery*, Luxembourg: Office for Official Publications of the EC and London: Kogan Page, 1997a.

European Commission, *The Single Market Review*, Subseries I: Impact on Manufacturing, No. 3: *Textiles and Clothing*, Luxembourg: Office for Official Publications of the EC and London: Kogan Page, 1997b.

European Commission, *The Single Market Review*, Subseries I: Impact on Manufacturing, No. 5: *Chemicals*, Luxembourg: Office for Official Publications of the EC and London: Kogan Page, 1997c.

European Commission, *The Single Market Review*, Subseries I: Impact on Manufacturing, No. 7: *Processed foodstuffs*, Luxembourg: Office for Official Publications of the EC and London: Kogan Page, 1997d.

European Commission, *The Single Market Review*, Subseries IV: Impact on trade and investment, No. 2: *Trade patterns inside the single market*, Luxembourg: Office for Official Publications of the EC and London: Kogan Page, 1997e.

European Commission, *The Single Market Review*, Subseries VI: Aggregate and regional impact, No. 1: *Regional Growth and convergence*, Luxembourg. Office for Official Publications of the EC and London: Kogan Page, 1997f.

European Commission, *The Single Market Review*, Subseries VI: Aggregate and regional impact, No. 2: *The cases of Greece, Spain, Ireland and Portugal*, Luxembourg: Office for Official Publications of the EC and London: Kogan Page, 1997g.

European Commission, *The Single Market Review*: *Results of the Business Survey*, Luxembourg: Office for Official Publications of the EC and London: Kogan Page, 1997h.

European Parliament, 'The Impact of the Internal Market Programme on European Economic Structure and Performance', *Economic Series* E-2, National Institute for Economic and Social Research, 1993.

Forward, N. and M. Clough, 'The Single European Act and Free Movement: Legal implications of the provisions for the completion of the internal market', *European Law Review*, 11, No. 6, 1986, 383–408.

Freeman, R.B., 'Are Your Wages Set in Beijing', *Journal of Economic Perspectives*, Vol 9, No. 3, Summer 1995, 15–32.

Gasiorek M., A. Smith and A. Venables, '1992: Trade and Welfare; A General Equilibrium Model', *CEPR Discussion Paper No. 672*, 1992.

Grubb, D. and W. Wells, 'Employment Regulation and Patterns of Work in EC Countries'. *OECD Economic Studies*, 21 (Winter), 1993, 7–56.

Grubel, H. and P. Lloyd, *'Intra-industry Trade: The Theory and Measurement of International Trade in Differentiated Products'*, London: Macmillan, 1975.

Guieu, P. and C. Bonnet, 'Completion of the Internal Market and Indirect Taxation', *Journal of Common Market Studies*, XXV, No. 3, 1987, 209–222.

Guillaume, Y., D. Meulders and R. Plasman, 'Simulation des impacts de l'achèvement du marché intérieur. Le cas de la Belgique', *Cahiers Economiques de Bruxelles*, Fourth Quarter 1989, 449–473.

Hammond, P.J. and J. Sempere, 'Limits to the Potential Gains from Economic Integration and Other Supply Side Policies', *Economic Journal*, September 1995.

Hart, R., *Working Time and Employment*, London: Allen and Unwin, 1987.

Haskel, J. and R. Jukes, 'Skilled and Unskilled Employment in UK Manufacturing over the 1980s', *International Journal of Manpower*, 16(2), 1995, 36–52.

Helpman, E., 'International Trade in the Presence of Product Differentiation, Economies of Scale, and Monopolistic Competition', *Journal of International Economics*, Vol. 11, 1981.

Helpman, E. and P. Krugman, *Market Structure and Foreign Trade*, Cambridge, Mass: MIT Press, 1985.

Henderson, D. and S. Sanford, 'A Regional Model of Import-Employment Substitution: The Case of Textiles', *Review of Regional Studies*, Spring 1991, 79–90.

Hendry D.F., A. Pagan and J. D. Sargan, 'Dynamic Specification' in Gilliches, Z. and M.D. Intriligator (eds), *Handbook of Econometrics*, Vol. II, Amsterdam: North Holland, 1984.

Hoeller, P. and M.-O. Louppe, 'The EC's Internal Market: Implementation, Economic Consequences, Unfinished Business', *OECD Economics Working Papers No. 147*, 1994.

Hogarth, T. and R. Lindley, 'Medium Term Employment Forecasts by EU Regions and Sectors of Industry, 1991-97: Occupation, Gender and Qualifications', Luxembourg: Office for Official Publications of the EC, 1994.

Institute for Employment Research, *Review of the Economy and Employment*, Coventry: IER, University of Warwick, 1987.

Jacquemin, A. and A. Sapir, 'International Trade and Integration of the European Community', *European Economic Review*, 32, 1988, 1439–1449.

Johnes, G. and T. Hyclak, 'Wage Inflation and Unemployment in Europe: The Regional Dimension'. *Regional Studies*, 23, No. 1, 1989, 19–26.

Killingsworth, M.R., 'A critical survey of neo-classical labour models', *Oxford Bulletin of Economics and Statistics,* August, 1970, 133–165.

Klau, F. and A. Mittelstadt, 'Labour Market Flexibility', *OECD Economic Studies*, 1986.

Krugman, P. and R. Lawrence, 'Trade, Jobs and Wages', *Scientific American,* April 1994.

Krugman, P. and A. J. Venables, 'Globalization and the Inequality of Nations', *Quarterly Journal of Economics*, Vol. CX, Issue 4, November 1995.

Larre, B., 'The Impact of Trade on Labour Markets: An Analysis by Industry', *The OECD Jobs Study Working Papers No. 6*, 1995.

Lawrence, R.Z. and M.J. Slaughter, 'International Trade and American Wages in the 1980s: Giant Sucking Sound or Small Hiccup?', *Brookings Papers: Microeconomics 2*, 1993.

Layard, R. and S. Nickell, 'Unemployment in Britain', *National Institute Economic Journal*, 1985.

Lee, K., M.H. Pesaran and R.G. Pierse, 'Aggregation Bias in Labour Demand Equations for the UK Economy'. in T. Barker and M. H. Pesaran (eds) *Disaggregation in Econometric Modelling*, London: Routledge, 1990.

Lee, K. and M.H. Pesaran, 'The Role of Sectoral Interactions in Wage Determination in the UK Economy', *The Economic Journal*, January 1993.

Lindley, R.M., 'Approaches to Assessing Employment Prospects', in R.M. Lindley (ed.) *Economic Change and Employment Policy*, London: Macmillan, 1980, 6–18.

Lindley, R.M., 'Input-Output Analysis and the Employment Effects Model', *Les Comptabilités en Travail*, Colloque International du Centre National de la Recherche Scientifique, Paris: CNRS, 1980, 59–68.

Lindley, R.M., 'Forecasting the Labour Market', Fourth International Symposium on Forecasting, London, July, 1984.

Maddala, G.S., '*Introduction to Econometrics* (2nd edition)', New York: Macmillan Publishing Company, 1992.

Markusen, J.R., 'The Boundaries of Multinational Enterprises and the Theory of International Trade', *Journal of Economic Perspectives*, Vol. 9, No. 2, Spring 1995, 169–189.

Marsden, D., *Pay and Employment in the New Europe*, Edward Elgar Publishing Ltd, UK, 1992.

Mayes, D.G. *et al.*, *The External Implications of Closer European Integration*, Hemel Hempstead: Harvester Wheatsheaf, 1993.

Oliveira Martins, J., 'Market Structure, Trade and Industry Wages', *OECD Economic Studies No. 22*, Spring 1994.

OECD, *Employment Outlook*, Paris: OECD, 1990.

OECD, *Employment Outlook*, Paris: OECD, 1993.

OECD, *Employment Outlook*, Paris: OECD, 1994.

OECD, *The OECD Jobs Study: Facts, Analysis and Strategies*, Paris, 1994a.

OECD, *The OECD Jobs Study: Evidence and Explanations. Part I – Labour Market Trends and Underlying Forces of Change*, Paris, 1994b.

OECD, *The OECD Jobs Study: Evidence and Explanations. Part II – The Adjustment Potential of the Labour Market*, Paris, 1994c.

OECD, 'The EC's Internal Market: Implementation, Economic Consequences, Unfinished Business', *Working Paper No. 147*, Paris, 1994d.

Pelkmans, J., 'The New Approach to Technical Harmonization and Standardization', *Journal of Common Market Studies*, XXV, No. 3, 1987, 249–269.

Pelkmans, J. and P. Robson, 'The Aspirations of the White Paper', *Journal of Common Market Studies*, XXV, No. 3, 1987, 181–192.

Petit, P. and T. Ward, 'The Implications for Employment in Europe of the Changing Pattern of Trade and Production in the Global Economy', European Commission Round Table on the Social Dialogue. Paris, July, 1995.

Pomfret, R., 'The Trade-diverting Bias of Preferential Trading Arrangements', *Journal of Common Market Studies*, XXV, No. 2, 1986, 109–117.

Poret, P., 'The Puzzle of Wage Moderation in the 1980s', *OECD Department of Economics and Statistics Working Paper No. 87*, Paris, 1990.

Revenga, A.L., 'Exporting jobs? The impact of import competition on employment and wages in US manufacturing', *The Quarterly Journal of Economics*, February 1992, 255–284.

Richardson, J.D., 'Income Inequality and Trade: How to Think, What to Conclude', *Journal of Economic Perspectives*, Vol. 9, No. 3, Summer 1995, 33–55.

Sachs, J.D. and H.J. Shatz, 'Trade and Jobs in US Manufacturing', *Brookings Papers on Economic Activity*, Vol. 1, 1994.

Sachs J.D. and H.J. Shatz, 'International Trade and Wage Inequality in the United States: Some New Results', First Draft, December 1995.

Saeger, S.S., 'Trade and De-industrialization: Myth and Reality in the OECD', Mimeo, *Harvard University*, November 1995.

Scharrer, H., 'Protectionism – A Necessary Price for Achieving the European Internal Market', *Intereconomics*, 22, No. 1, 1987, 9–13.

Schelzky, M., 'The Comparability of Vocational Training Qualifications Between Member States of the EU', *Vocational Training Bulletin*, 19, 1985, 48–50.

Symons, J.S.V. and R. Layard, 'Neoclassical Demand for Labour Functions for Six Major Economies'. *The Economic Journal,* 94, 1984, 788–799.

Turner, D., P. Richardson and S. Rauffet, 'The Role of Real and Nominal Rigidities in Macroeconomic Adjustment: A Comparative Study of the G3 Economies' *OECD Economic Studies*, 1993.

Venables, A. and A. Smith, *The cost of non-Europe: an assessment based on a formal model of imperfect competition and economies of scale*, Research on the Cost of Non-Europe, Luxembourg: Office for Official Publications of the EC, 1988.

Viner, J., *'The Customs Union Issue'*, New York: Carnegie Endowment for International Peace, 1950.

Waegenbaur, R., 'Free Movement in the Professions: The new EEC proposal on professional qualifications', *Common Market Law Review*, 23, No. 1, 1986, 91–109.

Walwei, U. and H. Werner, 'Employment and Social Dimensions of the Single European Market', *IAB Labour Market Research Topics*, No. 1, 1993.

Whitley, J.D., 'Comparative Simulation Analysis of the European Multi-country Models', *Journal of Forecasting*, 11, 1992, 423–458.

Windmuller, J.P. (ed.), *Collective Bargaining in Industrialized Market Economies: A Reappraisal*, Geneva: ILO, 1987.

Winters, L.A., 'Britain in Europe: a Survey of Quantitative Trade Studies', *Journal of Common Market Studies*, XXV, No. 4, 1987, 315–335.

Wood, A., *North-South Trade, Employment and Inequality*, Oxford University Press, 1994.

Wood, A., 'How Trade Hurt Unskilled Workers', *Journal of Economic Perspectives*, Vol. 9, No. 3, Summer 1995, 57–80.

Yannopoulos, G., 'The effect of the single market on the pattern of Japanese investment', *National Institute Economic Review*, 1990.

Zighera, J.A., 'Regional Dispersion of Industrial Labour Costs in the European Community (1988)' (revised version of paper to EALE Annual Conference, University of Warwick, 1992 available from Laédix, Université Paris X-Nanterre), 1993.